Management Accounting and Control Systems

Management Accounting and Control Systems
An Organizational and Behavioral Approach

Norman B. Macintosh
Queen's University, Kingston, Canada

JOHN WILEY & SONS
Chichester · New York · Brisbane · Toronto · Singapore

Copyright © 1994 by John Wiley & Sons Ltd,
          Baffins Lane, Chichester,
          West Sussex PO19 1UD, England

          Telephone (+44) 1243 779777

Reprinted September and October 1995; February and July 1996; February 1997

All rights reserved.

No part of this book may be reproduced by any means,
or transmitted, or translated into a machine language
without the written permission of the publisher.

Other Wiley Editorial Offices

John Wiley & Sons, Inc., 605 Third Avenue,
New York, NY 10158-0012, USA

Jacaranda Wiley Ltd, 33 Park Road, Milton,
Queensland 4064, Australia

John Wiley & Sons (Canada) Ltd, 22 Worcester Road,
Rexdale, Ontario M9W 1L1, Canada

John Wiley & Sons (SEA) Pte Ltd, 37 Jalan Pemimpin #05-04,
Block B, Union Industrial Building, Singapore 129809

Library of Congress Cataloging-in-Publication Data

Macintosh, Norman B.
  Management accounting and control systems : an organisational and behavioral approach / Norman B. Macintosh
     p. cm.
  Includes bibliographical references and index.
  ISBN 0-471-94409-2 (cloth)—ISBN 0-471-94411-4 (pbk.)
  1. Managerial accounting.   2. Management information systems.
  3. Organizational behavior.   I. Title.
  HF5657.4.M27   1994
  658.15'11—dc20   94-31614
                                                                  CIP

British Library Cataloguing in Publication Data

A catalogue record for this book is available from the British Library

ISBN 0-471-94409-2 (cloth)
     0-471-94411-4 (paper)

Typeset in 10/12pt Times by Photo-graphics, Honiton, Devon
Printed and bound in Great Britain by Bookcraft (Bath) Ltd

To my brothers William, Donald (in memory), Wallace, Kenneth (in memory), Douglas and our father William (in memory) with love and gratitude.

Contents

Preface		ix
1	Introduction	1
2	Case Studies	11
3	Universal Theories	29
4	Macromodels	51
5	Case Illustrations	75
6	Strategy and Control	87
7	Technology and Interdependency	111
8	Uncertainty, Scorekeeping, and Control	125
9	Interpretivist Models	149
10	A Structuration Framework	169
11	Structuration in Action	183
12	Conventional Critiques	197
13	A Radical Critique	221
14	Employee Accounting Systems	245
15	Conclusions	257
References		261
Name Index		273
Subject Index		277

Preface

Almost as soon as I began to study what was then called management control systems, I was deeply impressed with the possibilities of bringing developments in organizational behavior and business policy to bear on the pressing issues of this important subject. This was in the late 1960s and there were few, if any, systematic ways in circulation to understand these systems. Robert Anthony's three-tiered (strategic, managerial, and operational) framework was the standby and I learned the subject by analyzing individual case studies.

The central idea at the time was that control systems had a lot more to do with the motivation of managers and employees than it did with providing accurate, neutral, factual accounting data to rational, utility-maximizing decision makers. While academics of the latter ilk (the rationalists) aimed to fine tune the accounting system so that it provided true opportunity costs to decision makers, the motivationalists thought it more important to design control systems that triggered the manager to do what was best for the organization even if the numbers were less than accurate.

Motivation was also the golden thread that ran through nearly all the prevailing theories in organizational behavior at the time. Not surprisingly, then, when research in behavioral accounting got off the ground, pioneering scholars imported concepts and research methods from organizational behavior (just as the rational decision makers brought them in from neoclassical economics). In consequence, a variety of concepts based mainly on micro-level behavior at the individual and small group level, particularly those used by the human relations school, were welded onto management accounting and control issues. It was an exciting time.

The next major development involved the so-called contingency models of management accounting and control systems. Tagging along behind trends in organizational behavior again, some academics developed frameworks indicating how *impersonal* forces (environment, technology, task uncertainty, interdependences) shaped key characteristics of these systems. At the same time, other scholars dug deep into the literature on cognitive psychology and personal traits to construct models of how individuals process accounting information. As with the human relations school, the main concern of these two developments was to figure out how accounting systems could be designed to better regulate the status quo to make it more efficient and effective.

Over the same time period, slowly but surely the institutional support system for behavioral accounting research came into being. The American Accounting Association (AAA) formally acknowledged behavioral science and accounting in a 1974 supplement to *The Accounting Review* as a legitimate research activity. Later on, the Accounting, Behavior, and Organizations (ABO) special interest section of AAA came into being. Behavioral accounting conferences were organized and formal sessions at the AAA annual meeting were allocated to behavioral accounting research. And, perhaps the most significant development proved to be the emergence of the *Accounting, Organizations and Society (AOS)* journal started in 1975 by Anthony Hopwood in the UK. *AOS* slowly but surely got better and better and became a legitimate and respected home for behavioral accounting research at a time when the main US journals, dominated by the rationalists, paid only lip service to such "soft" research and gave center stage to information economics, efficient markets, positivistic events studies, and economic agency theory. Even so, the necessary institutional support for behavioral accounting came into place to make it possible for academics like me to carve out a respectable academic career in accounting based on behavioral research.

During this period, I decided to write a book about these developments. My aim was to bring the surprisingly substantial body of research about the people side (the social software as I labelled it) to the attention of academics, students, and practitioners working in the areas of management accounting, controllership, and management information systems. Luckily John Wiley, a highly respectable and energetic publisher, took on my book, *The Social Software of Accounting and Information Systems*, and published it in 1985. And to my great surprise, it was very well received in behavioral accounting circles. I shall always cherish the review by Chris Argyris, the founding father of behavioral accounting, who wrote in a review of three behavioral accounting books

> This book is a delight in its careful description of problems, its analysis of how research can be used to understand the problems, and its methodically stated set of conclusions and advice to the practitioners. The analysis made by Macintosh of the strengths, gaps, and inconsistencies of the existing research plus the models that he presents should serve as a basis for further research by scholars.

More recently, some of my colleagues around the world and my publisher have suggested I revise my book and bring it up to date. As much as I was tempted to try, it seemed to me that so much more had been added in the past decade to the corpus of behavioral accounting literature that to do justice to it would require at least eight or nine hundred pages. Moreover, I believed that most of the important microbehavioral pieces have been put in place for at least a decade and that this type of work had reached the mature (if not the decline) stage of the research life cycle. In the interim, however, the organizational sociological approach had proceeded apace. So I decided, for better or worse, to concentrate on the latter, thinking (hoping) that I might be able to make a better contribution to behavioral accounting by bringing some of these newer developments to the attention of practitioners, students, and academics, than by revising my social software book.

In developing the ideas in this book about an organizational sociology perspective of management accounting and control systems I have been lucky enough to work closely with several outstanding scholars. Dick Daft worked with me on the ideas in Chapter 7, Bob Scapens on the ideas in Chapters 9 and 10, and Trevor Hopper on the ideas in Chapter 13. Not only have they been gracious but also tough-minded in keeping me from making huge blunders in my thinking as I stumbled into new territories; they have also been great friends.

More recently, Dan Thornton, who came to Queen's this past year, was an invaluable source of intellectual support when I was stumped by agency theory (Chapter 3) and the theoretical aspects of ABC (Chapter 12). Bob Kaplan and Ken Merchant also commented helpfully with some of the ideas in Chapter 12, while Bob Simons was kind enough to critique Chapter 7. And always, somewhere behind the scenes, gently and patiently guiding my thinking from afar, have been people like Richard Boland, Wai Fong Chua, David Cooper, Anthony Hopwood, Larry Gordon, Sten Jönnson, Peter Miller, David Otley, and Tony Tinker. But my mention of these persons in no way implies that they endorse my views.

I have also been lucky to be at Queen's University where the tradition of the great liberal scholar working at his or her own pace on problems and issues of his or her choice is still part of the ethos. In this regard Don Akenson, a world-renowned historian, the editor of McGill-Queen's Press, and my writing guru, was kind enough to read an earlier draft and encouraged me to finish (although as an avowed antipostmodernist he sniffs at the ideas in Chapters 13 and 14). The Queen's Critical Theory Study Group seminars were also a great source of inspiration and comfort. Over the years, the research program at the School of Business has been highly supportive and generous in providing me with research assistants, visiting summer research scholars, teaching relief, and research stipends, while the individual scholar grant I received from the Social Science and Humanities Research Council (Canada) was extremely helpful. And the secretarial help, particularly Mary Brown and Linda Freeman, has been superb.

I should like to acknowledge the invaluable help of students in our Ph.D. program who, over the years, kept me on my toes when working on some of the sophisticated notions in critical social theory, as well as my MBA and undergraduate students who have been the guinea pigs for most of the material in the book. Finally, I must thank Tom Johnson, who provided the inspiration for the ideas in Chapter 14 about real employee empowerment and who took time out to help me resolve last-minute doubts about pushing forward with some radical ideas.

Norman Belding Macintosh
Kingston, Canada

1
Introduction

Control may be the most contentious word of our time. Half the world thinks of control as coercion and oppression protesting that we should have less of it. The rest believe society is pretty much out of control and that we need more of it. Either way, and regardless of one's political stance, control is a phenomenon that requires careful study if we are to make sense out of our world. This book is about one very important and specific sort of control, *management accounting and control systems*.

In fact, it is not too great an exaggeration to say that *management accounting and control systems* are so important and ubiquitous today that if accountants and information people wrapped up their systems and took them home, the whole process of producing society's material goods and services along with the governance of the social order would grind to a standstill. Banks would close, factories would produce goods at random, supermarkets would be out of many products and overstocked on others, police would arrest and release the wrong people, and military people would not have a clue which way to point their missiles. In this immense organism we call society, these systems are its central nervous system.

Even in the early part of this century, management accounting and control systems were recognized as absolutely essential to the affairs of large corporations such as du Pont, General Motors, Sears, Standard Oil of New Jersey, and Bethlehem Steel. As Alfred Sloan, chief executive officer, (CEO) of General Motors at that time, put it in his memoirs

> Financial method is so refined today that it may seem routine; yet this method—the financial model as some call it—by organizing and presenting the significant facts about what is going on in and around a business, is one of the chief bases for strategic business decisions. At all times, and particularly in times of crisis, or of contraction or expansion from whatever cause, it is of the essence in the running of a business (Sloan 1965, p. 118).

The globalization of these same corporations, along with newcomers and their counterparts in other capitalist countries, has made financial controls even more vital today. When operating a highly complex and multifaceted operation in over 150 countries around the globe, a common language is essential. That language, even more universal today than English, is accounting and finance.

Moreover, these systems are the major means by which a couple of dozen meganational enterprises virtually rule the world of the 1990s. Journalist Janet Lowe, in her carefully researched and disturbing book, *The Secret Empire*, convincingly documents how twenty-five multinational corporations, which developed out of the reach of public control, have become the new center of power of the world. Their combined financial resources—sales and assets—exceeds that of all but half a dozen or so nation states. The men at the desks at the top of these meganationals (there are virtually no women at the top) run their vast empires with highly sophisticated financial control systems.[1] Management accounting and control systems are an integral part of this staggering development.[2] These goliaths could not exist without them.

MANAGEMENT ACCOUNTING AND CONTROL SYSTEMS

This book is about management accounting and control systems. Sometimes they are referred to as *planning and control systems*; sometimes *management control systems*, and sometimes simply *control* systems. I call them *management accounting and control systems* to signal my primary concern with management accounting systems such as the annual operating budget for a division of a multinational conglomerate, or a standard cost system for a factory, or a case-mix costing system for a hospital. This is control in the small or tactical sense.

Management accounting is fairly easy to define in this narrow sense. It is "the process of identification, measurement, accumulation, analysis, preparation, interpretation, and communication of information that assists executives in fulfilling organizational objectives ... a formal mechanism for gathering and communicating data for the ends of *aiding* and *coordinating* collective decisions in light of the overall goals or objectives of an organization" (Horngren and Sundem 1990, p. 4).

In addition, however, management accounting is also about control in its broad sense. So the term control is added to signal that the book also deals with other related administrative devices which organizations use to control their managers and employees. Strategic planning systems, standard operating rules and procedures, as well as informal controls such as charismatic leadership and the fostering of a clan-like atmosphere are examples. This is control in the large.

My premise is that management accounting systems are only part, albeit usually a very important part, of the entire spectrum of control mechanisms used to motivate, monitor, measure, and sanction the actions of managers and employees in the organizations. So, to fully understand the workings of management accounting systems, it is necessary to see them in relation to the entire array of control mechanisms used by organizations. My aim is for the reader to develop a strategic perspective of management accounting and control systems in this larger context.

AN ORGANIZATIONAL SOCIOLOGY APPROACH

The book, importantly, does not deal with the "chug and plug" procedural and technical aspects of accounting. Nor does it discuss the principles and rules for calculating management accounting entries and preparing reports. Nor does it cover the application of quantitative techniques and spreadsheets to the work of management accountants.[3] And it does not focus on the influence of social psychological factors such as participation, leadership, group dynamics, and personality. There are already many excellent books covering these aspects of management accounting.[4]

Rather, this book takes an organizational sociology approach to the study of management accounting and control systems. This approach is more akin to organization theory, which studies organizations in the round, than to organizational behavior, which tends to focus at the micro-organizational level. There are two reasons for taking the macro approach.[5] First, a few behavioral management accounting books already focus mainly on macro-level control and organization.[6] Second, organizations are the prominent social institutions of our time; we are born, raised, educated, employed, retired, and even die in them. Management accounting and control systems are pervasive in organizations. For people who are managers of these systems (or potential managers) a macro approach can provide significant insight and understanding to help them design better systems.

Organizational sociologists study organizations, including their management accounting and control systems, from the perspective of a sociologist. Sociology involves the scientific analysis of a social institution, such as an industrial enterprise, as a functioning whole, including how it relates to the rest of society. More specifically, sociology involves the systematic study of the development, structure, interaction, and *collective* behavior of organized groups of human beings. So the sociologist aims to comprehend individuals, such as managers, as *social* beings who can be understood from their social context and the society of which they are part.

My aim for this book, importantly, is not to undermine the micro approach, but rather to complement it with the macro developments in organizational sociology. Since control—coercion, compliance, and cooperation—is at the heart of organizational sociology, it seemed that studying management accounting from this vantage point could be of considerable help to accountants and financial executives in developing a strategic and global perspective of the way management accounting and control systems work.

This book also advocates the use of models and theories to help analyze management accounting and control systems, problems, and issues. Models and theories are tools for understanding these complex systems. A model is a simplified (or cleaned-up) representation which identifies a few important dimensions or variables of the real thing and packages it into a comprehensive conceptual net. A theory is a description that explains how the dimensions and variables are casually related or intertwined. Models and theories help us to see and understand things we might otherwise overlook.

Instead of learning by trial and error or making do with personal experience in several organizations, models and theories can help accountants diagnose and explain the workings of a system in a more systematic and sophisticated way. The result should be greater effectiveness. Thus, in a very real sense, an organizational sociology approach can make management accountants and financial executives more competent and more influential. After all, accounting and financial discourse is one of the most pervasive and important voices in today's world; a strategic and global perspective, in addition to technical and micro perspectives, seems crucial.

MULTIPLE PARADIGMS

Accounting theories tend to line up into distinctive brands or paradigms about organizations and the social world. In fact, there are many to choose from and this book includes several.[7] But models and theories come in different forms: structural functionalist, interpretivist, radical structuralist, radical humanist, and postmodernist. This book presents a cross section of frameworks in each form along with some genealogical background on the historical evolution and development of some of their central concepts.

Structural Functionalist

The most common paradigm, sometimes called the structural functionalist (or rational–contingency) paradigm, assumes that an organization's social system consists of concrete, empirical phenomena which exist independently of the managers and employees who work there. So functionalists apply the methodologies used in the natural sciences to analyze the way organizational control systems work. They describe the variables, set forth testable hypotheses, collect quantitative data, and undertake statistical analysis. As with the particle physicist, they assume themselves to be neutral, objective, and value-free observers of the management accounting and control systems under investigation. Scientific positivism is *de rigueur*.

For the most part, functionalists pretty much take power and political arrangements as given and so treat the existing power structure as unproblematic.[8] Their concern is to make management accounting and control systems function more effectively. So they search for workable solutions to practical problems which will help accountants and managers achieve a better alignment of these systems with other organizational elements and forces. Their aim is pragmatic, to achieve better regulation and control of the status quo.

This book includes several frameworks based on the functionalist paradigm. Chapter 4 describes three conceptual schemes of this kind: a mechanistic–organic model, a command–market theory, and a cybernetic systems framework. The ideas in Chapter 5, which present a couple of frameworks linking corporate strategy to control systems, also adopt functionalist premises. Chapters 7 and 8 include several models showing how different technologies and interdependencies determine the kind of uncertainty in the work done in organizational components and how these variables dictate the appropriate style and characteristics for control and scorekeeping. And Chapter 10 includes structural functionalist thinking in an all-encompassing theory of management accounting and control systems. The structural functionalist approach, for better or worse, remains the dominant paradigm for most management accounting and control systems thinking.

Interpretivist

Another important but less well-known approach is the interpretivist (or subjective interactionist) paradigm.[9] It differs from the structural functionalist one in two major respects. First, the focus is not only on getting the organization to run more smoothly, but also to produce rich and deep understandings of how managers and employees in organizations understand, think about, interact with, and use management accounting and control systems. Interactionists write thick narratives about the way accountants and managers make sense of these systems.

The other major difference is that interactionists do not believe in the existence of any one single, objective, concrete organizational reality. Instead, they contend that each organizational participant interprets the situation in his or her own way and, more importantly, that their understandings become very real to them since they act towards events and situations on the basis of these personal meanings.[10] Moreover, in doing so, they interact with other managers and employees who also bring with them their private meanings. Relying on personal reflections and interactions, managers and employees socially construct and transform the meaning of accounting and control systems as they go along, always paying close attention to how others see them and the situation at hand.

For the subjective interactionist the meaning of an accounting system does not stem predominantly from social structures which exist independently from accountants and managers. Nor are they seen to be the result of the manager's psychological makeup. Although the manager must use language to construct meanings, the words used do not have fixed, stable meanings. Instead, they take on meaning only within the context of the circumstances and the people involved in the situation. The organizational world is seen to be socially constructed, dialogic, and hermeneutical.

Several frameworks in the book rely on the interpretivist paradigm. Chapter 8 introduces the notion of social tests for assessment and scorekeeping and outlines a model of management control systems whereby charisma and tradition discipline managers and employees. In Chapter 9, the subjective interactionist paradigm comes to the forefront in a couple of frameworks which demonstrate the way managers use management accounting and control systems to construct social reality in organizations. Chapters 10 and 11 incorporate this perspective into a comprehensive analytical framework of management accounting and control systems.

Radical Structuralist

The radical structuralist approach, as with the structural functionalist paradigm, assumes that social systems have a concrete and real ontological existence. It parts company with functionalism, however, by positing that any social organization exists in a state of dynamic tension; two basic opposing forces or principles are locked in a dialectic contradiction whereby they operate in terms of but also run counter to each other. Secondary contradictions are also present, but they arise as a consequence of the primary contradiction. The antinomy between the two forces is seen as an intrinsic aspect of any organization and basic to its system integration.

All social systems are seen to be inherently changeful. At any time, one of its two fundamental forces has the upper hand, but only temporarily, and can be overturned by the other. Radical structuralists pay particular attention to the way those with power (usually a small select circle of elites) hold the rest in check by means of their control over and use of power resources including management accounting and control systems. The radical structuralist aims to develop the possibilities for effecting a radical overhaul of any such coercive status quo.

Radical structuralists try to uncover the role management accounting and control systems play in supporting and maintaining a mode of organization whereby a small minority of executives and managers rule and exploit the rest of the employees scattered across the organizational landscape.[11] Importantly, these dynamics are more readily discerned in crisis situations than during periods of surface stability when underlying structural contradictions and tensions are camouflaged. The labor process framework in Chapter 3 and the historical–dialectical model in Chapter 4 are of this type.

Radical Humanist

The radical humanist paradigm assumes a subjective social world but, in contrast to the interpretivist position, adopts a radical change position.[12] Radical humanists aim for a people-oriented vision of management accounting and control systems practices whereby humanistic ideals and values come before organizational purpose. Like radical structuralists, they see management accounting and control systems as vehicles for elites such as upper echelons to take advantage of other managers and employees. They see this suppression stemming not from concrete, real, and independent forces and institutions such as the state, capitalism or bureaucracy, but from the self-laid traps which organizational participants construct themselves.

Radical humanists contend that managers and employees socially construct organizational control structures and processes, but later, through a process of reification, come to experience them as outside forces rather than as extensions of themselves. They argue that relations of power in organizations acquire their force just as much from subordinates' enactment of the prevailing structures of control and domination as they do from superordinates' initiatives. In consequence, managers and employees are prone to take on a false consciousness about the coercive nature of their working life. They become happy slaves or misinterpreters of the basic source of their suppression, themselves.

Radical humanists, importantly, have both enlightenment and emancipatory aims. The enlightenment goal is for agents to see through their self-induced fettered existence; while the emancipatory aim calls for them to regain power and responsibility for their self-constructed social world. It is possible, radical humanists believe, for managers and employees to realize that the social structures which discipline and control them are merely language games (albeit extremely serious ones) whereby socially constructed narratives and metaphors dictate the terms of social actions and interactions. Thus, the possibility opens up for the actors to understand these dynamics and restore power to themselves. People are not for organizations; organizations are for people.

Chapter 14 outlines a proposal along radical humanist lines. It argues for empowering employees by letting them take control of the design and running of management accounting and control systems in order to rehierarchize organizational goals in such a way as to put their needs for employment, stability, and interpersonal relationships at the top, instead of a hierarchy where the market price of the company's stock and cost-cutting come ahead of employee imperatives. Humanistic ideals, not materialistic goals, would reign.

Postmodernist

The final perspective goes under the heading postmodernist. Although almost unknown in accounting circles, postmodernist thought has profoundly influenced the methods and theories of a variety of disciplines in the human sciences as a fresh and vital development.[13] Postmodernists suggest the world has emerged from the modern epoch (1650–1970) and coalesced into the postmodern epoch, which constitutes yet another novel, but not final, major era in the history of humankind. Highly sophisticated technology, mass media, and a homogenized global marketplace dominate the contemporary world, making it qualitatively different from the modern epoch. New concepts, theories, and methodologies are needed to understand the postmodern world.[14]

So postmodernists abandon the dominant methods and theories of modernity (such as structural and radical functionalist ideas) in favor of poststructuralist tools. These include genealogical historical analyses of the human sciences and their related practices; deconstructivist interrogation of ruling ideas and canonical texts; dedoxification of scientific methods; and strategies for defining and constructing one's own individuality. Poststructural tools are used to create legitimate spaces for discourses and groups previously marginalized by modernist perspectives with their hegemonic tendencies.

But perhaps the distinguishing feature of the poststructuralist approach is its negation of structuralist claims that more or less permanent, subsurface structures, such as the invisible hand of the market or the forces of capitalism, running deep and strong give the true meaning of the surface world, which merely copies them. For poststructuralists, such an idea is simply a politically loaded language game. Instead they see the sociocultural landscape as a collage of unanchored, free-floating, and competing discourses. Chapter 13 presents a poststructuralist critique of management accounting and control systems and Chapter 14 draws on some poststructural ideas to sketch out and argue for a truly employee-empowered corporate world with its own employee accounting system.

MODUS OPERANDI

This book outlines nearly twenty frameworks for investigating and understanding management accounting and control systems. The frameworks have been carefully selected in order to offer a cross section from the five perspectives described above. Each was selected because it offered a unique and valuable way to look at these systems. Such a wide sweep brings with it the advantage of broadening the horizons well beyond the traditional approach of information for decision and control. Thus, it holds the obvious advantage of opening up fresh vistas and providing new insights, but alongside may come the feeling that such an expansion has been won at the cost of importing into our knowledge base a certain measure of confusion and disagreement about the best approach.

One way to handle this feeling is to adopt the view that no one paradigm is best. This means realizing there are different possible and self-contained traditions and ways of making sense of the organizational ramifications of management accounting and control systems and that we can judge each according to its own standards and purpose. Such a relativistic stance may not sit well with those who believe there must be one best perspective in an absolute sense, or with those who believe that accounting knowledge rests on some incorrigible and certain foundational scheme which we can discover. While such idealism is to be admired, the reality is that accounting, as with most bodies of knowledge, has yet to achieve such a state. Even physics, the crown prince of the sciences, lacks a general theory to unify the quantum theory of particles with the concepts of relativity that Einstein formulated about planetary systems and galaxies. So it is not surprising that accounting has competing and conflicting theories.

Instead of looking for the ultimate paradigm, we aim more modestly to bring to the attention of management accounting practitioners, a number of paradigms which look at different dimensions and aspects of these systems and to illustrate how to use them in a practical way. And we do this with a variety of case histories, which breathe life into what might otherwise seem dry and inert. We show how to put frameworks into action for analyzing, diagnosing, and prescribing for real management accounting and control systems problems and issues.

Thus, our modus operandi differs from both the traditional case method and the conventional theoretical approach. The case method treats each situation as unique with issues and problems which can be identified and sorted out diagnostically by drawing on intuition and insight derived from an in-depth analysis of the situation at hand. In contrast, the theoretical approach focuses heavily on the knowledge content of the subject in order to get across the theoretical concepts and the elegance of their interrelationships within the theory. Often oversimplified examples, such as a children's lemonade stand, are provided to illustrate the theory but little concern is shown for practical applications. The aim is to acquire a deep understanding of theories for appropriate application in the future. If the traditional case method is like a rich and fascinating story about a problem looking for a solution, the conventional theoretical approach is like an elegant and powerful solution looking for a problem.

This book veers away from any such theoretical–practical duality. Instead, it aims to show that both are important, that theory and practice are inextricably intertwined. Theory is always there in practice, regardless of how implicit or commonsensical it may appear. Likewise, practice is always present in the theory, no matter how thinly disguised. In this book, the theories are used to sensitize understanding of practical situations to vital but easily overlooked aspects of the situation at hand. And the case histories are used to bring to life what otherwise might come across as inert abstractions.

But it is also a slippery slope. The frameworks tend to focus on a few variables at the expense of excluding others that may also be vital. Moreover, there is no guarantee that different persons, even though they rely on the same framework, will reach an identical understanding. The frameworks do not inevitably generate correct answers, as would be the case for a die with three dots on each of its faces. There is still a great deal of individual skill and judgement that goes into the analysis. The richly painted case analyses presented here are idiosyncratic and are not deemed to be definitive solutions. Rather, they are offered as illustrations of one way to interpret the case situation using the syntax, concepts, and relationships included in the particular framework at hand. Nevertheless they are considerate and serious attempts to use the frameworks in a valuable way.

SUMMARY

This book deals with management accounting and control systems from an organizational sociology perspective. It presents a cross section of frameworks from five different paradigms: structural functionalist, interpretivist, radical structuralist, radical humanist, and postmodernist. It introduces a series of actual case studies in order to bring the frameworks to life and to show how they can be used to analyze, diagnose, and treat problems and issues in management accounting and control systems. Today these systems play a huge role in the management of society's major institutions and corporations, both private and public sector ones. We begin our journey into their organizational aspects by detailing three case histories.

NOTES

1. For an inside look at how these meganationals are run with financial controls, see Sampson's (1974) exposé of the rise and fall of International Telephone and Telegraph Company (ITT).
2. The morality of this kind of concentration is addressed in some detail in Chapter 13.
3. See Scapens (1985), Kaplan and Atkinson (1989), and Ezzamel (1992) for excellent books regarding the application of quantitative techniques to management accounting.

4. See Parker, Ferris, and Otley (1989), Macintosh (1985), and Emmanuel, Otley, and Merchant (1992) for coverage of the micro-behavioral approach to behavioral accounting.
5. See Daft (1992) Chapter 1 for an overview and description of the organization theory approach and purpose.
6. The exceptions are books by Puxty (1993) and Roslender (1992); both are excellent.
7. See Hopper and Powell (1985) for an overview of the various paradigms and related accounting studies; and Chua (1986) for a review of their epistemological and ontological assumptions.
8. Not all functionalists ignore politics and change. But those that do consider power and global goals see them as up for grabs in a pluralistic (and changeful) setting where all have a chance to become part of the "dominant coalition." Thus, the manager is considered to be either an astute or a naive politician.
9. See Chua (1988) and Covaleski and Dirsmith (1990) for useful overviews of the interpretivist paradigm in accounting.
10. A striking example supporting interactionist beliefs comes from Perrow (1970) who reports the case of two juvenile delinquent institutions. In one case, the authorities saw the situation in terms of a lack of authority and discipline in each of the delinquent's lives. So they set up a highly autocratic, military-like organization and meted out large doses of discipline. In the second case, the authorities saw each delinquent as a unique case study to be treated and understood in his or her own right. This institution featured highly organic and peer-type person-to-person relationships between counsellors and detainees. The reality in each case was socially constructed but, importantly, resulted in totally different control structures and processes.
11. See Tinker (1980) and Tinker and Neimark (1987) for a seminal accounting study along these lines.
12. See Puxty (1993) Chapter 6 for a detailed (if partial) review of this paradigm.
13. These include law, philosophy, architecture, history, communications and information studies, geography, visual arts, and women's studies.
14. See Sarup (1993) and Best and Kellner (1991) for an excellent introductory coverage of postmodernity and poststructural developments.

2
Case Studies

This chapter describes three case studies of management accounting and control systems: Wedgwood Potteries, Empire Glass, and Johnson & Johnson. They illustrate in some detail the subject matter of this book. Each case represents an instance of a well-thought-out and effective control system both in how it is designed, its structure, and how it works, its process. In addition, each represents a prototype of the state of the art in management accounting and control in three different eras: the beginning of the industrial revolution, the post World War II period, and today's late twentieth century capitalism. Thus, they also reveal the evolution that has taken place in the accounting and control systems over the past two centuries.

The Wedgwood Potteries case describes a 1770s management accounting system, one of the first at the dawning of the industrial revolution. Empire Glass documents the almost textbook design and use of financial controls in the early 1960s in the bottle-making division of a packaging industry giant. While the Johnson & Johnson situation shows how a very successful multinational health care conglomerate uses its sophisticated management accounting and control system to manage its sundry subsidiaries spread around the globe. We begin with the remarkable story about the achievements of Josiah Wedgwood and the role played by his "costing worksheets."

WEDGWOOD POTTERIES[1]

Josiah Wedgwood sat in his small, meagerly furnished office, staring at the spreadsheets he had been working over for nearly a month. It was a rainy day in August 1772, the year a major recession swept through Europe causing sales of fine pottery to plummet. Wedgwood believed if he could calculate the costs for his various pottery products, it would help his firm survive the depression without laying off any of his carefully trained workforce.

Yet something was amiss. Wedgwood had worked out in great detail the costs of labour, clay, and other materials for each product line. He had also figured in the costs of overseers, clerks, incidentals, and other indirect expenses and estimated their loading on specific products. He then meticulously developed the cost for each product down to a halfpenny. Next, he calculated the profit margins for each piece and every batch of pottery using selling prices. Yet in the aggregate, they did not agree with the overall costs and profits included in the financial accounting reports prepared by his head clerk. He went over his calculations again and again until he convinced himself they were roughly accurate.

For some time Wedgwood had harbored an uneasy feeling about the head clerk's propriety. He was suspicious, for instance, about the cash accounting. It always arrived a couple of months late, yet came up spot on whenever the head clerk produced a bank reconciliation and cash report. He decided to launch an inquiry into the affairs of the counting house. His hunch proved accurate. The difference in Wedgwood's figures was not due to any miscalculations on his part. The head clerk, in cahoots with his underlings, had been stealing cash and fixing the books for some time. But that was not all.

A detailed investigation revealed a distressing state of affairs. As Wedgwood described the situation in a letter to his London marketing partner, Bentley: "The Plan of our House in Newport Street is rather unfavorable to virtue and good order in young men," the housekeeper was "frolicking with the cashier," the head clerk was "ill with the foul disease" and had "long been in a course of extravagance and dissipation far beyond anything he has from us (in a lawful way) would be able to support." Embezzlement, blackmail, extravagance, and dissipation were the order of the day.

Wedgwood acted without hesitation. He dismissed the head clerk replacing him with a trusted employee from his own office. He sacked the cashier and most of the other employees. He set up a new system of weekly cash reporting and bank reconciliations. He put into effect a policy that customers' accounts must be settled weekly and that all receipts be taken to Bentley for depositing. He soon "put things to right in the counting house."

Wedgwood's cost accounting system had rescued the company from internal destruction. In addition, however, it helped stave off and overcome the potential ravages of a severe economic depression. Wedgwood used his costing system in a systematic and detailed way, to analyze the effect of possible product price cuts, calculate the cost savings accruing from scheduling longer production runs of specific products, determine optimum wage scales, set sound hiring policies, and calculate the most efficient assignment of workers to the production process. As many of his competitors went to the wall of bankruptcy, Wedgwood's pottery weathered the storm. Once again his costing system came to the rescue.

As the depression eased and sales recovered, Wedgwood continued to use his new management accounting system to maximum advantage as a powerful tool in managing his business. Cost accounting and control proved an essential ingredient in Wedgwood's success in the subsequent period of high demand, rising prices, and spectacular profits. The man who single-handedly invented his own detailed cost accounting became a legend in his time for this and other innovations in the techniques of management control.

Strategic Planning

Wedgwood started his pottery operation near Burslem, Staffordshire, England in 1759 with only a couple of leased sheds, two kilns, a small amount of family financial backing, and one employee—a cousin. The operation grew rapidly. Within two years it was a typical potbank operating under the traditional guild system with a dozen men and boys, a rudimentary division of duties, a relaxed master-apprentice hierarchy, no formal rules, and a small output of undistinguished pottery sold to the local gentry. Burslem alone could count 150 such potbanks.

For Wedgwood, however, this was only a beginning. A small, lame, God-fearing man, he harbored much greater ambitions. He foresaw a golden opportunity in the spectacular rise in demand for fine pottery precipitated by the general increase in the English economy, the spectacular rise in the populace's standard of living, the rapid growth in tea and coffee drinking habits, and the penchant in the colonies for imported fineries from the motherland. Wedgwood reckoned he could monopolize this burgeoning market by manufacturing large quantities of very high quality earthenware *and* selling it at a price lower than his competitors.

With this strategy clearly in mind, Wedgwood carried out literally thousands of experiments with various clays and kiln techniques. Eventually his remarkable persistence paid in spades. He developed the formulae for what was to become a renowned label, the *Queen's Ware* line, so named because it graced the table of the Queen of England. Armed with a product of superior quality suitable for the upper class, he sensed that if he could mass produce it he could cash in handsomely on the rapidly growing middle class market. Wedgwood never forgot that he was in business to make money.

Yet one big obstacle stood in his path. Transportation in England, particularly around Staffordshire, remained primitive. This made it extremely difficult to get cheap raw materials in and, even more distressing, to get the finished product out to markets in London, the rest of England, and overseas. Getting finished pottery to Bentley's London marketing arm was a treacherous undertaking. Typically, a potter lost one-third of every shipment by packhorse over the muddy ruts that passed for roads between Staffordshire and London. Breakage and theft were common. In consequence, Wedgwood organized the potters in his region to petition to Parliament for a turnpike and a canal system linking with existing rivers.

The petition proved successful. When the canal was completed Wedgwood built his dream factory, Etruria, alongside the canal. Modelling it after his friend Boulton's metal manufactory, he constructed five separate buildings in an arc alongside the canal. He situated the raw material house at one end on the water and built the packing and shipping house on the other end, also on the water. With the wheels, pottery-making equipment and kilns in the middle, production flowed smoothly along the arc. Clay and materials arrived at one end and fine pottery emerged at the other.

Control and Discipline: The Missing Link

Affairs were definitely on the upswing. Wedgwood had his dream factory, a reliable transportation system, a unique product, a brilliant competitive strategy, and a solid marketing operation run by his partner in London. But one more stumbling block proved to be his biggest obstacle—turning an obstinate, slovenly, intemperate band of employees into a disciplined productive work force.

The English working class at the time consisted of peasants and serfs who had been driven off the farms and pressed into unfamiliar nonagricultural work. On the whole they tended to be illiterate, ignorant, and lazy. Drinking and carousing on the job was the norm. Work habits were slipshod and wasteful. As was their custom, they would work for a few days and then take off with their earnings for long bouts of debauchery and gambling. They also participated in the frequent fetes, fairs, and holy days which featured cockfighting, bullbaiting, and visits to brothels. Every Monday was a Saint Monday, a holiday for merrymaking, gambling, drunkenness, gluttony, and carousing. All of this was a source of great distress for the ambitious and religious Wedgwood. Something was urgently needed to tame and discipline his recalcitrant work force.

Happily, Wedgwood's run of good fortune continued. John Wesley, the founder of the Methodist movement (a very severe brand of Protestantism) had just moved to Staffordshire. He and Wedgwood became fast friends. Wesley's inspired preaching soon attracted larger and larger congregations. He specifically railed at the working class, chastising them that their savage, drunken, licentious, and gambling ways were sinful in the eyes of God, exhorting them to give up their evil ways in favor of a pious life, and warning them that unless they uplifted themselves, they would never reach the kingdom of heaven where they could be with Christ for eternity. Over time, Wesley's influence proved remarkable. Little by little Wedgwood's workers began to give up their egregious vices to live sober and respectable lives.

At the same time, Wedgwood instituted a series of controls at the factory. For starters he wrote and had printed a book of potter's instructions. It described in fine detail the specific rules that were to be followed for each task in the production process. Such meticulous spelling out of duties and procedures was essential to making quality pottery. A little dirt, a marred piece of clay, some small error at any stage of production spoiled the entire batch of pottery. The instructions also contained detailed descriptions of proper decorum while at work. There was to be no drinking, gambling, swearing, or debauchery at Etruria.

In order that there be no mistake about the instructions and rules, Wedgwood spelled out a system of stiff fines for specific misdemeanors and disobedience. For example, a worker could lose 10 percent of his weekly wages for bringing ale into the factory, for writing obscenities on the wall, or for gambling on the premises. Habitual violators and anyone abusing an overlooker were sacked on the spot. Wedgwood also kept an eye out for any 'natural' leaders in the work force and the moment they challenged him or his potter's instructions, he got rid of them.

However, the pottery soon grew too large for Wedgwood to keep a personal watch over the work force. He solved this problem by installing a layer of overseers, clerks, and supervisors to make sure that his instructions and rules were carried out to the word. Wedgwood also designed a career structure whereby the more obedient and better workers could advance to more highly skilled jobs and subsequently join the ranks of supervisors. Some historians credit Wedgwood with inventing hierarchical supervision and establishing the traditions of the overseer class. In any event, the idea of a cadre of supervision proved so successful it was soon copied in other potteries and industries.

Wedgwood's next control was a primitive clocking-in system. Employees were issued tickets with their name on them. They put these in a box when reporting for work in the morning and when returning after lunch. The potter, instead of doing the rounds, merely recorded times of arrival on a board, checked the box, and investigated any absences. This was an early instance of management by exception.

Wedgwood followed up his instructions in person. He could be found almost any day stumping through his five sheds telling, showing, and scolding his workers. He was well known for raging over a speck of dirt, smashing substandard batches of pottery, and writing on the workbenches in chalk, "This won't do for Josiah Wedgwood." He also constantly harassed his clerks and overseers to be diligent in the extreme when watching the workers, making sure that no dirt got into the clay and that the potters followed his instructions to the letter.

Unlike most owners at the time, Wedgwood also paid close attention to his employees' quality of life at home as well as on the job. He made sure that his shopfloor conditions were clean, dust-free, and well ventilated and better than other Staffordshire manufactories. He also offered a wage scale that was the highest in the area. He went out of his way to encourage and praise individuals for hard work and good effort in addition to chastising them for shoddy output, thus maintaining a judicious balance between sticks and carrots.

And he took a keen personal interest in the spirituality and health of his employees and their families.

Wedgwood's human resources plan, as it might be called today, paid in spades. Although he demanded more from his employees, worked them for longer hours, and supervised them more closely than was usual in other potbanks, the workers believed him to be a kind and good employer. Wedgwood asked and got much more from his employees who worked a great deal more diligently and productively than did workers in competitive firms. These measures were not purely altruistic. Competitors frequently tried unsuccessfully to hire away his best employees, not least because they hoped to steal Wedgwood's secret formulae for fine pottery.

Summary

Wedgwood's superior management methods enabled him not only to survive a severe economic depression when numbers of other potters went bankrupt, but also to outstrip his competition during the boom years. Much of this success can be attributed to his innovative experiment in cost accounting. While others were making crude guesses as to their costs on the basis of intuition and ignoring relevant details in their account and waste books, Wedgwood made substantial use of his cost accounting system. It taught him about economies of scale, informed him about profit margins of his various pattern lines, guided him in wage decisions and pricing policies, and exposed dishonest employees. He considered it, along with his well-guarded formula for fine pottery, as one of his most valuable competitive weapons.

What is remarkable about the Wedgwood saga is that he put in place a sophisticated, effective, and smoothly functioning control system. In addition to his cost accounting system, this included his detailed potter's instructions; personal observation, criticism, and directives; hierarchical surveillance by a cadre of overseers and supervisors; a system of positive and negative sanctions; co-optation of the best workers into the supervisory cadres; concern for employees' quality of working life; social control in the form of Wesley's Christian preaching; and a clear strategic plan. It would not be wrong to say that Wedgwood had literally and single-handedly invented what we call today management planning and control systems.

The results proved spectacular. By 1775, Wedgwood employed over 150 workers in five specialized shops whose work was supervised by a fully trained cadre of overlookers and clerks. And by 1790, what was once a modest potbank now boasted a diligent and dedicated workforce of more than two hundred men, women, and children performing highly specialized tasks under vigilant and systematic supervision. Not surprisingly, he outstripped his competition in quality of pottery, efficiency of operations, and discipline of work force. His beloved Etruria factory became a showcase for the nation. When he died in 1795, Wedgwood's estate exceeded £500,000, a staggering fortune in those days.

EMPIRE GLASS

We move on nearly two hundred years to the early 1960s for a look at the financial control system at the Empire Glass Company (EG). EG is a national diversified container company manufacturing and marketing glass, metal, plastic, and paper containers of all sorts.[2] The company is organized by product divisions with operating responsibility delegated to the divisions, each of which is run by a divisional vice president, supported by functional line managers and several staff officers, who report to the executive vice president at headquarters.

Headquarters (HQ) consists of a small group of executives and a handful of clerks and three financial officers: the controller, the chief accountant, and the treasurer. The controller's office employs only the controller and one assistant. These HQ executives rely almost exclusively on EG's management accounting and control system to run the company. We now look in on the glass division and one of its seven glass plants where James Hunt is the plant manager.

It is early in November 1963 and Mr. Hunt sits in his office poring over the October batch of profit planning and control reports (PPCRs). These contain highly detailed financial information about almost every aspect of the plant's operations as well as comparative data for the previous month, the previous month a year ago, and the year to date. Importantly, the annual bonus for the management team at each plant is tied to attaining 90 percent of budgeted profit for the year. Sales from Hunt's plant for the year are running 3 percent below target. So it is going to be necessary to reduce costs substantially for the next six or seven weeks in order to reach or exceed the profit target for the year. Hunt is confident this can be accomplished.

The management accounting reports consist of five PPCRs which cover every aspect of a plant's operations. PPCR #1 contains thirty-nine line items summarizing the plant's financial performance including gross and net sales, fixed, discretionary, and variable costs, as well as a host of statistics such as profit/volume ratio, manufacturing efficiency indexes, return on capital, and capital turnover. Five other PPCRs provide the details behind the data in PPCR #1. The reporting system also includes PPCRs that contain the same information consolidated at the divisional level and contain financial information on almost every nook and cranny of each plant as well as of the division as a whole.

Under EG's incentive scheme plant management team members can earn a bonus of between 25 and 30 percent of their annual salaries for achieving 90 percent of the plant's budgeted profit for the year. Once this is accomplished, it is also possible to earn another bonus of up to 25 percent of salary for meeting cost reduction and efficiency targets. Further, managers can earn up to 20 percent more by exceeding the plant profit target. The bonus scheme has been in effect for several years.

When asked by the case reporter if there might be slack in some accounts which could be used to reduce budgeted expenses when results fall below target levels, Hunt responds

> No, we never put anything in the budget that is unknown or guessed at. We have to be able to back up every single figure in the budget. We have to budget our costs at standard assuming that we can operate at standard. We know we won't all the time. There will be errors and failures, but we are never allowed to budget for them (Dalton and Lawrence 1971, p. 143).

After carefully scrutinizing the October financial reports, Hunt schedules a special meeting of all line and staff management people and supervisors. The purpose of the meeting, he tells the case reporter, is to get things straightened out.

> The thing we have to do now is kick those accounts in the pants that are not making the savings they planned to make... The thing to do now is to get everyone together and excited about the possibility of doing it. We know how it can be done. Those decisions have already been made.

> It's not unattainable even though I realize we are asking an awful lot from these men [the work force].
>
> You see we are in a position now where just a few thousand dollars one way or the other can make a big difference in the amount of bonus the men get. There is some real money on the line. It can come either from a sales increase or an expense decrease, but the big chunk has to come out of an expense decrease.
>
> We never fight about the budget. It is simply a tool. All we want to know is what is going on. Then we can get to work and fix it. There are never any disagreements about the budget itself. Our purpose this afternoon is to pinpoint those areas where savings can be made, where there is a little bit of slack, and then get to work and pick up the slack (ibid., p. 153).

Hunt opens the meeting by expressing his disappointment with the October financial results and pinpointing those areas with good results and those with poor numbers. The plant accountant then calls on each individual with poor numbers to explain why. Hunt exhorts each in turn to do better with statements like "that's not good enough", or "get things straightened out now", or "do better for the rest of the year". He also makes it perfectly clear that he wants everyone to communicate to him as soon as problems come up with meeting their part of the profit budget.

The plant accountant then reviews in detail the budgeted targets for November, discussing for each account what specific savings in November and December can be expected. Hunt closes the meeting with a no-holds-barred pep talk.

> There are just a couple of things I want to say before we break up. First, we have got to stop making stupid errors in shipping. Joe, (foreman of shipping) you have absolutely got to get after those people to straighten them out. Second, I think it should be clear, fellows, that we can't break any more promises. Sales is our bread and butter. If we don't get those orders out in time we'll have no one but ourselves to blame for missing our budget. So I just hope it is clear that production control is running the show for the rest of the year.
>
> Third, the big push is on now! We sit around here expecting these problems to solve themselves, but they don't. It ought to be clear to all of you that no problem gets solved until it's spotted. Damn it, I just don't want any more dewy-eyed estimates about performance for the rest of the year. If something is going sour we want to hear about it. And there's no reason for not hearing about it. (Pounds the table, then his voice falls and a smile begins to form.) It can mean a nice penny in your pocket if you can keep up the good work. That's all I have got to say. Thank you very much.

These efforts pay off. The plant exceeds its profit budget in the year and beat its operating efficiency and cost reduction targets by a comfortable margin. The managers and supervisors, as in previous years, receive a sizable bonus.

Background

Empire Glass competes in an industry that is monopolized by a few companies. Quality across the industry is very high and price competition is virtually nonexistent (on occasion some companies have been charged by the government of collusion on prices). However, large customers, such as brewing companies, soft drink corporations, and large packing plants, play off the various glass companies if quality, delivery, or service standards are not met. Consequently, while glass firms compete on the basis of service and delivery, efficient low-cost production is essential to make profits.

EG's glass products division produces thousands of different sizes, shapes, colours, and decorations of glass products including food, soda, and beer bottles. Nearly all sales are made to order with a typical lead time of two to three weeks between delivery and order. EG's production process, originally a craft enterprise, is now almost fully automated due to its recent heavy capital investments in the most up-to-date production machinery. Quality is critical and a mistake anywhere in the production cycle (melting, filling the moulds, blowing, annealing, cooling, or coating) results in high rejection rates. Glassmaking, once the purview of highly skilled artisans, now comes under the discipline of the machine.

Most jobs in the plant require little skill. Output is machine driven and the workers, little more than machine tenders, have virtually no control over either the pace of work or the production methods. The higher skilled jobs include mould making, machine repairing, and machine changeover and setup. Employees are unionized with bargaining conducted at the national level. Wages are relatively high for industry in general. All employees below the supervisory level are paid on an hourly basis and do not participate in bonus schemes.

Planning and control at EG is elaborate, systematic, and formal. In May each year, the planning and control process starts when the glass products division vice president estimates sales, profits, and capital expenditure needs for the next year. The vice president also outlines planned capital spending requirements for the next five years. These data, along with estimates of long-term market and sales trends, are then submitted to headquarters.

Meanwhile, the HQ market research department is busy preparing detailed market predictions for the glass industry for the next year and two or three more years. These estimates are based on thorough reviews and consideration of all relevant economic factors for each product line. They submit these forecasts to the divisions for review, criticism, and adjustment.

At the same time each district sales manager prepares a forecast of sales for his or her district by estimating sales on a customer-by-customer basis. The divisional office amalgamates these estimates and then compares the total with the HQ market research department's top-down forecast. The district forecasts are then consolidated and reviewed by the divisional marketing manager but no changes are made in district sales managers' targets without their consent.

Headquarters also provides guidance at this stage if requested. But once the sales budget is approved by divisional marketing managers no one is relieved of responsibility for meeting the sales budget without the approval of division top management. At the same time, the divisional marketing managers cannot make changes to the sales budget without the explicit agreement of all parties responsible for meeting the budget.

This process is repeated at HQ and divisional levels until everyone agrees the sales budgets are sound. Only then are they approved by the divisional vice president and the HQ executive vice president and president. At this stage each level of management becomes fully responsible for its own share of budgeted sales. The sales budget, now fixed in stone, cannot be changed without the approval of HQ top management.

After the sales budget receives final approval, the divisional sales budgets are broken down and allocated to the various plants for manufacture and shipment. The plant sales budget is further broken down by month, price, volume, and end use. This information is then used by each plant to prepare their profit budgets. Standard variable cost information for each product and estimated fixed costs and overheads are used to estimate gross and net profit. Plant executives prepare budgets for each operation, cost center, and department in their plants.

Participation in preparing the plant budget is widespread. Each production section in the plant estimates the physical requirements of the sales plan (e.g., tons of raw material, labor hours, and machinery hours) necessary to produce budgeted sales. Costs are budgeted based on standards developed by industrial engineers at the plant and at the divisional headquarters. And estimated cost reductions (such as expected variances from standards due to improved methods and fixed cost improvement programs) are incorporated into the budget figures by the plant industrial engineer in conjunction with the plant manager, the assistant plant manager, and the plant supervisors. Such widespread involvement, it is felt, results in general commitment to meet targets.

Before submitting the plant budget to headquarters each plant receives an on-the-site visit during the summer by a group of HQ and divisional executives. Typically, this group includes the HQ controller and his assistant as well as members of the divisional manufacturing staff. During these visits, which last from half a day to a day, the budget is discussed in detail with the plant manager and his management team. The group also tours the plant paying particular attention to the details of the maintenance and capital replacement budgets. The purpose of these visits is to acquaint the upper echelon executive with the proposed plant budget and to guide the plant manager regarding corporate expectations concerning projected profit levels.

The divisional vice president also reviews each plant's budgets "keeping in mind the corporate headquarter's expectations". At this stage the plant managers are sometimes asked to cut their budgeted expenses. Then, in September, each plant submits its budget to headquarters for consolidation with those of other plants. When the executive vice president at headquarters and the divisional vice president are satisfied, they submit the consolidated budget to the president who accepts it as is or sends it back for modification. In December, the budget is presented to the Empire Glass board of directors for final approval.

During the year, HQ and divisional offices use the PPCRs to monitor closely each plant's performance. On the sixth working day of each month the plants wire key operating variances to headquarters. These are summarized on a variance analysis sheet and sent the next day to all key HQ executives. This sheet highlights variances in critical areas and helps headquarters take timely action. Along with the PPCR reports, plant managers also submit their most current estimates of key operating variances for the coming month and the next quarter. This exercise is believed to force plant managers to look at where their programs are going rather than run the plants using a day-by-day fire-fighting mentality.

The actual results come in on the eighth working day of each month. HQ executives then review each PPCR on a management-by-exception basis, looking only at numbers in excess of budget targets. Plant managers include written commentaries for variances, explaining where they went off base. Net sales, gross margins, price and mix changes, and manufacturing costs receive particularly close attention. HQ and divisional executives also watch fixed cost and capital expenditure items closely to keep a careful eye on whether or not the plants carry out their planned capital expenditure and maintenance programs. Plant managers must submit written explanations if these programs are not carried out or if they exceed budget allowances.

The financial controls allow the company to identify trouble spots in advance. Depending on the magnitude of the problem, headquarters might ask for daily reports from the plants. In some cases they dispatch HQ or divisional staff experts into a particular plant to help resolve the problem. Although plant managers are not required to take the experts' advice, they are expected to accept gratefully their help.

While the operating budgets are fixed in stone in December, changes can be made if early in the year sales decline and if plant managers can convince headquarters that the change is permanent. No changes are permitted, however, if sales decline unexpectedly towards the end of the year as in Hunt's situation described above. In this case, the HQ controller asks the plant managers to review their profit budgets with the plant staff looking for places where reduction and expense cuts will do the least harm including expenditures that can be eliminated or delayed until the subsequent year. When sales exceed budget, the plants are not allowed to keep the extra profits that accrue.

Sales and plant relations, while typical in terms of cooperation and conflict, also have a unique side to them. In theory, sales has the final say; but in practice it is the other way around since the plants control production, quality control, scheduling, and delivery. Nevertheless, if an important customer wants a rush order, the plant tries to accommodate them even if it upsets production schedules by disrupting runs already in process, thus increasing costs.

Importantly, both the plants and the sales organization have a common interest through the budgeting and bonus systems in seeking sales. The plants have a material interest in the customer's welfare since plant profit depends on sales as well as cost efficiency. Sales managers also earn a substantial bonus at year end if budgeted sales levels are met. Thus, the motivational force to meet the budget is substantial throughout the organization. And it is very unusual for either plant or sales management personnel not to meet the annual targets and not get their bonuses.

Several other controls operate alongside the planning and budgeting system. For all employees above the level of assistant supervisor detailed formal job descriptions exist. And industrial engineering personnel measure and set standards for all jobs up to supervisor level. As well, standards are developed by divisional engineers for each machine when the equipment is first used. These standards are subsequently used to compare performance across plants; they are also used in the preparation of costing data in the budgets.

Industrial engineers also perform job studies throughout the year aimed at cost reduction. Each month, headquarters prepares and distributes bar charts comparing and ranking divisions and plants across the company on manufacturing efficiency. And, since quality is critical, all output is inspected both electronically and visually by inspectors. These controls complement the PPCR system.

Managers throughout the company seem highly satisfied with the management planning and control system. HQ executives believe that the system's focus on planning is very valuable, especially when so many people up the line and down in the field units are involved. The field people, in turn, value the management prerogatives the control system offers them. They do not see it as a straitjacket. Rather, they feel it gives them almost total freedom to carry out the planned programs. Moreover, managers and supervisors at the plant level are well aware of, informed about, and respond in a positive fashion to the plans, budgets, standards, and targets. In recent years, the glass division had increased its share of market slightly and was one of the industry leaders in profit performance.

JOHNSON & JOHNSON[3]

Background

We move on again but this time only a couple of decades to look at the management accounting and control system at Johnson & Johnson (J&J) a global conglomerate that develops, manufactures and markets a wide variety of health care products. Products ranged from Band-Aids and disposable diapers to highly sophisticated lab equipment and surgical devices, as well as hospital management systems. Sales in the late 1980s approached $10 billion and the firm employed over 75 000 people in seventy-five countries. In recent years surveys have consistently rated J&J as one of the best managed companies on the Fortune 500 list. The reason for this lies in large part with the company's commitment to and sophisticated use of its unique management accounting and control system.

The headquarters executive group play a key role in this system. The group consists of twelve members including the chief executive officer (CEO), the president, the chief financial officer, the vice president of administration and eight executive committee members (ECMs). Each of J&J's 155 subsidiaries reports to headquarters through one of these ECMs. Remarkably, this small group of executives at headquarters in New Jersey are able to run a complex, giant empire spread around the globe.

J&J's decentralized organizational philosophy is a key feature of the management control system. Each subsidiary is treated as an autonomous, independent, integral business in its own right. The subsidiary president, in most instances a citizen of the country involved, takes full responsibility for the subsidiary's strategy and operations. Unlike most large, complex organizations, J&J does not have a strategic planning staff at headquarters nor does it have any explicit global corporate strategy outside the strategic business plans for its subsidiaries. Against this background we look in on the action at the Codman & Shurtleff (C&S) subsidiary as its executives work on the June update of the operating budget.

The June Update at C&S

It is a warm Boston spring day in May. C&S's executive team are sitting around a solid oak table in the executive boardroom poring over the sheaf of financial reports and plans in front of them. President Roy Black called the meeting to review the preliminary figures assembled by the vice president of information and control for the annual June update of the annual operating budget.

The June update involves reestimating and revising all budget estimates, right down to the lowest-level expense center, for the rest of the year. These estimates and the actual results to date are used to update the forecast for the rest of the year. The numbers indicate that C&S will be $2 million short of its profit target for the year. The June update is due at J&J's headquarters in New Jersey in a week.

The profit shortfall is due to a mixture of events that were unforeseen the previous September when the profit budget was put together. Actual sales and estimates for the rest of the year are running nicely ahead of budget due in part to the earlier than expected introduction of a new product. Actual expense and estimates for the rest of the year, however, are running over budgeted targets largely because of three unanticipated events: higher than predicted start-up costs of a recently combined manufacturing facility, an unfavorable product mix variance, and a larger than anticipated drop in US currency that pushed costs up on specialty instruments purchased from a J&J European subsidiary.

Roy Black begins the meeting by asking his vice president of marketing if he could "give something" (Anthony, Dearden, and Govindarajan 1992, pp. 485–486). The latter shakes his head and replies:

> I've been working with my people looking at price and mix. At the moment, we can't realistically get more price. Most of the mix variance for the balance of the year will be due to

increased sales of products that we are handling under the new distribution agreement. The mix for the remainder of the year may change, but with 2,700 active products in the catalogue, I don't want to move too far from our original projections. My expenses are cut right to the bone. Further cuts will mean letting staff go.

Roy Black nods and turns to his vice president of business development.

You and I should meet to review our research and development priorities. I know that Herb Stolzer (J&J's headquarters executive committee liaison member for C&S) will want to spend time reviewing the status of our programs. I think we should be sure that we have cut back to reflect our spending to date. I wouldn't be surprised if we could find another $400,000 without jeopardizing our long-term programs.

He then closes the meeting.

Well, it seems our work is cut out for us. The rest of you keep working on this. Excluding R&D, we need at least another $500,000 before we start drawing down our contingency fund. Let's meet here tomorrow at two o'clock and see where we stand.

Mr. Black along with the vice president of new development, the vice president of marketing, and the vice president of information and control meet again that evening to review all R&D projects included in the current year's budget. They review the priority and progress of all major projects searching for projects that can be deferred or terminated because of changing market conditions. Roy Black then asks the vice president of business development to meet with his staff the next morning to scrutinize the forty odd active R&D projects for potential savings to the tune of $400 000.

The next afternoon he chairs another meeting of his executive team. The vice president of product development reports that his group has come up with $300 000 of savings and are working on the other $100 000. The vice president of marketing, after some queries from Black regarding inventory levels allows that there might be savings by providing faster turnaround on the core of critical products and by putting high specialty items on a ninety-day made-to-order delivery basis. The vice president of human resources reports that an early retirement initiative will eliminate fourteen jobs in the personnel department. Roy Black then asks his executives to go back to their departments and chip 2 percent off their operating expenses.

In total, the projected savings, including the remaining $100 000 from R&D, will reduce the projected profit shortfall to $200 000. This is sufficient Roy Black believes to allow C&S to draw down on its contingency fund for the rest. Each subsidiary has a contingency expense account built into its annual operating budget. The amount depends on negotiations with the corporate office and is based on the perceived uncertainty in achieving the budget profit target. The June update is complete and will be sent to C&S's executive committee member for approval. The above decisions and actions are motivated in large part by J&J's management accounting and control system which is described next.

The Business Plan

The business plan is the overall strategic framework for the subsidiary's competitive stance. It consists of three parts: a mission statement, a detailed competitive marketing plan, and a simple financial forecast. The mission statement defines in general terms the focus for the subsidiary's products and markets. C&S's mission statement, for example, states its mandate as a primary focus on neurospinal surgery worldwide. All new R&D products as well as any business acquisitions are screened carefully to see if they fit under this umbrella.

The heart of the business plan is the marketing plan. It describes in detail how the subsidiary intends to compete in each major segment of its business. It also includes a concise summary of all major competitors' strategies along with an estimate of their sales and profits. The marketing plan then becomes the basis for each department head in the subsidiary to develop a plan for his or her area of responsibility.

The financial part of the business plan consists of only four numbers: sales volume, net revenue, net income, and a return-on-investment index. It also includes a qualitative description of how these figures are to be achieved. A unique feature of the business plan is that once the four numbers are approved they remain fixed for the next five-year period. For example, if 1989 was the first year of the five-year period, the financial forecast for the four numbers is made for the years 1996 and for 2001. These numbers then remain fixed until 1994 when a new forecast will be made for the years 2001 and 2006. The mission statement, the marketing plan, the departmental plans, and the financial plan are combined into the business plan and bound in a formal document.

The logic behind fixing the financial forecasts for a five-year period is to make sure managers do not get totally caught up in the daily routines and tactical problems of running a complex business. The business plan exercise forces them not to lose sight of long-run and strategic opportunities and helps them to maintain a dialogue between qualitative and quantitative plans. When formulating the business plan, managers are constantly made aware of previous as well as current financial forecasts. Likewise, when making financial forecasts, they are constantly aware of strategic and marketing plans. In order to further reinforce this back-and-forth process, a subsidiary cannot get capital expenditure funds for future opportunities unless they are identified in the business plan.

Company executives believe that locking-in the financial forecasts in the business plan for each five-year period is the cornerstone of J&J's management accounting and control system. Since the same two planning years are analyzed repeatedly over the five-year period by managers, many different perspectives throughout the subsidiary are brought to bear on business possibilities, problems, and issues. And managers are forced to repeatedly reconsider how the competitive environment has changed since the initial forecasts were made and what action is needed to compete effectively according to the plan. This process forces managers to continuously articulate strategic qualitative plans with financial results and targets in the annual profit budget.

The timetable for preparing the business plan is as follows. During the first half of the year, the business plan is discussed, debated, and adjusted by the subsidiary's top management group over a five-month period. Then in June, they meet with the subsidiary's executive committee member for an on-the-site, no-holds-barred, let-the-hair-down meeting. During this gruelling two-or-three-day session, the executive committee member challenges all aspects of the plan, exposes its basic assumptions, airs concerns about the particulars of the marketing strategy, questions the validity of the financial forecasts, and presses the subsidiary managers to spell out specifically how they expect to meet the numbers. The meeting ends when the executive committee member declares the plan to be sound and workable.

The subsidiary president then summarizes the business plan in a two-page report and submits it to headquarters. In September, the HQ executive committee meets to debate and approve each subsidiary's business plan, formally presented by the subsidiary president. These meetings feature frank, hard-hitting discussions and debate and adjustments to the business plan are not uncommon at this stage. Finally, when the HQ executive committee is satisfied, the business plan is approved.

Profit Budget

The approved business plan becomes the guiding framework for preparation of the annual profit budget for the coming year. Throughout each subsidiary managers down to department level prepare detailed financial estimates for their particular responsibility center. They also prepare a second-year forecast, but in less detail. In putting together the profit budget, managers keep a close eye on the financial numbers in the business plan and on the second-year forecast in the previous year's profit budget. Thus, the current year's profit budget is integrated with the thinking expressed in the business plan and in the prior year's profit budget.

Subsidiaries compile the first draft of the profit budget late in the summer. The various managers in each subsidiary are brought into the process as early as possible to complete their part. Each department prepares its own forecast, where applicable, of sales revenues, operating expenses, and capital expenditures for the coming year. Revenues and expenses as well as key balance sheet accounts are also broken out on a month-by-month basis. Production cost estimates are based on standard costs. R&D expenses pose a special problem since a subsidiary's project list is usually too long to support it in its entirety. R&D expense targets have to be determined by ranking projects in terms of the expected sales, expenses, and production capacity requirements.

All this information is consolidated and compared with the second-year forecast in the previous year's profit budget. If the first consolidation indicates a shortfall relative to that forecast, special budget presentations are held where each department manager is asked to remove any budget slack and meet the earlier targets. The second-year forecast is also useful in hindsight as an indicator of how well the managers plan and perform. This widespread participation, company officials believe, helps to gain commitment on the part of managers throughout the subsidiary to meet the targets in the profit budget.

When the subsidiary's top management group are satisfied they meet in October to discuss the profit budget with their executive committee member. The latter reviews it in detail looking for any differences in the new profit budget from the second-year forecast in the previous year's profit budget or from the financial figures in the current strategic business plan.

During this meeting, the amount of the subsidiary's contingency expense allowance to be included in the profit budget is negotiated and finalized. This allowance acts as a cushion against unpredictable events such as currency devaluations, inflation fluctuations, and supply price changes. Subsidiaries are allowed to draw down on their contingency fund when actual profits fall below budget targets and a reasonable explanation of the shortfall is forthcoming. The idea behind the contingency fund account is to mitigate any propensity by subsidiary managers to deliberately build slack into their budgets. When all agree on the profit plan in November, the executive committee member presents the subsidiary's profit budget to the HQ executive committee for final approval.

Each month starting in January, actual performance relative to the targets in the profit budget is monitored closely both at headquarters and in the subsidiary. Each week the subsidiaries submit a report to headquarters on sales performance which the designated executive committee member reviews. And the subsidiary president submits a monthly management report to headquarters highlighting key financial accounts and comparisons with the previous month and with budget targets. This report includes a written explanation of any significant variances. Clearly, all parties keep a close eye throughout the entire year on progress towards the profit target.

The profit budget is revised three times during the year. The March revision consists of an update by the subsidiary's executive committee member who presents its most recent estimate of the sales and profit for the current year to the HQ executive committee. This information is based on estimates provided by the subsidiary president. The more extensive June update was described earlier. Finally, the November revision, presented to the HQ executive committee at the same time as the profit budget for the coming year, features a close look at results for the first ten months and a revised profit budget for the last two months. The November revision also requires updates from all departments in the subsidiary, but not in the same detail as for the June revision.

Reward System

J&J's unique salary and bonus system is an integral part of the control system and merits special mention. The HQ executive committee decides the salary and bonus for each subsidiary president and its other top executives. But the subsidiary president has complete discretion in salary and bonus matters for all other employees in the subsidiary. These decisions, importantly, are not tied to any predetermined formula such as achievement of financial targets in the profit budget or the business plan. The logic is that these are bound to be inaccurate anyhow given the diversity and unpredictability of changes in the health care industry. Rather, initiatives taken to increase market share, efforts to introduce new products, and steps taken to develop long-term competitive advantages, along with energy put into increasing the current year's sales and profits, count heavily. These subjective assessments far outweigh objective criteria.

Reflections on the Control System

The case reporter asked several J&J executives for their opinion of the company's management accounting and control system. They unanimously expressed great pride in it and saw it as one of the corporation's distinctive competences. They believed it to be a big factor in getting the competitive edge over rival firms. Roy Black, who had been with J&J for twenty-five years in various line and staff positions at headquarters and in subsidiaries, responded that the control system's big plus is that it kept him and his managers aware throughout the entire year of the long-term strategic plan as well as the importance of achieving the annual profit. He put it this way:

> We should always be thinking about such issues, but it is tough when you are constantly fighting fires. The Johnson & Johnson system forces us to stop and really look at where we have been and where we are going.
>
> We know where the problems are. We face them every day. But these meetings force us to think about how we should respond and to look at both the upside and downside of changes in the business. They really get our creative juices flowing.

Some of our managers complain. They say that we are planning and budgeting all the time and that every little change means that they have to go back and rebudget the year and the second-year forecasts. There is also some concern that the financial focus may make us less innovative. But we try to manage this business for the long term. We avoid at all costs actions that will hurt us long term. I believe that Herb Stolzer is in complete agreement on that issue.

It is important to understand what decentralized management is all about. It is unequivocal accountability for what you do. And the Johnson & Johnson system provides that very well (Anthony, Dearden, and Govindarajan 1992, p. 495).

CONCLUSION

This chapter described the management accounting and control systems in three organizations. The aim was to clearly identify the subject matter of this book. The cases also illustrate the evolution of these systems over the past two centuries. At the dawn of the industrial revolution, Josiah Wedgwood single-handedly invented his own cost accounting system along with a host of other valuable controls. He used it to sharpen his intuition when making strategic pricing, production, and hiring decisions and to supplement his many other control devices. Two hundred years later at Empire Glass, we saw a sophisticated and highly detailed management accounting system that covered every nook and cranny in the organization. In this management accountant's paradise a handful of executives at headquarters and in the divisional offices simply ran the company by the numbers.

A couple of decades later at Johnson & Johnson, we witnessed another metamorphosis in management accounting and control. An elaborate strategic planning system was grafted on to and intertwined with the traditional profit budget system. Managers at all levels in the subsidiaries were constantly interacting with and adjusting strategic business plans as well as profit budgets, while a dozen or so executives at headquarters kept up a close and vigilant surveillance of the subsidiaries plans and results.

The job of top management at J&J differed markedly from the work of Josiah Wedgwood. Wedgwood formulated the business strategy, researched and developed the product, designed and organized the factory, and kept a close personal watch over the workers. At J&J, in stark contrast, the top executive's job is almost purely control. J&J's elegant and sophisticated management accounting and control systems allows them to rule a vast health care conglomerate operating around the globe with hundreds of thousands of products and employees. For top executives, control at a distance has replaced the actual doing of business.

Enterprises like J&J have emerged today as the dominant form of organization. These goliaths have command over resources in excess of those of all but a dozen nation states. Such an eventuality would have been unthinkable in Josiah Wedgwood's day. There are, of course, many ways to account for the miracle of the meganational conglomerate form of economic power. Some would point to their vast storehouses of proprietary scientific and technical knowledge. Others might attribute a good deal of weight to their marketing skills. Still others would point to random events and lucky breaks. Be that as it may, it does not seem too great an exaggeration to say that without concomitant advances in control systems, such enterprises could no more exist today than could we send a spaceship to the planet Saturn without the miraculous developments in space technology. Today's sophisticated and elegant management accounting and control systems play a huge role in this scheme of things.

While each of these three case histories represents an instance of what we would consider sound management control systems, there are important differences in both their design and implementation. In the next three chapters we look at theoretical frameworks which go some way towards explaining such variations.

NOTES

1. The source material for this synopsis is reported in McKendrick (1961, 1970) and Langton (1984). The Wedgwood case has been the focus of more than 50 000 articles and papers.
2. The case study is reported in Dalton and Lawrence (1971) and Anthony, Dearden, and Govindarajan (1992).
3. The source material for this case history is reported in Anthony, Dearden, and Bedford (1989) and in Anthony, Dearden, and Govindarajan (1992).

3
Universal Theories

The previous chapter described in some detail three case situations focusing on the management accounting and control systems in these organizations. The reason for doing so was to identify clearly the subject matter of this book and to demonstrate the vital role these systems can play in organizations as well as to illustrate their richness and complexity. Yet no useful body of knowledge can rest on the assumption that each case is unique and can be understood primarily on the basis of experience and intuition. In consequence, most fields of knowledge about organizations started out with the alternative premise that there are essential and universal elements and general principles that are applicable to all complex systems. Management accounting is no exception. This chapter looks at three major universal paradigms.

The first of these, *agency theory*, depicts the organization as a web of two-person contracts between owner and employees where the owner (or principal) uses accounting information to control the employee-manager who acts as the owner's agent. Agency theory, situated solidly in the structural functionalist paradigm, treats accounting systems and agents as commodities in order to bring the analytical techniques of neoclassical economics to bear on problems of control such as employee sloth and deceit. In sharp contrast, the second framework, the *informational nerve center* picture, sees the manager as immersed in collecting, storing and disseminating formal and particularly informal information during a frantic and relentless round of brief encounters. The nerve centre framework, following the subjective interpretivist paradigm, relies on sociopsychological role theory to make sense of the manager's chaotic life. Finally, the *labor process* view, following the radical structuralist tradition, situates the manager in a contradictory position as both the victim and user of management accounting and control systems. It applies notions of deskilling and cheapening of wage labor to raise important issues about accounting and control systems. While these three paradigms are competing ways of thinking about the role of management accounting and control systems, each has something vital to say about them.

AGENCY THEORY

The agency theory view of management accounting usually assumes a world of two-person explicit or implicit contracts between owner and employee in which both parties behave in a rational utilitarian fashion motivated solely by self interest. It depicts the agency relationship as a contract under which the owner (or principal) delegates decision-making authority to the manager (agent) who then performs services on behalf of the owner. Given that agents are utility maximizers, it seems the agent will not always take actions that are in the principal's best interests. The owner, however, can limit such aberrant behavior by incurring auditing, accounting, and monitoring costs and by establishing, also at a cost, an appropriate incentive scheme (Jensen and Meckling 1976).

Most proponents of agency theory do not really believe the organizational world consists mainly of bilateral contractual relationships. Rather, they hold that this portrayal can be used to advantage to shed light on and develop insights into the way accounting and information systems can be used by owners to control employees and managers. Importantly, by positing a contractual world, agency theorists can proceed by relying solely on neoclassical economic theory and the techniques of information economics. Agency theory is built around the key ideas of self interest, adverse selection, moral hazard, signalling, incentives, information asymmetry, and most pervasively, the contract.

Self-Interest

Agency theory has its roots planted firmly in neoclassical economics, which has always made self-interest its basic platform. The self-interest theme stems from the influential writings of Hobbes, Smith, and Spencer, three giants of social philosophy. Thomas Hobbes (1588–1679), living in an era of tumultuous political, economic, and spiritual upheaval, pictured human beings as wholly selfish and, when living in the state of nature, leading a solitary, poor, nasty, brutish, and short life. The individual's happiness came from getting what one desires. One's desires were deemed to be goods, the converse of bads, what one did not want. Moreover, Hobbes argued, people are fundamentally equal in terms of strength, ability, and hopes for attaining these goods but since everyone is motivated by and is equal in the attributes needed for satisfying self-interest, everyone's hopes and happiness are constantly threatened. The individual's life, it seemed, was permanent and constant misery.

In contemplating this miserable state of affairs, Hobbes hit on a way out. The key, he surmised, lay in enlightened self-interest. The individual must create some sovereign, central power (king, parliament, state, whatever) and agree to be governed by it. The sovereign's job is to remove the threats to the individual inherent in the state of nature. Importantly, the sovereign acquires the right of power and command from "man's" consent, not from some divine being. And from command comes law, justice, morality, and civilization. In return, the enlightened but still self-interested individual must not only consent to the rules of the sovereign, but also take on the duty of obedience. Command, consent, and duty emerged as the essential ingredients of Hobbes' social contract.

A century later, the self-interest theme received a large boost from Adam Smith (1723–1790) who, in his monumental treatise, *Inquiry into the Nature and Causes of the Wealth of Nations*, worked out an intriguing and elegant theory of the political economy of self-interested effort guided by the invisible hand of the marketplace. Smith was greatly influenced by the French physiocratic school of economics, who believed that wealth was not merely possessions of land, gold, silver, livestock, castles, and so on, but also arose from labor on land. The physiocrats also held that wealth must be circulated through the nation (just as blood flows through a healthy body) and not be hoarded in counting houses and treasuries. Smith held, against the physiocrats, that labor created wealth not just on the land, but wherever it performed, particularly in the manufactory. Moreover, he concluded, it worked in a miraculous fashion when specialized into discrete tasks. Specialized labor, not nature, which was already in abundance, was the source of a nation's wealth.

Labor, however, Smith held, must be free to circulate so as to perform wherever the individual might detect some personal advantage in meeting the needs of others. There was no need of any central direction for this circulation, because the laws of the market (not the laws of nature or a sovereign) automatically work things out. Self-interest directs labor towards those needs for which others willingly pay. For example, the factory owner, driven into action by other people's self-love, makes pins to take advantage of his specialized abilities to make a profit, not out of benevolence towards humankind. But if the owner is able to command an exorbitant price and impressive profits, capital will circulate as others will rush in, also make pins, charge less, and take away business. Similarly, if the owner pays workers too little they will circulate elsewhere to earn a better wage. In both instances, it is competition which produces the unexpected result, social harmony.

Competition for labor, moreover, not only regulates its price, but also directs it towards the quantity of goods society desires. If society wants more zippers and fewer pins than currently produced, customers will scramble for zippers while the pin business will be overstocked. As the demand for zippers rises, zipper prices will increase as will the owner's profits. Conversely, the price for pins will fall and profits disappear for the pin maker who then puts the workers out on the street. These workers, however, are snapped up by zipper makers whose business is booming. Quantities of pins on the market fall while quantities of zippers rise, just what society wants. Self-interest creates competition, which keeps self-interest in check. But best of all, Smith emphasized, these laws of the market have no need of any central regulator since they work of their own accord. The market acts as its own chaperon.

Smith also reasoned that capital as well as labor must circulate freely. The profits, which accrued to the shrewd and industrious factory owner, should be pumped back into the economy and put to use in more machinery. This in turn would call for more labor, so the price of labor would rise until higher wages for the working class led to better conditions for their children. More babies and children would live and a greater proportion would enter the work force causing wages to fall again. The owner's profits, conversely, would rise and fall with the supply of wage labor. Thus, the system would automatically regulate the supply and demand for capital. Self-interested specialization of labor and the accumulation of capital would move society inexorably towards a Valhalla of material wealth and happiness.

A century later, self-interest received another boost from Herbert Spencer (1820–1903), a much acclaimed philosopher during his lifetime but hardly read today. Spencer linked economic theory to Darwin's revelations regarding the law of survival of the fittest in the natural world. Spencer argued that the individual was uniform, discrete, and ultimate and that society existed for the benefit of its members rather than, as tradition had held, its members existing for the benefit of society. The endowment of the individual with a self-regulating mechanism ensured that when each pursued his or her self-interest and private wants, the result would be the greatest possible satisfaction of everyone's wants. Rational, self-interested utilitarianism, unfettered in any way by state or sovereign, ensured the survival of society and propelled humankind along its evolutionary path towards abundance, peace, and happiness for all. Darwin's laws of nature not only prevailed over beast, bird, fish, insect, and fauna, but also over humanity.

This self-interest theme remains today as the cornerstone of neoclassical economics. So it is not surprising people see it as the foundation of the agency theory perspective of the manager. In the first instance, the manager consents to carry out certain duties as the owner's agent. Then, the owner and the manager (explicitly or implicitly) enter into a social contract whereby the duties and obligations are worked out during a bargaining process. Such arrangements provide ample opportunity, as we shall see next, for the manager to pursue self-interest with guile given the problems of adverse selection, moral hazard signalling, incentive schemes, and asymmetric information.

Adverse Selection

The adverse selection, or hidden information, problem arises when an owner puts out a contract to the market for managers. The owner sets a specified price for designated effort and output based on a probability distribution of the payoff associated with his or her prior experience with managers. While the managers are privy to private information regarding their own abilities to perform the contract, the owners have never observed the managers in action so they cannot ascertain their quality. Thus, a situation of asymmetric information arises in the market for managers.

In these circumstances, an opportunity arises for the less able managers to pursue self-interest through lack of candor and honesty as the owner is unable, at low cost, to distinguish between the managers who respond to the bid. Moreover, managers whose ability to command a higher price on the market will not regard the offer as sufficient. Nor can the latter establish that they are indeed superior as well as honest in representing their abilities and therefore entitled to a higher price, since the inferior managers, acting opportunistically, can make the same representations by withholding private information and lying about their abilities. As a result, the superior managers withdraw from the bidding. In consequence, the owner ends up paying a premium for the wrong managers thus incurring the adverse selection cost. At the end of the day the low quality managers, the lemons, crowd out the high quality managers, the plums.

In order to circumvent this situation, the owner is obliged to consider several alternatives, each of which comes with an impending cost. For example, the owner might incur the cost of obtaining detailed information regarding the managers who submit bids. Alternatively, the owner might offer a premium price over and above the going market price. Or, the owner might purchase insurance against the risk of contracting with inferior managers. When the problem is framed this way, agency theory calls for the owner to calculate the costs and benefits of the various alternatives and, using the logic of information economics and Bayesian statistical techniques, work out a Pareto optimal solution for both parties.

Moral Hazard

The adverse selection problem concerns situations where the owners, on one side of the market for managers, before settling on a contract, cannot observe at low cost the quality of managers on the other side of the market. Along somewhat similar lines, the moral hazard (or hidden action) problem refers to situations where owners cannot observe the actions and efforts of the managers *after* they have been hired and are now under contract. The term moral hazard is borrowed from the insurance industry and describing it in that setting can be helpful.

In the insurance market for bicycle theft, for example, the probability of theft may be affected by the effort and actions bicycle owners take. For the sake of simplicity, assume all owners live in the same neighborhood and that all owners lock their bicycles, so the insurance company is not bothered by the adverse selection problem. A bicycle is more likely to be stolen from those owners who do not lock their bicycle or use only a flimsy lock, than for those who carefully do so. But owners who take out full insurance may change their locking behavior, which affects the probability of theft.

The insurance company, recognizing this, takes into account the incentives owners have to take proper care of their bicycles. At one extreme, if no insurance is available, owners have the maximum incentive. While at the other extreme, where full insurance is available at low cost, they have the minimum incentive to take care. In the former case, owners bear the full cost of their actions and will invest in caretaking until the marginal cost of more care just exceeds its marginal benefit. In the latter case, owners have no pecuniary incentive whatsoever to lock up their bicycles securely. So it seems that too much insurance leads to inadequate care, while too little insurance leaves owners bearing most of the risk. This lack of incentive to take care is moral hazard.

Insurance companies, however, are well aware of these risks and the owner's trade-off decision. When the company can observe the amount of care taken by the insured, they base their rates on these observations. But when they cannot observe the relevant actions, the insurers will not offer complete coverage but instead include a deductible amount the insured has to pay on any claim. This way the insured always has *some* incentive to take care. Moreover, even if the insurance company could verify the amount of care taken *after* the fact of theft, it still would not offer complete insurance when it cannot observe the level of care taken *before* theft for if they did, owners would rationally choose to take less care. Overcoming the moral hazard problem leaves some of the risk in the hands of the insured, at the cost of not maximizing market opportunities, thus paradoxically tampering with the law of market equilibrium.

The moral hazard problem in the firm arises in a similar fashion. The owner and the manager enter into a contractual arrangement which calls for a certain amount of input and effort on behalf of the manager. Yet, as in the bicycle example, the owner cannot directly observe the input and effort exerted. In consequence, the owner must rely on some observable output measure (profit, cost of production, sales or billings, etc.) as a basis for administering the contract. This arrangement, however, can result in a reduction of the manager's incentive to take care and to supply maximum input and effort. In terms of agency theory, this is the incentive to shirk.

The result is a loss of welfare for both owner and manager. On the one side, the owner obviously suffers because the manager shirks on inputs and effort. On the other side, the manager suffers since the owner (as with the insurance company) will arrange the contract so the manager shares some of the risk for the output. Moral hazard involves a cost for both parties. But agency theory does not rest there; the costs of adverse selection and moral hazard can be minimized through signalling.

Signalling

We saw how asymmetric information caused problems in the market for managers and resulted in adverse selection costs. So the superior managers (assuming they must bear some of the agency costs) have an incentive to try to convey to owners the fact that they are honest and superior. They would like to take actions that signal their abilities. Two such signals could be their level of education and their work experience. Let us use education, say an MBA degree, for illustration.

Superior managers might decide to acquire an MBA "sheepskin" as a signal of their competence, but they cannot do so costlessly. Attending university involves out-of-pocket costs, plus the costs of study efforts, plus the opportunity-lost costs of temporarily withdrawing from the market. Now it may or may not be the case that the MBA education increases the manager's productivity. Either way, however, there is a sheepskin cost to the manager as well as to owner and, importantly, to society at large, especially when university education is government subsidized.

Agency theory may be helpful at this point. By putting the various opportunity costs to all parties (inferior managers, superior managers, owners, and society) into cost–payoff equilibrium models and making assumptions regarding the various parties' utilities and risks, the agency theorist can indicate under which conditions signalling leads to efficiencies (or inefficiencies) for the various parties. While it is beyond our scope here to go into details, agency theory modelling can show when equilibrium with signalling is preferred to equilibrium without it. But each situation must be analyzed on its own merits and assumptions. Signalling can make things better, but it can also make things worse.

Incentive Schemes

Incentive schemes also play an important role in agency theory. Given adverse selection, moral hazard, and signalling, how can I get the managers to work hard on my behalf? That is the central problem for owners. One obvious answer is to give the manager a lump-sum fee independent of output. But then the manager might have little incentive for hard work. A better solution might be to link the payment or some part of it in some way to the output produced. In this case the owner will try to determine exactly how sensitive the payment should be to the output produced. Economists call it the efficient incentive scheme.

An efficient incentive scheme is one which ensures that the utility the manager gets from the job is at least as great as the utility available elsewhere in the market for managers. This is called the participation constraint, where the utility the manager gets from the job must be at least as great as the utility available elsewhere. Second, the owner wants to induce the manager to choose an effort level that yields the owner the greatest surplus, given that the manager is willing to work for the owner. This is the unconstrained maximization problem and is solved by choosing the effort level to make the marginal product equal the marginal cost. Finally, the owner has to determine how much to pay the manager to achieve that effort. This is the incentive compatibility constraint, which simply states that the utility to the manager from choosing the determined effort level must be greater than the utility of any other choice of effort.

A few examples may help to clarify these incentive issues. For starters, the owner could simply rent the business to the manager for a specified price so that the manager gets all the profit produced after paying the rent. This, however, could result in too much risk for the manager. Alternatively, the owner could pay the manager a constant salary per unit of effort (where the salary rate is equal to the marginal profit of the manager at the optimal effort level) plus an amount which makes the manager indifferent between managing elsewhere and managing for the owner. This would satisfy the participant constraint. Another arrangement could be to fix the contract so that the manager, who makes the effort decision, is the residual claimant to the profit. A scheme where both parties get a fixed percentage of the profit does not lead to an efficient incentive scheme. (Either way, the owner wants a scheme whereby the manager shares in the uncertain output in a way that he or she also shares the risk.) But the effort exerted by the manager on behalf of the owner is also conditioned by private, or asymmetric, information.

Asymmetric Information

Asymmetric information is the most fundamental concept in agency theory since it gives rise to all the other problems: adverse selection, moral hazard, signalling, and incentives. Asymmetrical information refers to important information to which the manager (but not the owner) is privy. A special case of asymmetrical information is impacted information, information regarding deep knowledge about one or more of the factors of production. It is very difficult and costly for the party without impacted information to achieve a state of information parity.[1] According to agency theory, managers will opportunistically exploit asymmetrical and impacted information to its fullest before, during, and after exerting effort. Driven by self-interest, they can use their private information to shirk (shrink selfishly) from responsibility and duty; to consume excessive perquisites to which they are not entitled (steal and cheat); to manipulate signals (lie or distort); and to hide their actions, private information, and information-impacted special knowledge from the owner (sneakily hoard essential knowledge). There is no place for either altruistic or naive behavior in this scheme.

Normally it is impossible for the owner to perfectly observe the effort of the manager. At best the owners may observe some signal of effort such as output. Yet even here, output also depends, in addition to the manager's effort, on factors outside the latter's control: competitor's moves, economic upturns, changes in national monetary policies, even the weather. This kind of "noise" means that a payment from the owner to the manager based on output alone will not be equivalent to a payment based on effort alone. Effort and output are always less than perfectly correlated.

The owner again faces the information asymmetry problem. The manager can select the effort level but the owner cannot observe it perfectly or costlessly. So the owner has to guess the effort from the observable output then design the optimal incentive scheme to reflect this inference problem. This involves calculating an incentive scheme that shares the risks and provides the appropriate incentives. Framing the owner–manager relationship this way, the agency theorist can apply the techniques of Bayesian decision theory, linear and dynamic programming, and equilibrium theory to various situations and calculate an arrangement where neither party can improve his or her welfare at the expense of the other. This Pareto optimal solution becomes the basis for the contract between the owner and the manager. Management accounting and control systems provide crucial information to owners for fixing, instituting, and monitoring the contract.

The Contract

The idea of the contract as the primary unit of analysis can be traced back over the years to other similar concepts. Hobbes, as we saw above, developed his idea of the social contract—consent to let some sovereign rule and duty of the individual to obey—as the fundamental bonding agent in society. A century later, Adam Smith replaced consent and duty with market exchange as the basic unit of social interaction arguing that on-the-spot exchanges were all society needed to regulate social relations. But by the end of the nineteenth century, however, industrial societies bore little resemblance to either Hobbes' rule by a sovereign or Smith's world of atomistic exchanges.

Instead, the social landscape featured large hierarchical organizations and institutions. Giant corporations, large-scale government, denominational specific churches, powerful unions, public utilities, military establishments, educational institutions, and professional associations dominated the social landscape. These hierarchies, which operated as collective action centers of economic power, exercised considerable influence over the coordination of production and the distribution of wealth and the life of the individual. This new institution-dominated world bore little resemblance to the suppositions of prevailing neoclassical economic theory. Economists scrambled back to their drawing boards to put into place a revisionist version of market economics called transaction cost economics.

Transactions, the revisionists argued, involve more than just on-the-spot exchanges of material wealth. They also include long (even indefinite) time commitments, general (even vague) obligations, and promises of forbearance (agreement for nonaction or nonexchange). Moreover, transactions take place within and between hierarchies as well as in the marketplace. And many transactions involve the exchange of legal, moral and power conditions such as the right of a superior to command a subordinate. The transaction, it seemed, is a more fundamental unit of social interaction than is the market exchange.

This realization proved to be a milestone. It meant that economists could explain both the internal workings of hierarchies, previously only a black box in economic theory, *and* the functioning of the market using the same conceptual apparatus—transaction cost economics. Transactions, they surmised, incur costs, over and above the "natural" price for a commodity, costs which the parties to the transaction must bear, such as the costs of negotiating, monitoring, administrating, insuring, and even litigating in the case of nonperformance. And, since costs are a key phenomenon in economic theory, they can be analyzed using the economist's bag of tools. There was no need to muddy the theoretical waters with theories from sociology, social philosophy, psychology, or organizational theory.

These developments later on gave rise to what we know today as agency theory. For agency theorists, the contractual arrangement for a transaction in a market for the sale/purchase of a commodity is exactly the same as a contractual arrangement in a firm (hierarchy) whereby one person (the principal) engages another person (the agent) to perform the service of making decisions on the principal's behalf. Importantly, agency theorists also transform the hierarchy, where the owners (superiors) have authoritarian directive and disciplinary legal and moral power over agents (subordinates), into a peer or team relationship where all the players are on an equal footing and one team member "is the *centralized contractual agent in a team productive process*" (Alchian and Demsetz 1972).

Limitations

While agency theory offers an elegant, even alluring, way to think about management accounting and control systems and to pinpoint some worrisome issues, it is not without its limitations. One concern arises from its concentration on problems encountered by the owner when the manager relies on asymmetrical information to cheat and shirk. Yet it is just as reasonable to assume that the owner would also have private access to crucial information, such as the condition of the firm's production technology and its market competition, and would use it to advantage in negotiating an enforceable contract, suboptional for the manager. Asymmetrical information is not a one-way street. That is to say, in order for agency theory to work, the markets for both managers and owners must feature pure competition where typically there are many competitors in these markets, each of whom have negligible effects on prices. In today's world of large oligopolistic organizations and high unemployment rates, this seems a heroic assumption.[2]

A related but more problematical concern with agency theory centers on its treatment of the owner–manager relations of power. While transaction cost economists gave full recognition to power relations in the hierarchy whereby fiat and command play a crucial role in coordinating production, agency theory assumes power relations between owners, managers, and employees that are far from asymmetrical. Since employees can order the owner to pay them money or quit, thus terminating the contract as easily as can the employer, the argument goes "...long term contracts are not an essential attribute of the firm. Nor are 'authoritarian', 'dictatorial' or 'fiat' attributes relevant to the conception of the firm or its efficiency" (Alchian and Demsetz 1972, p. 783). So the owner, instead of being the boss, all of a sudden becomes merely the team captain; simply the team member who acts as its central contractual agent and thus entitled to be the residual claimant to the team's net output. Agency theory renders opaque the asymmetric power situation in the hierarchy.

Moreover, the owner-now-captain "...has no power of fiat, no authority, no disciplinary action differently from ordinary contracting between two people" (ibid., p. 783). One gets designated captain because of one's special abilities to monitor other team members for shirking but will not shirk because he or she is the residual claimant. These supplementary moves are absolutely necessary to make agency theory work and they put owner and employee on an equal power footing. Issues of domination and power are central to most social and organization theories. But agency theory, having conveniently set them aside, can explain the workings of the firm on the basis of markets for managers and information.

Nevertheless, agency theory offers insights into some of the tough issues and difficult problems involved in the design of such systems. So it is not surprising that a great amount of intellectual effort by accounting academics has been exerted in working out the details.[3] Yet, this rarefied air of economic reductionism seems a long distance from the realities of managers' working lives. Next we turn to the closeup observations of real managers in real organizations.

THE MANAGER AS NERVE CENTER

Our second paradigm, the manager as nerve center, also foregrounds the role of information as a central theme. But instead of portraying a disturbing picture of sloth, deceit, and mendaciousness motivated by self-interest, the picture is almost completely opposite. In sharp contrast with the agency theory portrait, or the labor process view of relentless deskilling and degrading at lower work levels, day after day the dedicated manager is seen to put in exceedingly long and hard hours, immersed in an unrelenting stream of unconnected rounds of activities. Managers use intuition and a sixth sense, not rational calculations, to hurriedly make on-the-spot decisions, and spend most of the time gathering, storing, moving, sharing, and processing huge quantities of informal and formal information.

This view emerged unexpectedly a couple of decades ago as a result of a now classical field study in which the researchers followed managers around observing, recording, and asking them what they actually did (Mintzberg 1972, 1975). The big surprise proved to be that the picture to emerge did not conform at all to the conventional managerial literature, which portrayed the work of managers as neatly working out a blueprint for the organization, making sure it is properly staffed, then using budgets and reports as feedback on progress to clearly defined ends. Instead of methodically planning organizational objectives and systematically achieving them, the managers lived in what appeared to be chaos.

So, in stark contrast to some orderly, highly controlled world, the managers encountered about 250 separate incidents per week, which on average lasted less than ten minutes. They also kept dozens of longer-term projects orbiting simultaneously, periodically checking progress, juggling them a little, then relaunching them; and, importantly, they seemed to abhor reflective tasks such as planning. The managers saw their working life as disorder, brevity, and discontinuity instead of orderly progress towards predetermined contractual goals or rational calculation of how to shirk and cheat the owner. Yet the managers reportedly thrived on this chaotic existence.

Managerial Roles

In order to make some sense out of this apparent chaos the researchers pigeonholed each incident into one of three major managerial roles: decisional, interpersonal, and informational. This proved to be an insightful step. Before long, a clearer but newer picture of managerial work emerged. Decision roles took considerably less time than either of the other two. This may come as a surprise to those who teach accounting stressing the decision-making role. This is not to say that decisional roles (entrepreneurship, disturbance handling, resource allocation, and negotiation) are not important; they are critical. Managers have the authority and responsibility to make decisions; indeed they are required to make them. Yet, surprisingly, managers in the sample spent only a fraction of their time on them.

Interpersonal roles took more time. In fact, figurehead chores, such as greeting visitors, taking important customers to lunch, attending weddings, and presenting gold watches to retiring employees, consumed about 12 percent of the working day. The formal authority vested in them called for the managers to spend a great deal of time in personal contact with other people, including clients, suppliers, peers, media, government officials, union representatives, professional association contacts, and trade association officers. Managers also spent a lot of time with subordinates, hiring, coaching, encouraging, reprimanding, and even firing them. Though they do not involve decision making, these interpersonal roles are important; they can be performed only by the manager.

Informational roles, surprisingly, took more time than either decision-making or interpersonal roles. Rather than behaving like wise owls in a tree, the managers acted more like trout in a mountain stream, poking briefly here, flitting there, investigating this, nibbling that, turning suddenly and darting into a new pool to repeat the process. Information seemed to be their basic source of energy. A recent field study undertaken to learn more about managers' use of management accounting systems found pretty much the same story (Bruns and McKinnon 1993). The managers, working in a variety of large manufacturing organizations and in a variety of jobs in the US and Canada, revealed that they hungered for better information to support their work and that informal sources dominated formal ones. As in the previous study, the managers were prone to develop their own personal information networks and systems. The study concluded

> Managers are hungry for information to support their work. They seek information from every source available to them. Informal sources of information—face-to-face meetings, observation,

telephone calls, and informal reports—dominate other sources of information for day-to-day needs and remain important for longer term needs. Unit data is the metric in which day-to-day management takes place, and financial information increases in importance and use as the management horizon lengthens (ibid., p. 94).

The realities of the manager's working life, it seems, have not changed much in two decades.

Managerial Information Processing

Managers apparently spend the majority of their time processing information, and the manner in which they gather, store, move, and share it should be of vital interest to accounting and information systems managers. Managers receive information from external and internal sources, and they give it out in their disseminator, spokesman, and strategy-maker roles. Three aspects of this information processing activity stand out.

For one thing, each day managers collect information from a wide variety of sources. They receive mounds of mail. They are flooded with reports. They get briefings from their superiors and subordinates. They go on observational tours. They scan professional journals. They talk on the telephone. They grill peers and subordinates. They even plant private sources in strategic places. In fact, they spend most of their time planting, cultivating, and harvesting information from private networks.

There are good reasons for this. To be effective, managers must be the nerve centers of their organizational units. In order to perform this role they must know, not necessarily everything, but at least more than their subordinates. This is true in formal organizations and informal organizations, such as street-corner gangs. The leader is the person at the center of the information flow, not necessarily the best technical expert or the toughest street fighter, but the one who perpetually monitors the environment for information and disseminates it within the organization. Subordinates must be informed, peers must be appraised, and bosses must be briefed. Managers also speak for the unit by passing information to outsiders as required. For leaders, then, information is their most valuable asset. It is a precious commodity; and managing their private information network is their most important skill. It must be carefully husbanded, skillfully exchanged, and judiciously passed around. Information is the manager's major lever to survival.

Another outstanding feature of the way managers process information is their predilection for soft, verbal, detailed, and current information. They reach out eagerly for every scrap, including speculation, hearsay, rumour, and gossip. They also show a strong preference for verbal communication and seem to thrive on telephones, meetings, encounters in the hall or washroom, and coffee-break chats. In fact, the managers in the survey spent nearly 80 percent of their working time in verbal activities; and they like their information fresh, hot, and spicy.

But managers also find information in management accounting systems, such as the annual operating budget, valuable for enacting managerial roles including decisional and interpersonal roles in addition to informational ones. A recent study found that one group of managers seemed to use budgets strategically as an aid in monitoring the environment to keep on top of trends and events and for allocating resources to areas of opportunity and need within the organization (Macintosh and Williams 1992). The study also found that high-performing departments are headed up by managers who are more active in performing managerial roles and are more involved in, and responsive to, information in budgeting systems than are their counterparts in the low-performing units.

The final striking feature of the managerial information processing is the way they use information to make decisions. Managers are constantly gathering odds and ends of data from their network. These bits and pieces are tucked away in the brain until one new piece suddenly acts as a trigger for all of it to come together, like a bolt from the blue, and form into an important message. This is not to say that managers are unthinking bundles of nerves reacting synaptically to each piece of information. Instead, it seems as if they have plans and models tucked away inside their heads to be updated as new information appears. Nevertheless, when this ah-hah strikes, the managers react as if charged by an electric current. They think nothing of interrupting meetings, rescheduling workdays, cancelling long-standing engagements, in order to follow the new insights.

So managers see their work as information processing. It is not aloofly orchestrating a master plan. It is not cunningly shirking while pursuing self-interest within the constraints of a contractual arrangement with the owner. And it is not deliberately deskilling and downgrading the labor force as the subjectified pawn of capitalist interest. Rather, it is frantically monitoring, storing, and disseminating information, especially the soft and current variety, then using it to make intuitive, off-the-cuff decisions. This style, it should be underlined, suits the nature of managerial work. It also matches managers' thinking processes. Just as war is too important to leave to the generals, information processing is too important to leave to the accountants. In any event, this picture of managers derived by actually observing what managers do, seems quite remote from the ethereal realm of economic theory.

THE LABOR PROCESS PARADIGM

A third universal view of management accounting and control systems emerged as part of the labor process critique of the treatment of workers in capitalist enterprises. This perspective focuses on the "sad, horrible, heart-breaking way the vast majority of my fellow countrymen and women as well as their counterparts around the world have been obligated to spend their working lives."[4] From the labor process view, management accounting systems play a crucial role in the institutionalized subordination and control of wage labor by capitalist interests.

The overriding concern of labor process theorists centers on the historical fact that the control of work has passed from the hands of the individual worker during precapitalist times to the capitalist owner class during early capitalism and subsequently into the hands of the managerial class working on behalf of owners. Since the labor process perspective draws heavily on the works of nineteenth century political economists such as Engels (1820–1895) and Marx (1818–1883) a brief review of some of their key concepts—commodification, alienation, surplus value, and class contradiction—seems essential in order to realize their pertinence today.[5]

Nineteenth Century Capitalism

Engels and Marx were first and foremost ardent and dedicated humanists. Both had witnessed firsthand in the mid-1800s how Adam Smith's wonderful world of the division of labor, the invisible hand of the market, and the newly released self-interested capacity of the factory, mine, and farm owners had transformed the social landscape.[6] But instead of a Valhalla, the result for the working class was a horrific life of overcrowding in filthy slums, abject poverty, child slave labor, sexual exploitation, dirt, and drunkenness. This savage indictment of the greed and hypocrisy of the bourgeoisie went against the humanistic aim of the Enlightenment project for the individual to attain dignity and unity with self, humankind, and nature.[7]

Commodification

Commodification, a central concept in the labor process perspective, refers to an object or thing outside the human being which satisfies some material want. Examples include a bushel of wheat, a tonne of coal, a side of beef, or a suit of clothes. These commodities obtain their value simply from what they fetch in a market transaction (the exchange value) rather than from their useful qualities (the use value) or from the amount of labor necessary to produce them (the labor value). The division of work, Engels and Marx observed, had turned working-class men, women, and children into commodities.

Commodification resulted from the specialization of labor within the production process. Specialization produced "man as a commodity, the human commodity, man in the form of a commodity...as a mentally and physically dehumanized being" (Marx 1944, p. 121). Moreover, "the worker becomes an ever cheaper commodity the more goods he creates. The devaluation of the human world increases in direct relation with the increases in value of the world of things" (ibid., p. 121). Under the capitalist system of production, like a machine or raw material, the worker was just a thing to be treated for whatever price it would fetch in the market for labor. No longer a whole human being with spiritual, aesthetic, and communal needs, the worker became merely another factor of production. Commodification of the worker meant that this total person had been swept away.

Alienation

According to Engels and Marx, commodification led inexorably to alienation, the estrangement of human beings from the product of their work, from other human beings, and, importantly, from oneself. Hegel (1770–1831), the great social philosopher of his time, saw alienation as a vital aspect of the human condition; the separation of body and soul, nature and intelligence, object and subject. He believed, however, that alienation could be transcended through the reason of the mind and intuition to reach a state of unified synthesis. Thus, it was possible for the individual to become truly human, far removed from animal origins and preoccupation with the problems of basic material existence. The individual could achieve an "Absolute" state of existence.

Marx, against Hegel, did not believe that such an unalienated condition could be achieved simply by reflection and pure thought, *logos*. Instead, he argued, alienation stems from society's way of producing its material wealth, its mode of production. And the main feature of the capitalist mode of production was a class-divided social order whereby the human beings of the underdog working class produced society's material needs by dint of their labor but in return were treated as commodities by the reigning bourgeoisie class. Furthermore, the bourgeoisie maintained workers at a mere subsistence level and were quick to appropriate the lion's share of the wealth created, especially the surplus wealth produced over and above its labor cost.

Commodification and class domination, Engels and Marx reasoned, meant that the individual's human existence had lost its unity and wholeness. With the specialization and deskilling of work, the individual became alienated from the product of his or her labor. Workers no longer produced a whole product so they never saw their work completed. Moreover, deprived of any intrinsic satisfaction with their work since the finished product belonged to the factory or mine owners, the worker was separated from the product of his or her own labor only to encounter it as an alien thing in the marketplace.

Further, the system made workers compete with each other for jobs and advancement in the market for labor. This splintered the social comradery of working conditions that had existed in the workshops of skilled artisans. And, since one class exploited the other, the human beings of each class became estranged from those of the other. Alienation, rooted in the material existence of the capitalist mode of production, could not be exorcized by thinking it away as Hegel had held.[8] Commodification, alienation, and class domination became the building blocks for the labor process humanistic critique of the inner workings of capitalism. At bottom, however, deskilling paved the way.

Deskilling and Control

The critical problem for the capitalist enterprise in the late eighteenth and early nineteenth century turned out to be control. The owner needed to develop an effective mechanism for keeping wage labor hard at work and so adapt it to the needs of capital. Deskilling of specialized labor proved to be a highly effective way of accomplishing this. Deskilling, it is important to recognize, is different from specialization. Specialization means being an expert in the entire process of production of a commodity, such as a carriage, a house, a shirt, or a bottle of wine. In contrast, deskilling refers to the systematic emptying of a traditional craft by carefully breaking it down into basic activities then assigning each one to a single worker along with a set of minute instructions.

Deskilling worked not only to make wage labor more efficient, as Adam Smith had shown, but even more importantly, it transferred control of the way work was performed from the individual worker into the hands of the owner and thus to the new managerial class. This enabled managers to treat labor like all other commodities in the production process with the aim of extracting ever increasing productivity from the individual worker. Appropriating control of the knowledge of the production process was vital to the capital accumulation process.

From a humanistic point of view, however, deskilling meant more than just the transfer of work knowledge to the managerial class. During the early stages of the industrial revolution, artisans of all kinds (weavers, china makers, brewers, mechanics, carpenters, instrument makers, even cigar makers) specialized in fabricating the entire product.[9] They spent long apprenticeships learning their craft, including studying the available scientific and technical knowledge underlying its practice. Weavers, for instance, not only worked with cloth but also studied biology, botany, entomology, mathematics, geometry, and music in order to develop a deep understanding of patterns and shapes. They even formed associations and societies to advance the scientific knowledge of the marvels and workings of nature. They enjoyed a cultured and intellectual existence. A century later, their offspring, now commodified, led a life of incredible poverty and degradation. Deskilling meant the obliteration of artisanship *and* the destruction of a whole civilized way of life for the craft worker.

Control of Work Process Knowledge

The transfer of the control of work from the laboring class to the owning class meant that the conception of *how* to do the job was uncoupled from its *execution*. As small firms developed into large enterprises in the late nineteenth and early twentieth centuries, their managers could systematically study work and apply science to the increasingly complex problems of the control of labor. This undertaking became the focus of the scientific management movement, spearheaded by the ardent and relentless Frederick Taylor who insisted that it was absolutely essential for management to dictate to the worker precisely how to perform each task. Management must assume the burden of "...gathering together all the traditional knowledge which in the past has been possessed by the workmen and then of classifying, tabulating, and reducing this knowledge to rules, laws, and formulae" (Taylor 1911, p. 36).

Scientific management brought to an end an era where owners derived their knowledge of the work process in a haphazard and hazy way. Taylor mandated, "All possible brain work should be removed from the shop and centered in the planning and layout department" (Braverman 1974, p. 118). Management could now fully lay out in advance the task of every worker and provide him or her with complete, detailed, written instructions regarding the exact means to be used for its completion. For the management scientist the science of work became the prerogative of management; it must never be left to the worker.

Control of Financial Information

As knowledge of the work process passed into the hands of the owners so did financial knowledge. Efficiency did not automatically accrue from technical advances. The abandonment of subcontracting, common in the mid-nineteenth century, in favor of direct control by hiring wage labor was not, in fact, more efficient, as some historians argue (Johnson and Kaplan 1987). Rather, it was done to transfer information about costs and profits from the workers to the factory owners. Owners soon developed costing systems and records which gave them detailed knowledge of internal costs and profits and allowed them to discipline labor in an intense drive for ever more efficiency. The ultimate aim, argues the labor process view, "was to cheapen the worker by decreasing his training and enlarging his output" (Braverman 1984, p. 118). Once again the worker was dispossessed, this time of the financial knowledge of the production process.

The advent of scientific management and the emergence of the large-scale capitalist enterprises also precipitated the rise of a new class of employees, clerical workers. The sales office handled travel to customers' factories, sales correspondence, order processing, commissions, sales analysis, advertising and promotion; while the accounting office looked after double-entry bookkeeping, financial statements, credit, collections, cash management, and shareholder records. Moreover, they were often paid double the wages on the factory floor.

As part of this development, the factory office expanded substantially. Previously, it consisted of the timekeeper and the supervisors' clerk who kept track of pay records, materials, work in progress, stages of production completion, and finished goods. To these basic duties were added production planning and scheduling, purchasing, engineering, design, and modern cost accounting systems in order to keep track of work on the shopfloor. The physical work process came to be mirrored in a stream of records and accounts, a vast paper empire which had a life of its own separate from the material world on the factory floor. The labor process was writ large in the factory office.

Sophisticated and detailed cost accounting systems, driven in large part by the scientific management movement, swept through factories and clerical offices across industrialized nations to become vital to the completeness of this writing. Employers already used historical cost accounting systems as a way to contain the powers of superintendents and overseers who were hired to control the work force. Engineers, following the principles of scientific management, using stopwatches, ergonomics, and close observation, redesigned labor processes and developed time standards for work operations rather than relying on historical experience. Such data were combined with accounting information to create full-scale standard costing systems including variance reporting. Each individual worker's output could now be compared daily with scientific norms. Cost accounting emerged as a "highly effective instrument of management control and the intensification of labor" (Hopper and Armstrong 1991, p. 420). This was not Adam Smith's *invisible hand*; but rather the very *visible fist* of owners and their managers.

The Ambiguous Status of Managers

These developments give rise to a new occupational group—managers. The status of the new managerial class, however, remains problematic within the labor process conception of a bipolar capitalist society. On the one hand, as with the working class, managers possess neither capital nor occupational independence but rather labor for the capitalist class. Thus, it should not surprise us that managers sometimes manipulate accounting controls, such as when divisional managers engage in activities which make them score well on profit budgets but which go against the best interests of the owners (Merchant 1985, 1987). From the labor process perspective such "manipulation of accounting controls . . .is thus more accurately portrayed as, in part, managerial resistance and coping with contradiction, rather than any failure of systems or corporate loyalty" (Hopper, Storey, and Willmott 1987, p. 452). Goal incongruent behavior is seen to be endemic to the manager's ambiguous position in society.[10]

On the other hand, owners hire managers to act on their behalf in directing and controlling wage labor. Furthermore, managers' remuneration runs significantly above that of the work force. So it seems fair to say that managers share with owners the surplus value produced by wage labor. Moreover, managers enjoy the prerogatives of fiat in that they hire, command, and fire workers. This seems to indicate that their interests lean more in the direction of owners than towards wage labor. So at the end of the day, labor process proponents conclude, managers contradictory actions with regard to accounting controls, which leave them sometimes advocating, sometimes resisting, arises from their ambiguous class position (Hopper 1990, p. 130).

Summary

The labor process perspective sees the role and historical development of management accounting and control systems not as an objective, neutral enterprise feeding information to decision makers, nor as a way for owners to counteract managers' shirking and cheating tendencies but rather in terms of a set of broader structural power inequalities endemic to capitalist societies. These systems developed during early capitalism to secure control over recalcitrant labor and later developed into "sophisticated control mechanisms designed to ensure the institutionalized subordination of labor to the needs of capital" (Puxty 1993, p. 78). Together with management science techniques they effected the usurpation of technical and financial knowledge of the labor process of production by owners and their manager agents from wage labor. According to labor process scholars, management accounting continues to play a crucial role in subordinating workers to contemporary capitalistic needs.

FURTHER CONSIDERATIONS

This chapter reviewed three prominent theories about management accounting and control systems. While each claims to be universal, each provides a strikingly different version of the workings of these systems. [see Table 1] Agency theory, taking the owner's side, depicts employees (agents) as using their asymmetrical information advantage to opportunistically cheat and shirk on the owner. Accounting and control systems are seen as a way to mitigate such behavior. In contrast, the labor process perspective is almost the antithesis of agency theory. Taking the underdog side of the exploited wage laborer, it depicts management accounting, buttressed with the techniques of management sciences, as a way for owners to deprive workers of the technical and financial knowledge of the production process,

Table 1 Universal Perspectives

Characteristics	Agency Theory	Nerve Center	Labor Process
Type of paradigm	Structural functionalist	Subjective interpretist	Radical structuralist
Intellectual forerunners	Hobbes, Smith, Spencer	Parsons, Mead, Merton	Engels, Marx, Braverman
Major thesis	Agents will use asymmetrical information to pursue self-interest with guile	Managers act as informational nerve centers whose main role is the collection, storing, and dissemination of information	Capitalistic interests appropriate the technical and financial information of the labor process to weaken and exploit wage labor
Central focus	Asymmetrical informational relations	Manager's information processing behavior and relations	Asymmetrical power relations
Main concepts	Moral hazard, signalling, asymmetric information, residual claim sharing	Decisional, interpersonal and informational roles, nerve center management	Commodification, alienation, deskilling, class contradiction, expropriation of surplus value
Advocacy position	Owners	Managers	Wage labor workers

treat them as commodities, and pressure them for even more productivity. Owners also use accounting systems as a way to legitimate grabbing the lion's share of any surplus value accruing from the enterprise. The nerve centre portrayal is neutral in this important debate. Taking the manager's position, it portrays them as involved in a hectic nonstop round of seemingly unrelated events, all the while gathering, storing, and disseminating information including that contained in formal accounting reports.

So we are left with three different portrayals, three different advocacy positions, three different views of the function of management accounting and control systems. Let us see how each performs in light of the case studies in Chapter 2.

Agency Theory

Agency theory receives some support. It could be said that Wedgwood's decision to pay his workers slightly more than the going rate constitutes a form of residual output sharing and thus mitigates the adverse selection problem since this practice would attract the better workers in the area. While in the Empire Glass (EG) case, since the owners cannot directly monitor the plant managers' efforts, the profit budget acts as an explicit contract for a specified output (profits and efficiency) and the bonus scheme works as a way of sharing some of the risk for output thus moderating the moral hazard problem. The Johnson & Johnson (J&J) case is less supportive of agency theory. Bonuses are deliberately uncoupled from budget performance and, since HQ executive committee members are intimately familiar with the subsidiary managers' abilities and comportment habits, the adverse selection problem is minimized. Furthermore, there is no indication of cheating and shirking on the part of the managers. On the contrary, they apparently put in long, difficult, and arduous hours on behalf of the company.

Labor Process

The labor process perspective performs somewhat better. In order to put his brilliant strategy to work, Wedgwood needed to suppress a recalcitrant and undisciplined work force. This he did by breaking up the traditional potting operation into separate tasks and assigning each to its own small building. This tactic served to deskill the traditional potter's craft. And by applying his own version of management science (his potter's instructions which spelled out in meticulous detail how to perform the various tasks as well as proper comportment on the job) he effectively took control of the knowledge of the production process. EG also illustrates deskilling in that the workers are no longer glassblowers but mere machine tenders. Moreover, they are programmed in accordance with the industrial engineers' scientific management work studies. Thus, the workers are deprived of both the technical and financial knowledge of the production process and in the end get treated as commodities.

The cases also demonstrate the ambiguous position of managers within the relations of capitalism. EG headquarters uses the profit planning and control reports to closely monitor the managers at the glass plants. Managers show some resistance in the form of making the numbers always come up right by the end of the year. At the same time, the glass plant managers advocate the use of the controls for disciplining the supervisors and the workers.

Similarly, J&J headquarters almost continuously monitors the subsidiary managers by means of the profit budget and the strategic business plan; whereas managers use them to watch over and discipline each department in the subsidiary.

Nerve Center

Turning to the nerve center framework, the case studies provide a good deal of support for its main propositions. Josiah Wedgwood seems an archetype of the nerve center manager, emerged in a relentless round of seemingly unrelated events he incessantly collected, processed, and disseminated information, including financial information. EG also confirms the picture of manager as nerve center. Plant Manager Hunt seemed to be constantly collecting, analyzing, and distributing information, especially data in the formal control system reports. While at J&J, HQ and subsidiary executives are literally awash with information, much of it from the formal management accounting and control system. In fact, it seems as if these systems are designed at J&J to ensure managers act as nerve centers.

Limitations

While each of the three universal theories receives some confirmation from the evidence in the case studies, not one is fully supported or seems to perform better than the others. The reason for this may be that each puts the spotlight on different facets of management accounting and control systems while ignoring or putting aside factors stressed in the others. For example, agency theory stresses the potential of agents to exploit owners due to their asymmetrical information advantage but ignores the latter's legal power of command which accrues from ownership. The labor process framework, in stark contrast, foregrounds the power of owners to exploit workers (agents) due to their control of the technical and financial information about the production process, as well as their powerful ownership position. On the other hand, it ignores the worker's possibilities of using private information for indolence and deceptive manoeuvring. And the nerve center picture ignores power positions altogether. So it is hard to prefer one over the others purely on its descriptive richness or its predictive ability; it seems to depend on one's advocacy position—owner, worker, or manager.

And universal theories have other important limitations. While they take us beyond the assumption that every case is unique, and they uncover elements essential to all management accounting systems, they do not provide much, if any, leverage on how the universal elements are capable of variation. Nor do they identify patterns to such variations. For example, the asymmetrical information problem is likely to loom larger at J&J than at EG. At EG the technology is well understood and uncertainty is low, whereas at J&J the technology is complex and uncertainty runs high. Similarly, the importance of the nerve centre role looms much larger at J&J, where uncertainty runs high, than it does at EG, where certainty is the order of the day.

Moreover, there are many important factors in the cases which are not included within the scope of these frameworks. Wedgwood, for example, relied more heavily for control on nonaccounting controls, such as the reforming influence of Wesley's Christian preaching, the clocking-in system, his supervisors and overseers, his potter's instructions, his daily inspection tours of the five work sheds, and his personal concern for the spiritual and physical well-being of his workers and their families. While at EG, group solidarity and the positive attitudes of the plant management group towards the company as well as for the formal accounting reports played an important part in the highly effective control we witnessed. And at J&J, the clear mandate of decentralized responsibility for business plans and their execution, along with a seasoned group of executives and managers weaned on the company's unique formal controls, played a vital role in the success of its control systems. These and other crucial factors are not addressed well by any of the three universal frameworks.

This suggests that the importance and nature of any universal elements of management accounting and control systems may be contingent on impersonal forces and factors surrounding the organization, such as environment, corporate history, strategy, technology, and interdependencies. If this is so, then organizations operating in similar environments and with similar strategies, technologies, and interdependencies should exhibit similar patterns in the way they design and make use of their management accounting systems. As it turns out, a lot of work has been done investigating this idea and, as a result, several valuable frameworks along these lines have been put in place. The next couple of chapters present several frameworks which show how systematic differences in such impersonal forces result in patterned variations in important characteristics of management accounting and control systems. This does not mean, however, that the universal frameworks are unhelpful; merely that by themselves they are not sufficient to capture all the important aspects of these systems.

NOTES

1. See Williamson (1973). A university is a case in point. Faculties in the various parts of the university—nuclear physics, genetic engineering, literary theory, law, electrical engineering, and so on—are privy to impacted information which is well beyond the ability of the president to comprehend, unless the president comes from that discipline.
2. See Perrow (1986) and Armstrong (1991) for detailed and trenchant critiques of agency theory as well as Scapens (1985) for basic exegesis of the mathematics involved in agency theory and an evenhanded synopsis of its limitations and possibilities.
3. See Baiman (1982, 1990) for detailed review articles of agency theory in management accounting.
4. Paul Sweezy, p. xii in the forward to Braverman (1974).
5. See Armstrong (1987, 1991); Hopper, Storey, and Willmott (1987); Hopper and Armstrong (1991); Roslender (1992); and Puxty (1993) for expositions of the labor process accounting paradigm.
6. Sir John Byng, touring the North Country in 1792 is reported to have exclaimed on seeing Manchester first hand, "Oh! What a dog's hole is Manchester," (Reported in Heilbronner 1980, p. 58.)
7. Children often proved better workers than grown-ups since they could be easily trained and controlled. It was not uncommon for half-starved, ragged, untaught children as young as seven or nine to be working fifteen hours a day six days a week. While a government commissioner described life after work this way: "I have seen wretchedness in some of its worse phases both here and upon the Continent, but until I visited the wynds of Glasgow I did not believe that so much crime, misery, and disease could exist in any civilized country. In the lower lodging-houses ten, twelve, sometimes twenty persons of both sexes, all ages and various degrees of nakedness, sleep indiscriminately huddled together upon the floor. These dwellings are usually so damp, filthy and ruinous, that no one could wish to keep his horse in one of them" (Reported in Engels 1987, p. 79). One need only go today to Cairo, Calcutta, or Karatchi to see how these slums have been exported to developing countries.
8. Similarly today, the university professor never (or only rarely) sees the product of his or her work—the completion of the student's education—and so is deprived of any intrinsic satisfaction

from the work of educating students. As well, the examination system creates a world of them and us, thus estranging professor and student.
9. Braverman (1974); see particularly Chapter 5, "The Primary Effects of Scientific Management."
10. See Macintosh (1994) for a detailed discussion of this issue within the widely diversified multinational firm.

4
Macromodels

Management accounting and control systems are pervasive in today's world. When looked at from close up, however, they seem to come in different forms and they seem to be used in different ways by different organizations. In consequence, suspecting that these variations might be related to the specific circumstances facing each organization, some researchers began to identify how patterned variations in these circumstances lead to systematic differences in key characteristics of management accounting and control systems. So they put aside the search for universal truths, such as "Participative budgeting is better than imposed budgets" and "Budget targets should be difficult but achievable." Instead, taking a contingency approach, they argued that an organization's performance depends on matching its control system characteristics with the constraints of its environment and other important contextual circumstances. These efforts bore fruit. Careful studies began to identify contextual dimensions that seemed to influence the way a particular organization would design its administrative systems. The findings were used to develop general frameworks that are useful for analyzing an organization's *domain*, that is to say, its environmental field of action, and for prescribing the appropriate administrative controls. This chapter outlines four frameworks of this kind.

The first one describes mechanistic and organic control and identifies the circumstances under which each is suitable. The second distinguishes between command and market controls and relates each to the type of information available to upper management. The third introduces two information processing strategies for coping with information requirements. These three frameworks fit squarely in the structural functionalist paradigm. They provide guidelines for achieving the best alignment of the organization's management accounting and control system with the type of environment and the kind of uncertainty facing the organization. The final framework, using a historical–dialectical approach, identifies five stages of organizational development and pinpoints the appropriate accounting and control system for each stage. This model, following the radical structuralist paradigm, brings into the picture the need for radical changes in the status quo of autonomy–direction relations in order to overcome or avert crisis situations.

MECHANISTIC AND ORGANIC CONTROL

The first framework stems from a milestone study by two Scottish organizational sociologists whose work caused so much excitement that senior civil servants had them put on a series of seminars around Britain for upper-echelon executives of industrial organizations (Burns and Stalker 1961). The hopes were that the insights from their research would help British industry regain its competitive edge, boost exports, and ameliorate the critical balance of payments conundrum faced by the UK.

The researchers built their framework using a biological metaphor. As they saw it, there were two *species* of firms. One type, living in a stable environment, reproduced its social technology in a mechanical way, just as a machine faithfully repeats its motions and functions. The other species, facing shifting conditions in its environment, constantly adjusted its administrative mechanisms just as some species change their shape, color, and physiology to suit their changing habitat. Adaptation, evolution, and survival are key concerns.

The insect hive provides the root metaphor for the mechanistic firm. In an ant hive, for example, each ant is born with a distinct role which it carries out automatically during its lifetime. Some are programmed to gather food and haul it back to the anthill. Some farm fungi and raise aphids as livestock. Some act as guards and attack any potential enemies. Some do the construction and housekeeping. And some look after the breeding and the babies. These roles are programmed into the ant's genes and are mechanically reproduced generation after generation. There is no question that each ant will faithfully carry out its instructions. The ant is a social creature in the extreme and a solitary ant on its own, lost and far from the hive, cannot survive. The social order of the hive is identical today to that of its ancestors millions of years ago.

The herd provides the root metaphor for the organic firm. The great caribou herds of North America exist by migrating with the seasons in search of food. Predators and weather patterns are its greatest enemies. Some fall prey to wolves, others drown while crossing swift-running rivers, and still others die of starvation or insect bites. Males compete with each other, literally head to head, for mating rights, with the losers deferring until the next rutting season. The herd is an organic unity and social roles—leader, sentinel, protector, and breeder—are passed around and exchanged as the need arises. It is survival of the fittest all the way.

The researchers paid particular attention to the relationship between a firm's external circumstances and its *social technology*, as they aptly labelled its management systems. External circumstances refer to markets served, production techniques developed, and scientific knowledge available. Social technology pertains to the flow of formal authority, lines of communication, division of tasks, arrangements for coordination of effort across the organization, and the required planning and control systems. Social technology, they concluded, is determined largely by external circumstances.

Mechanistic Control

As evidence for this conclusion the researchers observed that some of the firms they studied operated in relatively stable environments. These had a well-defined social technology. They divided tasks and problems into distinct slots. They defined precise duties for each function. They allocated power unambiguously amongst managers. They relied heavily on vertical hierarchical arrangements. And they coordinated the entire operation from the top, the only place where overall knowledge of the firm resided. Their management systems worked like a machine, constantly repeating the same motions and producing the same effects. The researchers labelled these kinds of management systems *mechanistic*.

The archetype of the mechanistic organization was a firm that produced and marketed viscose rayon filament yarn. The efficient production of rayon called for a highly predictable, precise and explicit production program. The requirements and tolerances for each stage of production program were set down in a book called the factory bible. Each department had a copy of the bible and used it each day to control part of the program.

The management system was also devised to keep production conditions stable. Each position in the hierarchy was specialized in terms of clearly defined parameters for authority, technique, and information requirements. No one could act outside the defined limits of their position. Reports, including expected and actual performance, were produced for every position and elaborately recorded in logbooks. All departures from stable conditions were immediately reported upward. Managers throughout the hierarchy made decisions within a tightly controlled framework of familiar program expectations.

The researchers also reported that the firm had a distinctly *authoritarian* character. Superiors issued firm commands, which subordinates obediently followed and took for granted as appropriate to the work situation. Authoritarianism at work, however, did not interfere with the relations off the job; the same people interacted within friendly social activities during which all were treated as equals and peers. The stable program conditions needed to manufacture rayon called for a mechanistic, authoritarian social order which all accepted as necessary and appropriate in the plant, but not away from work.

Organic Control

Other firms, in contrast, existing in relatively more fluid environments had a quite different social technology, which the researchers labelled *organic*. Scientific knowledge moved ahead relentlessly. Manufacturing equipment and techniques constantly improved. Tolerance levels for products became ever more demanding. And new products perpetually emerged in their market sectors. These environments seemed in a state of perpetual motion.

This turbulence meant that problems and actions could not be precisely defined or assigned to distinct specialist departments. Individual tasks could not be performed without knowledge of the goals and tasks of the entire organization. So a manager's methods, duties, and power arrangements had to be continually negotiated and redefined. They accomplished this through frequent and intensive interaction with other managers throughout the organization. Lateral, rather than vertical, relationships dominated. Formal definitions of hierarchy melted. Complete knowledge of the organization was no longer ascribed to top management. Environment and social technology never settled down.

An electrical engineering firm, one of the research sites, typified the organic firm. Upper management deliberately avoided assigning specific functions to particular people and they were loath to define lines of responsibility. And the makeup of the top management group fluctuated with prevailing circumstances. Managers met regularly and interacted frequently on an informal basis whenever they needed to find out what was going on or to decide what to do about a problem. All employees regardless of official rank could consult with top management or anyone else in the firm about issues and problems. All managers had an equal voice in making decisions. This organic social technology, the researchers concluded, matched the firm's unstable program conditions.

The researchers had identified two distinct types of bureaucratization, mechanistic and organic. An essential part of upper management's job, they concluded, is to interpret correctly both the degree and kind of instability in the market circumstances, manufacturing technology, and scientific knowledge facing their firms. Only then are they in a position to match the appropriate social technology with the environment. They found many firms in the sample had failed to make this analysis, consequently their form of bureaucracy proved inappropriate and ineffective.

In some instances, firms had successfully utilized the mechanistic form but had not noticed their external circumstances had shifted from stable to unstable program conditions. They struggled on with a mismatch of unstable program conditions and a mechanistic social technology. In other instances, upper management tried to utilize an organic social technology when their external circumstances were stable. One reason was that managers had come from firms where organic arrangements worked well, so they blindly assumed they would work in their new firm. The organic type of bureaucracy including *matrix management* was trendy at the time, so consulting firms urged nearly all their clients to adopt the new fashion. And in some firms, even when the environmental–social technology mismatch had been correctly diagnosed, politics and careerism forestalled a realignment. A sinecure in a mechanistic firm can readily disappear in an organic firm.

Implications for Management Accounting and Control

This work holds special interest for our purposes because it also theorized about the appropriate type of accounting and control system for each of the mechanistic and organic types of firms. In mechanistic firms, operating in a stable environment, overall knowledge is available only at the top of the hierarchy, resulting in a simple but forceful management control system. Superiors govern operations and work behaviour by issuing strong, clear commands. Unilateral instructions flow downward and become more explicit with each successive layer of bureaucracy. Upward-flowing information, in contrast, is successively filtered out on its way to the top. As this process continues it loses its impact, accuracy, and force. This unilateral, autocratic style of control is uniquely suited to the mechanistic management system. These ideas are depicted in Figure 1.

In organic systems, by contrast, the management control pattern is remarkably different. All-encompassing information is no longer available at the top and often critical knowledge exists only at the tentacles of the firm. So managers seek out and interpret for themselves the information needed to perform their part of the overall task. Much back-and-forth

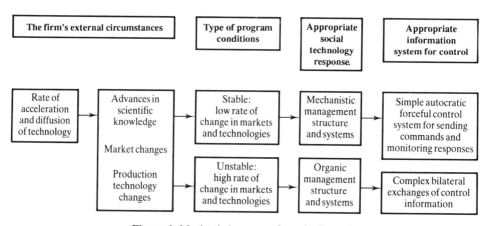

Figure 1 Mechanistic versus Organic Control

exchange of control information ensues. Communication patterns resemble lateral consultation more than vertical command. These systems seem almost to be self-designing.

In sum, external circumstances which shape management systems also influence the shape of the appropriate accounting and control systems. The organization of the accounting department, as well as the form of the system itself, should be consistent with the overall organizational design. In centralized mechanistic firms the management accounting and control function should be centralized and unilateral. In decentralized organic firms, horizontal flows of control information should be featured. A great deal more participation and discussion with line managers is required at all levels; and decentralization of responsibility to management accounting offices down the line is a necessary aspect of successful design in organic systems. Accounting and control systems need to match both the external circumstances and the social technology.

Before moving on to our next framework, it is important to note that most organizations include both forms of control. Other researchers, following upon the above study, noted that the type of program conditions could vary within organizations as well as between organizations.[1] Some parts of the same organization faced stable program conditions and a slow rate of acceleration in technology, while other parts faced rapidly changing program conditions and technology. Research and development departments, for example, tended to have the organic form, while production departments tended to follow the mechanistic one. In fact, purely mechanistic or organic firms are the exception rather than the rule. The important implication is that management accounting and control systems should be different in form and use within the same organization, depending on the type of program conditions facing the particular component.

COMMAND VERSUS MARKET CONTROL

Our second model of control and organizational context is anchored in some fundamental ideas from economic theory. For the economist, the key to modern life for both the individual and the organization is *rational choice* which, fired by the vigorous pursuit of *self-interest*, leads to *utility-maximizing* behavior. While many, if not most, economists recognize that rational utility maximizing is seldom achieved, they maintain that we can, nevertheless, be *intendedly rational* and so achieve *utility-satisfying* behavior (Cyert and March 1963). Either way, the key to optimal behavior for the economist lies with the kind and quality of information available to the decision maker.

Kenneth Arrow, a Nobel Prize winner in economics, used the economist's toolkit of ideas in his classic article to address the issue of control in large organizations (Arrow 1964). An organization ". . . is a group of individuals seeking to achieve some common goal, or, in different language, to maximize an objective function."[2] At the same time, however, each member also has a personal objective function, which may or may not coincide with the common goal. Thus the crucial problem for organizations is to solve the problem of ". . .how it can best keep its members in step with each other to maximize the organization's objective function" (ibid., p. 3). This, of course, is the *goal congruence problem* so familiar to management accounting and control practitioners, academics, and students.

Further, as the economist argues, the decision parameters for any individual are constrained by the external environment and by the decisions made by the other members. Some but not all of the observations about the world both internal and external to the organization get communicated to its members. So communication and information processing becomes a prime consideration in organizational design.

Managers' work is seen to consist of processing information. Managers spend most of their time sending and receiving signals from other managers and from the environment to make decisions based on their revised, current assessment of the various factors in play. In the jargon of the economist, they acquire information at a cost and use it to gain a benefit, such as better decision making, as long as the incremental gain accruing from additional processing exceeds its marginal cost.

Moreover, the process is dynamic. As decisions are made, they generate further information, which is transmitted in one form or another to the environment and to other managers. This in turn generates new decisions and more signals. At the same time, new signals are coming in, so managers are constantly revising their conditional probability distributions about the *state of nature* and calculating the effect of these new distributions on their *utility functions*. They behave in very *rational* ways.

As managers go about this process, they do so within the framework of *operating* and *enforcement* rules laid down by the organization. Operating rules instruct organizational members about how to act; they *tell* managers what to do. Enforcement rules compel them to act in accordance with the operating rules, they *persuade* them to do it. Operating and enforcement rules, the nervous system of any organization, come in two bureaucratic forms: the command organization and the market organization. Each of these is outlined in Figure 2.

Command Control

The command style of organization is almost identical to the mechanistic form in our first model. When the upper echelons have low-cost access to the sundry conditions prevalent throughout the organization, a centralized command organization can be employed to advantage. The key factor here is the availability of omniscient information to top management. The vitally needed coordination of individuals' efforts and interests is achieved via a structured, vertical hierarchy and general acceptance throughout the organization of the master plan made at the highest level. Control is exercised through specific detailed operating rules

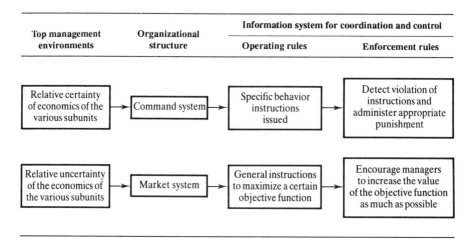

Figure 2 Command versus Market Control

followed up with punitive enforcement rules. The latter are designed to detect and report violations of operating rules to upper management who then have only to reward compliance or sanction deviations. Omniscient information lends itself to a command organization.

Management accounting and control systems are an integral aspect of both the operating and enforcement rules. Top management's plans (the operating rules) are expressed in great detail in the master budget for the entire organization and in the operating budget for each component. The controller's department, closely monitoring variances from the budget, detects any deviations from the master plan. Budget performance is then used as the basis for rewarding and sanctioning managers throughout the organization. The master budget masters the organization and its individual members.

But perfect transmission and assimilation of knowledge by top management is not always possible, especially in complex, widely diversified, multinational companies. The reason for this is that managers are information channels of decidedly limited capacity. Information, especially that circulating lower down in the organisation, is not necessarily transmitted to the next step up the hierarchy. Further, the absorption of every piece of information would lead rapidly to information overload on the part of sundry managers in the hierarchy.[3] After a certain level of information coming in, managers simply stop processing information.

But perhaps more importantly, in today's huge, worldwide organizations with their complex flow of products and services, lower-level managers will always know much more about their spheres of activities than will higher officials. The net effect is that the centralized command management system, so well suited to the context of omniscient upper management, becomes more and more cumbersome and eventually becomes a major impediment to organizational effort.

Market Control

A widespread response is for organizations to adopt a decentralized, market-based approach to managerial arrangements. This is accomplished by rearranging the enterprise into many small, separate, widespread but related organizational subunits. These units then trade among themselves and with their external markets as if they were independent, autonomous entities. The organization is transformed into a sort of a quasi-miniature free market economy featuring free-wheeling economic subunits.

The information requirements for operating in this mode are quite different than those of the command organization. The major operating aim for each subunit is to maximize the profits of its own activity center. The enforcement rules here are designed to encourage each manager to increase profits as much as possible in his or her sphere of responsibility. Economic performance, not obedience to orders, is paramount.

This creates new demands on the management accounting system for information about the economic performance of each autonomous subunit. Management accounting and control systems shift from budgets full of highly detailed information to reports that focus on general profitability indicators such as return on investment and residual income. Reports provide a kind of market performance information in that profits are seen to measure success in the marketplace.

With each subunit vigorously pursuing its own profit, however, the various parts of the organization may work against the common organization purpose. Goal incongruence behaviour might prevail. Goal incongruence exists when the actions taken by individual managers that are in their own best interests, as far as subunit profit maximization is concerned, are not in the best interests of the overall firm. As a result, the organization might even fly apart. Some mechanism is needed to hold the separate pieces together.

The answer, according to the economist, lies in market mechanisms. External market prices and internal transfer prices provide the necessary information with which to calculate profitability, that is to say, economic performance. And need for external and internal prices places new demands on the management accounting and control departments.

Regardless of the actual external price paid, the information used for internal reporting must be based on solid, longer-term market prices, not on distress prices. Where available, internal transfer prices should be based on external prices for the same goods and services. Even when these are not available, or seem inappropriate, they can be simulated. One way is to rely on engineering estimates of the cost structure of a hypothetical producer of the same product. Another way is to approximate them with the aid of mathematical programming techniques.

Such market-based information is of great value. It reveals to upper managers those component managers who are capable of prospering under competitive market conditions. By the same process the less-fit managers are exposed. The invisible hand of the marketplace reaches in and disciplines the managers throughout the organization in a marvellously efficient manner.

It is important to recognize that the economist's belief in the efficacy of the market to coordinate and control individuals in society stems from Herber Spencer's (1820–1903) evolutionary, utilitarian, and individualistic social philosophy (long since out of favor with social theorists) built on positivist scientific assumptions. For Spencer, each individual was "blessed with an automatic, self-regulating mechanism which operated so that the pursuit of self-interest and private wants would result in the greatest satisfaction of the wants of all" (Parsons 1937, p. 4). Society would evolve until it reached the final state where each individual (free from ignorance, superstition, and fear) would stand unfettered by state or church and act "rationally." All nonrational or nonutilitarian aspects of existence were pared away in Spencer's scheme. Utilitarian economic action explained all.

In sum, for the economist, the nature of the information available to top management determines the choice of either a command or a market organization. When omniscient information is available a command structure is appropriate. When this is not the case, a market system is required where real or quasi-market prices replace perfect information. Management accounting and control systems vary depending on which of these arrangements is adopted. Importantly, neither form is perfect. Each presents unique problems for management and for management accounting. In reality, many if not most of our large, complex firms employ both types, adopting command control for some parts and market control for others.

AN INFORMATION PROCESSING MODEL OF CONTROL

Our next model of context and control takes an *information processing* or *cybernetic* approach.[4] Researchers following this path focus on the decision making and informational process aspects of organizational action, which they believe are a function of the limitations of human ability to process information. This approach aims to improve decision making by improving data collection and retrieval; building and managing useful data bases; grafting computer-based models onto human information processing and decision making; and getting on-line, real-time information into the decision process. The fundamental idea is to get the right amount of information into the decision-making process at the right time. But what, we might ask, determines the right amount and the right timing for information?

The answer from the information processing camp is *task uncertainty*. When task uncertainty is great, that is to say when the task is poorly understood, more information needs to be processed among decision makers to get the work done than when task uncertainty is low. Uncertainty is the difference between the amount of information already available for task execution and the amount of information required for the job. So it follows that organizational design and control is a function of the information processing requirements necessary for task completion.

These information requirements in turn are seen as a function of three task characteristics: output diversity, input resources utilized, and the level of goal difficulty inherent in the task. The degree of diversity of output depends on the number of different products, services, and programs the organization produces and the number of markets, customers, and clients it serves. The variety of input resources utilized is measured mainly by the number of different work locations, the number of different specialists employed, and the number of different resource requirements (such as inventory and suppliers) needed to produce the outputs. The level of goal difficulty is a matter of the tolerances and quality needed in the inputs and outputs in order to meet efficiency and effectiveness expectations of customers and clients. The difference between the combined information requirements of outputs, inputs, and goal difficulty and the amount already possessed determines the level of task uncertainty.

Rules, Goals, and Hierarchies

Organizations can usually cope with task uncertainty by means of rules, goals, and hierarchies. Rules specify prescribed behavior in advance and become the organization's memory for handling routine work thus eliminating the need for further communication between the subunits involved. New and unusual situations, however, are not covered in the rules. New information must be processed through the formal hierarchy. But hierarchies have a finite capacity to process information and after a point become overloaded. So instead of using the hierarchy, it becomes more effective to delegate decisions to the point of action and set superordinate goals (or targets) to cover interdependencies among the various subunits. Eventually rules, hierarchies, and goal setting are unable to cope with new demands for information processing. When they fail, organizations look for new strategies to cope with their information processing needs.

Uncertainty-Reducing Strategies

When such a disparity exists between the amount of information that needs to be processed and the actual amount possessed, organizations respond by increasing the organization's capacity to process information, by decreasing the amount required, or by some combination of both. Two main strategic alternatives are available for decreasing and increasing the amount of information processing.

Slack Resources

The first strategic move for reducing the quantity of information required involves creating slack resources. In place of information, slack or excess resources are used to alleviate immediate problems. Excessive inventory levels, long delivery times, extra machine capacity, overtime, and staff departments are examples of areas where slack is created rather than increasing the amount of information processed. Slack acts like a sponge to absorb information.

Self-Contained Units

The other strategy to counter insufficient information processing is to create self-contained tasks. These could take the form of a shift from an interlocking responsibility for all products to responsibility for a segment only of the organization's total products. Airplane manufacturing firms, for example, have two main choices in their organizational arrangements. The first is to allocate responsibility to product engineers, design technicians, process engineers, fabricating, assembly, and test units, and so on, with each unit having responsibility for the entire aircraft. The second choice is to organize around self-contained units by airplane section such as a wing or the tail.

The creation of self-contained units reduces the amount of information processing needed in two ways. First, it reduces output diversity. Each self-contained unit deals only with *one* body section, whereas the functional response would require each unit to deal with *all* body sections. Second, creation of self-contained units reduces specialization. Rather than sharing process engineering across all aircraft section units, the engineers in each unit would be involved in process engineering as well as product design and quality engineering. This reduces the need for information processing *across* functional departments.

Vertical Information Processing

Two major strategies also exist for *increasing* an organization's capacity to process information. The first involves investing in vertical information systems. It is advantageous to develop new annual operating plans and budgets, rather than making incremental changes to the old ones, when, for example, uncertainty levels become intolerable. This entails collecting information and new plans at appropriate times and places instead of overloading the organizational hierarchy by forcing it to cope with a vast number of budget exceptions.

New Lateral Relations

The second strategy for increasing the capacity to process information involves the selective creation of new lateral relations. This move lowers the level of decision making to the point where the information is located, instead of transmitting it to high echelons who then make the decision. This strategy decentralizes decision making; but it does so without calling into existence new self-contained groups.

The creation of new lateral relations can be accomplished by several means. These include: direct contact between those who share a problem; establishment of new liaison positions, such as project or product administrations, to manage interdepartmental contacts; formation of task forces, such as product or project teams; or in the extreme case employment of dual reporting relations and matrix organizational structures.

New lateral relations, however, are not a free good. They lead to an increase in the time managers must devote to the processing of horizontal information. This cost in time can be offset by moving decisions to the level where the information is located, rather than to a centre high up in the organization. This has the added advantage of guaranteeing all relevant information is included in the decision processes. The result should be sounder decisions.

It is important to emphasize that these moves are not called for because of incompetent management but because the information processing and computational capacity is insufficient to deal with the complex and interdependent coordination requirements of an organization. This model of information and organizational design is summarized in Figure 3.

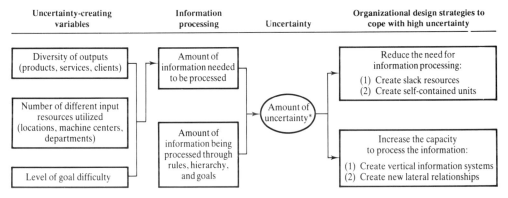

*The amount of uncertainty is the difference between the amount of information needed to be processed and the amount actually processed.

Figure 3 An Information Processing Model

A Case History

The case history of a firm which produced different assembled mechanical devices for the aircraft industry lends vividness to this theory.

> Work in this firm ranged from acquisition of raw materials to forging and stamping, to machining and the assembly of units in an orderly fashion. A conventional type of functional organizational structure was employed. Costs were under control; growth was adequate; and profits were good.
>
> In spite of these favourable conditions, management were dissatisfied with the prevailing state of affairs. Their impression was that their time was spent in dealing with short-term problems, 'fighting fires', to the neglect of longer-run tooling and capital investment programs. Of course, the lack of attention to longer-run problems eventually fed more fuel to the short-term 'fires'.
>
> In due time a management task force was formed to study the problem. The task force produced an historical analysis of the firm's development. Members of the task force were surprised when they documented the changes that had taken place in the previous decade. The product line had doubled. The number of individual parts produced had increased by more than half, as had retooling. The number of machining and assembly stations had expanded. Quality specifications and tolerance limits had risen significantly. The production department had acquired a new expertise in dealing with exotic metals, and in order to keep up with expanding volume the number of shifts had increased from one to three. The firm had increasingly met delivery dates, while at the same time reducing inventory levels. Remarkably, until formation of the task force, the magnitude of these changes had remained almost unnoticed; but the information processed had not kept pace with production (Galbraith 1973, pp. 67–68).

Viewed over a suitable time period, the firm had experienced substantial changes in product diversity and volume, the number of parts produced, and the requirements for retooling. Output diversity had increased significantly. Almost simultaneously, the number of input resources also increased, as witnessed by the expansion in the number of machining and assembly stations. At the same time, the goal levels required for quality, tolerance, delivery, and new metal technology had risen sharply. These changes in output diversity, input resources, and difficulty in maintaining the new goals meant that the amount of information in need of processing had grown by leaps and bounds. The result had been a large increase in uncertainty.

Once the problem had been diagnosed, management moved rapidly. They selected two of the prescribed strategic responses. First, they created more self-contained units, such as separate departments to concentrate on new products and processes. Second, they formed liaison units, including both line and staff people to handle the multifunction interdependence at low levels in the organization. In the model's terms, they brought into existence new self-contained units, and they created new lateral arrangements. These strategic alignments ameliorated the immediate problems, and management was able to concentrate on the long-run situation.

A major strength of the information processing model is its ability to frame the problem in terms of a repertoire of responses. So instead of redefining authority, responsibility, and accountability every time a storm of uncertainty passes through the environment, the problem can be resolved by judicious selection from a repertoire of strategies until the storm passes. This would avoid a series of substantial realignments from mechanistic to organic or from a command to a market structure, a phenomenon we see often, perhaps too often many argue, in our organizations.

The case studies in Chapter 2 also illustrate some of the ideas in the cybernetic model. As Wedgwood grew, the rules, supervisory networks, and quality goals were no longer sufficient to process the expanding required amount. In consequence, Wedgwood invented a vertical information system for highly detailed cost reporting, including accurate allocations of general and fixed costs. He also created self-contained units for each of the major phases of production and materials management.

Empire Glass headquarters relied on a highly detailed vertical information system in the form of specific plans and detailed budgets, and set up the plants as self-contained units. Given the maturity of the glass industry, the firm could ill afford slack resources. Nor did it need new lateral arrangements since both the sales organization and the plants had an important and shared stake in the form of a substantial bonus in sales, quality, and delivery targets. In the Johnson & Johnson situation, operations featured nearly 160 self-contained subsidiaries and intensive vertical information systems (the strategic business plan and the annual profit budget). All three companies judiciously selected from the model's repertoire of strategies to cope effectively with their information processing requirements.

A PHASES MODEL OF CONTROL

The final framework of organizational context and control is constructed using a *historical–dialectical* analysis.[5] The historical strand brings attention to the way organizations are shaped by their past and how they evolve through distinct phases over time. The dialectical strand emphasizes the inherent changefulness within any system due to a concealed but fundamental contradiction embedded in any system such as an organization.[6] The central idea is that systems exist in a state of dialectical tension and that they evolve through distinct historical phases as the contradiction is played out.

Dialectical Analysis

Contradiction in the dialectical sense refers, not to some logical contradiction,[7] but rather to the existence in any system of some mutually interacting but incompatible opposition between two fundamental elements or forces. The elements, locked in a relational struggle, exist in a state of *dynamic tension*. At any moment, one element gains the upper hand and represses the oppositional one. Thus the current condition of the system is seen to be the result of its most recent resolution of this struggle. Paradoxically, the essence of reality is the unity of opposites.

This current resolution, however, is only temporary. The seeds of its downfall are already planted inside the system. The reason being that the present state (or thesis) inevitably leads to excesses on the part of the temporarily superior element. Feeding off these excesses, the subordinate element rises up, overturns the hierarchy, and establishes itself in place of privilege but, importantly, at a level that subsumes the previous resolution. However, this new synthesis is subject to the same fate as the old resolution. A new antithesis forms in opposition to the synthesis, and the process repeats.

Thus, for the dialectician the essence of reality for any system is motion, restlessness, and mutability. Its innermost being is changefulness since it consists of the unstable coexistence and perpetual resolution of the two incompatible forces. The dialectic unfolds in a way that is integral to the system but at the same time destructive of its current state. This idea of contradiction helps us ascertain the dialectical logic to any system's historical tendencies and evolution.

An Example

The classic illustration of dialectic tension is Plato's Master–Slave contradiction. Two independent people stranded on a desert island see each other as a limit to their own freedom and power over the island's natural abundance. To settle this unstable state of affairs, a struggle takes place and one conquers and enslaves the other. Although, it may seem that the Master has the upper hand and the situation is stable, such is not the case. The situation will reverse due to the contradiction concealed in the relationship.

It is the Slave, who by dint of physical labor changes the natural (material) world, gets satisfaction from this, and so develops a true self-consciousness. The Master, not recognizing the Slave as a real person worthy of an ideal consciousness, debases himself to a lower form of existence; he has not seen his own universal self in the other. Further, the Master is now dependent on the Slave for his material existence. Spiritually and materially, the Master now occupies the lower position. Control has passed from Master to Slave who now enjoys the upper hand.

Autonomy and Direction: The Basic Contradiction

The basic contradiction in our historical–dialectical model of control in organizations is the unrelenting struggle between the forces for autonomy and freedom on one side and the forces for authority and order on the other. When authority gets the upper hand, autonomy is suppressed. This resolution of the dialectic tension, however, sows the seeds for its inevitable reversal when the excesses of authority lead to its overturning by the forces for autonomy. In turn, an overdose of autonomy brings on another crisis and authority eventually comes to the fore. And so it goes, as the organization's basic character swings back and forth, one time in favor of autonomy and the next time giving authority the nod.

The dialectic plays out as follows. When autonomy has the upper hand the outcome is innovation, creativity, and risk-taking behavior on the part of managers and employees. An overdose of autonomy, however, eventually leads to chaos, parochialism, and even anarchy. At this juncture, authority, which has been lying dormant, comes forward to gain the upper hand. The shift to authority fosters leadership, direction, and attention to global purpose. In the course of events, the surfeit of authority induces its own crisis of apathy, lethargy, and paralysis within the organization. The time has arrived for another reversal.

The model also contends that organizations tend to experience lengthy periods of evolutionary growth followed by short revolutionary periods. Each phase of evolution precipitates its own revolution and its ensuing dialectical crisis. If the firm survives the crisis, by making the correct organizational adjustment, then it moves into the next phase of prolonged evolution. During each evolutionary stage, only minor fine-tuning of its overall pattern of organization and control is necessary for sustained growth. Importantly, each evolutionary phase has a characteristically distinct and dominant management style as well as its own unique management problem. And as each phase effects the subsequent phase, it too is a result of its predecessor phase.

As the dialectic process unfolds, organizations that survive each crisis develop through five distinct phases: creativity, direction, delegation, cooperation, and coordination. Each phase is shaped by historical factors including its size, the growth rate of its industry, and its previous stages of evolution and crisis. Next we sketch out these five phases, summarized in Figure 4.

Phase 1: Creativity

Early on in its history every firm is small. It begins its historical journey because its founder created a new product, invented a new production method, or developed a new market. The organizational climate is informal and communication is face to face. Often the founder-owner, usually highly technically or entrepreneurially skilled, relies on charismatic leadership and personal relationships for coordination and control. Employees, feeling awe and respect for the founder and his or her innovation, perform faithfully and diligently. Creativity and charisma, essential for the firm to get up and running, provide the necessary discipline and control.

Success leads to growth. Longer runs and an increasing number of changeovers characterize the production process. New capital is raised to finance increases in working capital and to purchase new production equipment. The size of the work force expands. The founder, still relying on creativity and charisma to cope with the expansion, works harder and longer in order to keep things under control.

The founder also tends to be engrossed in selling the product and getting it through the plant as designed. Customers' demands, production mishaps, and technical breakdowns in the product consume more than the available hours in the day. Administrative practices and

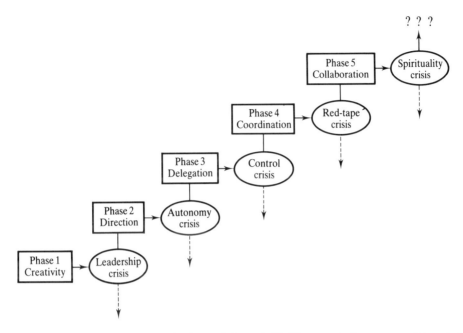

Figure 4 A Dialectic Model of Coordination and Control. (Solid arrows indicate successful evolution to the next phase; dashed arrows indicate failure to resolve the crisis. Adapted from Greiner (1972).)

management systems, such as accurate cost accounting systems, inventory controls, and capital expenditure analysis, are neglected or even disdained.

Moreover, the founder no longer knows each employee on a personal basis. Nor are new employees motivated by an intense loyalty to the founder or by an enrapturement with the original innovation. Administrative systems, including formal management accounting systems, are urgently needed to bring order to the now chaotic situation. The firm faces its first perilous turning point, the *leadership crisis*.

Strong leadership and direction is needed, particularly on the administrative end. Founders now perceive themselves to be forever bogged down in mundane, petty administrative matters. Moreover, they find the new conditions much less appealing than the crusading, swashbuckling days of old. Even though they understand the need for strong leadership and direction on the administrative side of the house, and they recognize that they themselves are unsuited in temperament for the administrative life, they are reluctant to delegate these activities to a capable manager. Yet survival is contingent on sound management practices and formal systems.

Phase 2: Direction

Firms that survive the leadership crisis are those that develop sound administrative practices which provide the vitally needed direction and authority. Production scheduling and inventory controls are formalized. Written communications replace face-to-face encounters. Standards and budgets are put in place. Cost accounting systems are introduced. Formal hierarchies and position titles are set up. The happy result is that order and routine begin to subdue chaos and uncertainty. Moreover, the new directive techniques provide a climate fostering efficient growth. The firm prospers, becomes more diverse and complex, and embarks on a sustained period of development.

The successful bureaucratization of the firm, as the dialectic predicts, precipitates its own crisis. The restriction of the prerogatives of lower-level employees produces a stagnant organization. They know a great deal more about manufacturing processes, products, distribution, and customers, than do the administrators. They experience the administrative mechanisms as cumbersome, centralized, and repressive. Yet, feeling they must follow procedures, they are reluctant to take the initiative for fear of reprisal from the administrative arm. A sort of paralysis creeps in as the inevitable *autonomy crisis* hits the organization.

The solution, although drastic, is clear enough. Lower-level managers need greater autonomy and freedom to act as they know best. Often this proves difficult for the administrators, after all they gained their success by being directive. Moreover, it can be troublesome for lower-level managers. Used to being bossed, they are unaccustomed to making their own decisions. Yet delegation of important decisions is inescapable if the firm is to overcome the autonomy crisis and reach the next era in its historical evolution, the *delegation phase*.

Phase 3: Delegation

Rather than stubbornly adhering to a dictatorial central bureaucracy, successful firms overcome the autonomy crisis through delegation. Responsibility and authority is given to the managers of the various field units: manufacturing facilities, sales districts, marketing departments, and engineering development and research offices. These components now run their own shows. Upper managers refrain from interference in the operations of the field units by restraining themselves to management by exception on the basis of strategic plans and operating budget reports from the field. Field units are treated as autonomous profit or marketing centers. Communication from the upper echelons is infrequent as they now concentrate on acquisitions, investment, and capital market considerations and monitor the field units by means of simple reports such as a profit budget.

Delegation usually does the trick. Released from the fetters of an overzealous central bureaucracy, these managers and their subordinates experience a heightened motivation. Taking full advantage of their newfound autonomy, they now react rapidly to customers needs, penetrate new markets, introduce advanced manufacturing technologies, and launch new products. Delegation proves to be the engine for a new and prolonged period of steady growth and progress.

As in previous phases, however, success breeds its own emergency, the *control crisis*. The autonomous managers, having developed their own fiefdoms, are reluctant to coordinate their operations with the other independent components. Opportunities to economize on purchasing, central services, technology, computing, specialized personnel, and R&D go unrealized. Moreover, arrogant parochial attitudes emerge as the various managers fiercely protect their territorial boundaries and prerogatives. The organization becomes Balkanized.

Sensing their loss of influence and direction over the diverse operating units, top management attempt to regain control for the center. Some firms opt for a return to the bureaucratic measures of the direction phase. In this case, failure is the usual outcome. The company is too diverse and complex to be run from the top. Others find and adopt a new social technology based on sophisticated integration mechanisms. The *coordination* phase is under way.

Phase 4: Coordination

The *coordination* phase features a realignment of duties throughout the organization. A new layer of management is inserted between headquarters and the field units. At the field level, the sundry operations such as various previously autonomous profit centers are merged into sector, industry, or product groups and treated as return-on-investment centers. The product group manager is given authority for nearly all phases of bringing the product to market: engineering, production, sales, marketing, and so on. Each product group is considered to be a separate autonomous business. This authority, however, is not granted without responsibility. The product group managers become responsible for achieving profit and return-on-investment targets.

At headquarters level, new staff offices are created and formal administrative systems are implemented to coordinate and control activities at product group level. Technical staff units—marketing, personnel, purchasing, engineering—are expected to provide expertise and assistance to the field. In some firms they must sell their ideas to the product groups; in others they have a direct line of authority over their counterparts in the product group. These units have the responsibility for the technical quality of their area of specialty. The result is dual lines of authority. The manager of engineering, for example, in the product group is responsible to both the HQ director of engineering and to the product group general manager.

As well, administrative staff offices at headquarters get busy developing various management systems in an all-out effort to effect more control over the product groups. A host of administrative control systems are put in place including strategic planning, long-range financial plans, operating budgets, capital investment budgeting, as well as standard operating procedures, rules, and policies. While the product groups are left with as much decision-making responsibility as possible, they are required to adhere to these control and coordination systems. They must now justify their actions and decisions to the watchdogs up the line.

In order to make the new arrangements stick, stock options and company-wide profit-sharing schemes are put in place. These act as inducements for the product group managers to respond positively to the new coordination and control systems and look beyond the needs of their own component. The realignment of the field operations, the institution of coordination and control systems, and the new global outlook by field managers lead to economies throughout the firm. The result is a more effective allocation of resources and another lengthy period of profitable expansion.

After a time, however, the various technical and administrative staff offices at headquarters take on a life of their own. They become an end in their own right rather than a means to assist and guide the line operations. The result for the field managers is a feeling of being overwhelmed by a proliferation of bureaucratic paper systems. They come to resent the unsolicited advice from staff specialists who are unfamiliar with local problems and conditions. While at headquarters, staff become increasingly frustrated by the parochial attitudes and ignorance of new technical developments on the part of the field product groups. They perceive the line managers as uncooperative and stubborn. Once again success has created the conditions for a new kind of crisis and yet another transmutation. Initiative, innovation, and problem solving grind to a halt as the *red-tape crisis* strikes.

The General Appliance Corporation (GAC) provides a prime example of an organization operating in the coordination phase.[8] GAC, an integrated manufacturer of home appliances of all kinds, strove to be the quality-products industry leader. It operated with four product divisions, four manufacturing divisions, and six central staff offices. GAC treated the product and manufacturing divisions as profit centers responsible for the design, engineering, assembling, and marketing of various home appliances. The product divisions purchased 20 percent of their parts from outside suppliers and 80 percent from the manufacturing divisions; the manufacturing divisions sold 25 percent of their output to outside companies.

The product and manufacturing divisions dealt with each other as if they were independent companies competing in the marketplace. Headquarters held each division responsible for a budgeted amount of profit and a target return on investment. The transfer price system for internal transactions relied on market-based prices wherever possible. The various staff offices had functional authority over their counterparts, who were also responsible to their divisional general manager, in the manufacturing and product divisions but they had no line authority over the divisional general managers.

The case describes an incident where the president, deeply concerned over customer and dealer complaints about product quality and a customer survey indicating that GAC's reputation as a quality leader had deteriorated, gave the production staff vice president of production unilateral decision power for six months to bring the quality of all products up to a satisfactory level. The case documents a transfer price dispute over a stove top which had been improved in quality as a result of the vice president's mandate.

The general manager of the chrome products manufacturing division passed on the extra cost (plus a normal profit) to the electric stove division. The electric stove division rejected the revised transfer price pointing out they had not requested the improved quality, the chrome products division argued that since they had been ordered to improve the quality by the staff vice president, they should not have to absorb the additional cost. Negotiations led nowhere and the electric stove division appealed to the finance office to arbitrate a transfer price.

The case illustrates how firms in the coordination phase attempt to get the best of both worlds by relying on market controls, including a sophisticated transfer price system, for the manufacturing and product divisions and a parallel command organization for the technical and administrative support side of the organization. The natural tension between the two systems came to a head over the stove-top transfer price. In terms of the dialectic model, this clash can be seen as a valuable situation whereby the autonomous market-oriented product division spontaneously attempted to resist nonprofitable quality improvements, while the technically oriented staff units took on the leadership role to ensure that GAC's products were of the highest quality, regardless of cost. Rather than seeing this as a dysfunction conflict, the model suggests it may be a natural and healthy confrontation in which two sides automatically provide the necessary checks and balances on each other. Coordination phase organizations strive to manage such tensions and the inevitable conflicts arising from coordination needs without precipitating a red-tape crisis.

Phase 5: Collaboration

Efforts to manage coordination phase tensions are frequently unsuccessful and the organization is overcome by a ponderous bureaucracy. Rule by red tape threatens to smother it. In order to surmount the new crisis successful organizations take steps to recover interpersonal relationships and to align organizational relationships more laterally than in the coordination phase. Sometimes cross-functional teams are put together to speed up complex problem solving. Or matrix management is introduced with functional, product, and line people coming together to debate, discuss, and make decisions with no party having more formal authority than the others. Social controls, including clan-like relationships, are frequently fostered in place of formal controls. Some companies in Europe even went so far as to restructure on a democratic parliamentary basis whereby the chief executive officer (CEO) and other top executives were elected by the employees, each of whom had one vote. Either way, the emphasis is on personal interaction and positive horizontal relations.

The consequence is that power and influence of headquarters staff experts melts away. Some regroup into consulting groups who now must sell their expertise to line people. Others are relocated into the line organization on a special assignment basis. Accounting and management information systems (MIS) departments strive to get real-time information to the right places. As these changes take place, the organizational ethos shifts from single career paths, specializing in one discipline, to a program of retraining and continuous education.

Conferences of like-minded managers and employees are held frequently. Sabbatical leave is granted so managers can catch up on leading-edge technology, knowledge, and developments at universities, government bodies, and international institutions such as the United Nations, the World Bank, and volunteer agencies in developing countries. Experimentation is encouraged and rewards are based on personal contributions to the problem at hand, rather than to conformance with administrative systems and edicts. The result is a gradual loosening and rejuvenation of the organization.

But pressures for innovation and the psychological stress of peer-group problem solving and decision making eventually take their toll. Managers feel mentally burned out. Japanese companies are notorious in this regard, literally forcing managers to spend most weeknights entertaining customers in long bouts of drinking, eating, and carousing and to attend weekend attitude training courses. In consequence, organizations must find ways of permitting a more balanced life-style for their employees. Some companies accomplish this through rest and recreation programs. Others install gymnasiums, squash and tennis courts, and arrange daily aerobics sessions and jogging groups. Some even introduce yoga classes and stress management programs.

While these measures help, organizations today face a new crisis. Managers and employees still sense a lack of spirituality. The specter of the economic and the material needs of society with its discourse of markets, competition, survival of the fittest, and wealth accumulation still prevails. In all likelihood, it will be some time before spirituality pushes economics and politics off center stage. Yet there are signs. The dawning recognition of the egregious ruination of the world's life support systems—air, water, farmland, lakes, forests, the ozone layer—will eventually force us to design and give pride of place to systems of accountability that feature environment and quality of life. This *spirituality crisis* is not limited to individual corporations, it invades all aspects of the living world.

GENERAL ELECTRIC: A CASE ILLUSTRATION

The history of the General Electric Company (GE) vividly demonstrates aspects of the historical–dialectical model of control.[9] In 1878, a small group of electricity enthusiasts formed the Edison General Electric Company to exploit the commercial opportunity presenting itself in electricity, electric lighting, and electric motors. For the next seventy years,

GE operated pretty much as a direction phase company with a traditional functional organizational structure featuring research, engineering, manufacturing, marketing, and financial divisions. Direction and strong leadership took GE to over $2 billion in sales by 1950.

But in the early 1950s GE's top management sensed the functional organizational structure was stifling innovation, flexibility, entrepreneurship, and managerial initiative throughout the company. In consequence, in a dramatic move, they reorganized GE into the delegation mode by decentralizing responsibility and authority for operations to nearly one hundred product departments, treating them as profit centers. Entrepreneurship, flexibility, and aggressive marketing flourished in these new departments and GE, capitalizing on the favorable economic conditions in the next two decades, experienced explosive growth in sales.

By 1970, GE featured 160 profit product departments reporting to headquarters through about forty divisional general managers, who in turn reported through one of ten vice presidents or group executives, who reported to one of three top executives at headquarters. Growth in profitability, however, lagged behind the spectacular boom in sales. Moreover, opportunities to economize by sharing R&D, technology, and administrative functions went unrealized. GE was ripe for coordination.

GE overlaid the department, division, and group hierarchy with a layer of strategic business units (SBUs), each of which represented as far as possible a stand-alone business. Headquarters also put into effect a strategic planning system which called for each SBU to prepare a strategic plan for its business. The information which emerged from the strategic planning process allowed top management to identify wasteful overlaps, initiate cross-businesses coordination, and allocate resources to areas of competitive advantage and away from areas with limited potential.

In 1977, top management added yet another layer of administration to the existing organizational structure by installing five sector vice presidents just below the headquarter's office. Each of the five sectors—consumer products and services, power systems, technical systems and materials including aircraft and aerospace operations, industrial products and components, and international operations—represented a separate industry. The sector executive became Mr. General Electric and acted as CEO and GE's institutional leader for that particular industry. A very important part of the job for these executives was to prepare a detailed strategic plan for their sector, reviewed by headquarters to pinpoint GE's major areas of competitive advantage, allocate resources, identify duplication and wasteful overlapping activities, and above all to exploit potential synergies across sectors and SBUs. In consequence, staff planning offices sprang up throughout GE.

Not surprisingly, as the historical–dialectical model predicts, the inevitable red-tape crisis hit GE in the early 1980s. Recognizing this threat, newly appointed CEO Jack Welch, previously SBU head of technical systems and materials and later chief executive of GE's huge consumer products and service sector, embarked on a massive restructuring and downsizing exercise. He culled the headquarters planning department to eight people from thirty, reestablished responsibility for strategy to line operating managers, eliminated layers of administrative managers, pruned product lines, closed down marginal and nonproductive plants, and introduced advanced manufacturing equipment in the remaining ones. He also eliminated the five sector offices, and reorganized GE's 150 separate businesses into fifteen lines of business, lumping them into three circles or groups.

When the smoke cleared towards the end of the decade, GE had divested 125 businesses and eliminated a staggering total of over 300,000 employees. For better or worse, Welch had "taken out the layers," "pulled out the weeds," and "scraped off the rust" of the GE family of businesses.[10] But the inexorable, massive downsizing left GE with a dispassionate "survivor mentality" cadre of middle and lower-level managers and a "running scared" work force.[11] A *spirituality crisis* was in full swing.

In the face of this pivotal moment in GE's history, Welch embarked on a long-term "spiritual revitalization" program designed to shift managers' and workers' focus "from cost-cutting to the murky realm of human values...pushing soft values because he sees them as the only way to maintain the pace of GE's productivity drive...by inspiring the remaining workers to produce more" (Tichey and Sherman 1993, p. 240). Welch aimed to do this by instilling throughout GE a new "spiritual credo" containing values which he hoped all employees would not only understand but, more importantly, would feel passionately about.

The new spiritual credo contained several major injunctions. For instance, all employees were warned that GE would not ensure job security, only satisfied customers could do that. Also, they were enjoined to develop a high level of emotional commitment to the new values. Moreover, the credo demanded that every business component achieve number one or two status globally in the design, technology, and quality of its chosen product line. In order to accomplish this, hierarchical grouping labels would be erased, cross-functional barriers would be removed, and the distinction between domestic and foreign operations would be eliminated. GE would become an organization without boundaries.

In order to institutionalize the new credo, Welch, started at the top of the organization and set up the Corporate Executive Council (CEC) composed of GE's thirty highest-ranking business unit chiefs and senior headquarters staff directors. CEC met quarterly to discuss the most important problems facing GE at that time. In order to encourage a wide open debating style, meetings were held in a small amphitheatre along the lines of a miniature Roman coliseum. A topic would be teed up and combatants took each other on until they reached a consensus of sorts. Unlike the old days, everyone at the meeting received the details of each others' financial results.

This open atmosphere, Welch believed, induced participants to share information, generate solutions to each others' problems, and most importantly, break down parochial barriers so that good ideas mined from one part of GE were shipped to the rest of the company. By 1988, CEC executives had not only come to understand and embrace the new credo, but also they became an effective force for pushing the shared values further down in the organization. Similar councils were organized at the next level in each of the thirteen major business components. Finally, in order to get employees and the work force to "walk the new talk", Welch set Operation Work Out in motion.

Operation Work Out consisted of a series of meetings patterned after "good old Yankee" town meetings. Groups of thirty to a hundred hourly and salaried workers within a particular business met off site for two or three days to discuss common problems. In order to promote candor, bosses were locked out during discussion times. Initially, employees spent a lot of time airing complaints and griping, but later the meetings took on a more constructive mode as participants started to identify and define work-related problems *and* to develop concrete proposals. The bosses then returned to the meeting and were required to make on-the-spot, public decisions regarding the proposals, particularly for problems which could be fixed readily. Operation Work Out not only empowered lower-level employees but also served to expose any managers and workers who were not walking the talk.

Welch summed up the new spiritual credo this way:

> The only way I see to get more productivity is by getting people involved and excited about their jobs. You can't afford to have anyone walk through a gate of a factory, or into an office, who's not giving 120%. I don't mean running and sweating, but working smarter. It's a matter of understanding the customer's needs instead of just making something and putting it into a box. It's a matter of seeing the importance of your role in the total process...The point of Work-Out is to give people better jobs. When people see that their ideas count, their dignity is raised. Instead of feeling numb, like robots, they feel important. They *are* important...With Work-Out and boundarylessness, we're trying to differentiate GE competitively by raising as much intellectual and creative capital from our work force as we possibly can.
>
> Trust is enormously powerful in a corporation. People won't do their best unless they believe they'll be treated fairly–that there's no cronyism and everybody has a real shot. The only way I know to create that kind of trust is by laying out your values and then walking the talk. You've got to do what you say you'll do, consistently, over time. It doesn't mean everybody has to agree. I have a great relationship with Bill Bywater, president of the International Union of Electronic Workers. I would trust him with my wallet, but he knows I'll fight him to the death in certain areas, and vice versa. He knows where I stand. I know where he stands. We don't always agree—but we trust each other.
>
> That's what boundary-less is: An open, trusting, sharing of ideas. A willingness to listen, debate, and then take the best ideas and get on with it. If this company is to achieve its goals, we've all got to become boundary-less. Boundaries are crazy. We're not that far along with boundarylessness yet. It's a big, big idea, but I don't think it has enough far on it yet. We've got to keep repeating it, letting everybody know all the time that when they're doing things right, it's boundaryless. It's going to take a couple of more years to get people to the point where the idea of boundarylessness just becomes natural (Tichey and Sherman 1993, pp. 248–249).

In order to ensure that the new values would stick, Welch revamped GE's management compensation scheme. For the top four hundred executives, take-home pay would consist of 50 percent bonuses while the next top four thousand managers could receive one-quarter of their compensation in the form of performance incentive bonuses. Most of these executives and managers also got stock options. Two main criteria were used for awarding bonuses and stock options: financial performance *and* demonstrated adherence to the new, shared values. The latter was measured on the basis of ratings by peers, superiors, and subordinates, as well as self-ratings. Thusly, an individual's accomplishment rating included a pragmatic financial measurement as well as a subjective assessment of his or her emotional commitment to the new spiritual credo.

These recent developments at GE can be read as a striking example of one organization's attempt to develop a blueprint for coping with the spirituality crisis. Welch's revitalization efforts—pushing soft values, promoting a boundary-less organization, creating cross-functional situations like CEC and Operation Work Out, assessing commitment to the new credo, encouraging a clan-like atmosphere, and diffusing the power of HQ staff experts—mirror the prescriptions of the historical–dialectical model for coping with the spirituality crisis. It remains to be seen whether or not Welch's gambit, with its injunction to "control your own destiny or someone else will," can pull GE through its current crisis. In spite of its emphasis on employee empowerment, the new credo's materialistic injunction—produce or else—seems to outweigh its spiritual precepts. Either way, the GE experience provides a striking example of one organization's attempt to develop a blueprint for coping with the tensions of the collaboration phase.

SUMMARY

This chapter presented four frameworks for investigating management control issues and problems. The first model, relying on a biological metaphor, saw the problem of organizational survival as one of adapting social technologies to the changes taking place in its environment. The next model relied on classical economic notions to develop two caricatures of control and coordination: command-based and market-based organizations. The third model built on the cybernetic view of organizations as information processing entities and outlined different strategies for coping with the need to reduce uncertainty when information needs exceed the organization's available processing capabilities. The final model relied on a historical–dialectical framework to identify coordination and control needs during five phases of organizational evolution.

Each approach contains its own biases, strengths, and limitations. Moreover, they tend to ignore agency, that is the actions and decisions of key actors who play important roles in the survival and growth of organizations. So they put aside the actions of great leaders such as Ghandi, Churchill, Gorbachev and the likes of Sloan, Iacocca, and Welch. The role of human agency will be addressed in Chapters 10 and 11, which outline a scheme that includes *both* structures and agency. For now, it has been sufficient to concentrate on organizational context and the problems of coordination and control from an organization-in-the-round perspective. The next chapter puts our frameworks into action in the context of two detailed case studies.

NOTES

1. See Hall (1962) for a pioneering study along these lines.
2. Arrow (1964), p. 3. Arrow explains his perspective this way: "My point of view is rationalistic and derives, with appropriate changes, from the logic of choice as it has been developed in the pure economic theory of prices and the mathematics of optimization."
3. See Schick, Gordon, and Haka (1990) for an excellent review and critique of the information overload phenomenon.
4. This model is adapted from Galbraith's (1973, 1977) pioneering work. See also Amey (1979, 1986) for a detailed and technical application of cybernetics to management accounting and control systems.
5. This framework is adapted from Greiner's (1972) model of organizational evolution and revolution.
6. "*Dialectics* come from the Greek term for dialogue, whereby a residue of give-and-take and of relentless questioning continues to inform the context. At the core of all dialectics, we find a continuation of incessant querying and an active engagement with the resistant stuff of knowledge. Dialectical inquiry is epitomized by Plato in the person and style of Socrates" (Heilbronner 1980, p. 31).
7. A logical contradiction refers to the situation where two statements contradict each other as is the case if one statement holds "A is true" but another statement says "not A is true."
8. See Anthony, Dearden, and Govindarajan (1992), p. 255 for a description of GAC.
9. This brief history relies on the case studies General Electric Company (A) and (B) in Anthony, Dearden, and Bedford (1989); "Background Note on Management Systems: 1981" *Harvard Business School;* "GE's new billion-dollar small businesses," *Business Week* December 19, 1977; and "Why Jack Welch is Changing GE," T.J. Lueck, *New York Times*, 1985.

10. Tichey and Sherman (1993), p. 8. The details of this "spiritual" revival are described in some detail, albeit in a highly protagonistic way, by Tichey and Sherman (1993) who provide an inside look at GE from the point of view of a long-time GE consultant and business professor and a senior writer for *Fortune* magazine.
11. See "An American Workplace, After the Deluge," Peter T. Kilborn, *The New York Times*, September 5, 1993.

5
Case Illustrations

The previous chapter introduced the idea that organizational structure and management systems might vary according to the type of environmental conditions faced, the availability or lack of omniscient information, the degree of uncertainty prevailing, and an organization's historical evolution. The contribution to control system design of this way of thinking is different from approaching management accounting and control as a problem of technical design. Rather it seeks a congruence in the characteristics of these systems with those of the organizational setting including characteristics such as environment, technology, available knowledge, type of work, uncertainty, and the state of the dominant crisis of its dialectic tension. The general thesis is that factors which shape context and organizational structure may also account for important differences in our management accounting and control systems.

Conceptual apparatuses which link impersonal forces with organizational processes would seem to hold great potential for explaining some of the major variations we see in the design and use of management accounting and control systems. When we see how these systems can be rearranged in various settings so that they are no longer random, we may develop a clearer notion of how system designers should proceed. This proposition is demonstrated next in some detail by using the frameworks in Chapter 3 to analyze two case studies of organizational design and control: a national consumer finance company and a high-technology computer company.

TRANSAMERICA FINANCE COMPANY

Our first case situation is relatively straightforward. It concerns the management control system at Transamerica Finance Company (TFC), a large consumer finance company.[1]

Case Information

TFC was one of the largest finance companies in the nation with hundreds of branch offices across the country. Its primary business was the acceptance of conditional sales contracts from customers who had purchased consumer goods. Usually customers could not get loans from chartered banks and so turned to commercial loan companies for finances. The company paid the retailer while the customer paid TFC in monthly instalments. It operated nationally and had several regional offices located in major cities. Branch offices were located in most cities and towns including downtown areas and suburb shopping centers. The branch offices varied in size from six to fifty personnel depending on the market served and the amount of money loaned. Branch offices received money for loans from the Central Office located in a large city in the center of the nation and at the end of each working day all branch bank accounts were closed out to a central office bank account. Branches, however, reported through one of fifteen district offices to the central office.

The corporate central office functioned as the hub for strategy, policy-making and administrative systems. Central office established specific policies and procedures regarding every aspect of branch operations, including loans, collections, record-keeping, branch control and reporting methods, personnel policies, and detailed job descriptions for all branch positions and office administration procedures. These standard operating procedures, designed to ensure consistent branch operations across the nation, were contained in procedures manuals located in each branch. They emphasized that branch managers should aggressively seek new profitable accounts, make sure payments were received on time, and run a neat, tidy, and efficient office with all records and reports continually up to date.

On the third working day of the month each branch submitted to the central office, as well as to its district office, a report for the previous month. The report contained the essential statistics about branch operations, including the number and dollar value of all loans, details of collections, branch expenses by line item category, and a detailed schedule which stressed overdue accounts and delinquency rates. These reports were sent to the central office for computer analysis of branch operations including a multiple regression analysis which analyzed office expenses and personnel levels according to the number of branch transactions for the month. The computer also compiled this information into a district report, which compared all branches in the district and ranked branch performance for each district. These reports were provided to each branch so they could compare themselves with other branches in their district. Branch managers, very sensitive about their relative performance, eagerly awaited this report each month. Interestingly, budgets and long-range plans were noticeable by their absence.

The district managers closely monitored the monthly reports, scrutinizing them for anything that appeared out of line, especially new loans and accounts which were over thirty days overdue. If a branch seemed out of line, the district manager and a staff member paid a quick visit to get to the bottom of the problem; and, if warranted, a staff expert was assigned to the branch until performance reached a satisfactory level. In addition, each district manager conducted a thorough, on-hand inspection of each branch at least once a year. During the visits, which were made on a surprise basis, the district manager and his staff performed a comprehensive audit of all loan and collection records as well as inspection of employee performance evaluation and pay rates. At the end of this visit the district manager conducted the annual performance evaluation review of the branch and assistant managers. The review included setting objectives for the next year regarding any aspect of branch operations needing attention.

More recently, TFC had installed an on-line computer system which connected each branch to a large computer at the central office. Branch transactions were entered each day, and a summary of operations by branch, district, and for the entire company was submitted early the next morning to the president. Branch managers could also call for an update on their branch operations for the month to date.

Branch managers operated their branches with some autonomy, hiring and training their own staff personnel and taking full responsibility for loans, collections, and expenses. They could earn an additional 30 percent of their annual salary in the form of a bonus based on branch performance. In addition, company-wide contests were a regular feature where the best-performing managers won, say, a free trip for two to Europe. Branch managers were highly satisfied with these arrangements including the controls and reports, and particularly the autonomy and responsibility that they perceived with the job.

Analysis

At first blush, the management accounting and control system at TFC seems excessively comprehensive. Standard operating procedures cover all aspects of branch operations. Every nook and cranny is monitored, reported on, and scrutinized on the spot. Surprise visits and spot checks are the order of the day; and branch performance is reported daily to the president's office. It would seem to be a prime illustration of the very kind of oppressive and punitive control system described by some human relations accounting researchers.

Analysis of TFC's accounting information, and control systems with the models in the previous chapter, however, yields a different perspective. In terms of the mechanistic and organic control model, it is apparent that TFC has a well-defined social technology. Tasks are divided into distinct slots. Precise duties are defined for each branch position. Power and authority arrangements are unambiguously allocated to corporate, district, and branch managers. Heavy reliance is placed on vertical hierarchic arrangements; and operations are coordinated from the top, the only place where overall knowledge of the firm resides. The simple but forceful management control system reinforces this pattern. Instructions flow downward through the hierarchy, becoming more explicit at each successive layer in a unilateral and autocratic control style. TFC seems almost the archetypical mechanistic firm.

This social technology, nevertheless, is well suited to the firm's circumstances. Markets are well defined. Procedures for operating a branch are well understood. Scientific knowledge is minimal. With the exception of the short-term money market, the environment is stable. TFC, like most consumer financial companies, had a precarious financial structure consisting mainly of short-term money market notes, some long-term debt, and very little equity. It turns a profit on the margin between short-term borrowings and interest rates charged to customers. Consequently, it is absolutely essential that the branches are tightly controlled and responsive to calling in or expanding loans depending upon short-term fluctuations in the money market. TFC's mechanistic social technology is well matched to its operating environment.

According to the command–market model of uncertainty and control systems, TFC mirrors the centralized prototype. Top management have at their disposal, at a low cost, omniscient information about nearly every circumstance anywhere in the organization. Centralized coordination is achieved through specific and detailed operating rules contained in the standard operating procedures. These are complemented by enforcement rules which detect and report violations from the procedures to upper echelons. The branches are monitored closely, and visits by district managers follow quickly if anything untoward is detected. So uncertainty is low; in fact, it cannot be tolerated because of the precarious capitalization. The centralized management system fits the top management environment perfectly.

Analysis of the TFC situation using the ideas in the information processing model of control confirms the above conclusions. Output diversity is low, goals are reasonably achievable, rules are in abundance, and reporting channels are more than adequate to handle information processing requirements. The branch offices are self-contained and highly homogeneous. Consequently, the amount of information to be processed is relatively small and most if not all is available to top management. Formal control information is simple, low in quantity, and handled readily with a straightforward but forceful vertical information system. TFC has made an appropriate response to its uncertainty-creating variables.

Finally, TFC is readily situated in the dialectical model of organizational development as an archetypical direction phase organization. Sound administrative practices provide close and tight control over every activity at the branches. The formal hierarchy is clear-cut, job responsibilities are unambiguous, and direction and authority prevail. The highly directive lean and mean management control system ensures that order and routine prevail, the necessary conditions for efficient operations in a highly competitive environment. The vitally needed direction and authority are institutionalized. Moreover, TFC has taken steps to fend off any autonomy crisis. Branch managers experience a great deal of freedom to run their operations according to the standard operating procedures while healthy performance bonuses and frequent interbranch competitions with exotic prizes keep branch managers from stagnating. TFC's organization functions like a well-oiled machine.

APOLLO COMPUTERS[2]

Our second case study traces the evolution of the organizational design and control through a corporate takeover and reorganization of Apollo Computers (AC). With headquarters in Monterey, California, AC was a relatively small and successful company specializing in sophisticated computer applications. Its products included computers for advanced weaponry and space explorations; microprocessors for trains, trucks, and ships; and integrated data base management systems. AC also conducted highly sophisticated research in sundry areas like bubble memory chips, DNA engineering and molecular rectifiers, and crystal lattice computers. At the time of the takeover, sales had grown to nearly $300 million annually from $10 million in 1978.

Case Information

AC's organization structure before the takeover was essentially functional. Different executives were assigned responsibility for R&D, engineering, manufacturing, marketing, and finance. Interfunctional problems and issues went up the hierarchy for resolution and decisions were passed down. While top management attempted to coordinate the various functions from on high. Recently these arrangements seemed to be working less than optimally. Decisions on differences of opinions at lower levels moved very slowly up and down the vertical hierarchy. R&D efforts were inefficiently allocated among product lines. Attractive market opportunities were lost when other companies, quicker off the mark, jumped in to steal business from under AC's nose.

The manager of organizational development had already begun to identify these problems. She concluded that the present organizational design was ill suited for exploiting new product opportunities. The different functional people simply were not working together. Many reasons were suggested for this. The managers did not know who held the responsibility and authority to develop new products. Upper management did not pay enough attention to this activity. Delays occurred as information travelled up and down the proper functional channels. R&D efforts appeared to be disproportionately allocated among product lines due to the personal influence of a few product specialists. Further, squabbles seemed continually to erupt among the functional managers over designs and production schedules. These disputes had to be sent up the line to top management and so were delayed, postponed, or even left unresolved.

These symptoms underlined the difficulties encountered by the various functional groups as they wrestled with new product development. In a few instances, however, engineers, marketing managers, and manufacturing personnel had formed informal work groups and successfully worked out common problems for specific products. As well, a partial division of executive responsibilities by product groups had been established as functional jurisdictions began to melt. Further, the controller had instituted monthly product line profit and loss statements and a few small, informal groups, composed of engineers, plant people, and marketers had sprung up to work out common problems. These people participated in company rest and recreation activities such as aerobic sessions, computer chess competitions, and 'Cal-Mex' cooking seminars. During these nonwork get-togethers informal groups discussed specific product problems.

Analysis

AC's situation can be understood using the frameworks described in the last chapter. According to the mechanistic–organic model, AC operated in an unstable environment. There is ample evidence for this conclusion. Engineering technology changed rapidly. Products required a high degree of engineering competence. AC custom-manufactured products according to the needs and whims of customers. And the market for the products had grown rapidly. Program conditions were characterized by a high rate of change in both markets and technologies.

Turning to AC's social technology, we see evidence that they were employing a mechanistic management structure. The organization was essentially functional with different executives assigned responsibility for manufacturing, marketing, engineering, and finance. Vertical hierarchical arrangements predominated for issuing orders. Conflicts went up the hierarchy for resolution and decisions were passed down. Top management attempted to coordinate the various functional units from on high. The top management structure, it would appear, was mechanistic.

According to the model, AC had a serious mismatch in that a mechanistic structure does not suit an unstable environment. In fact, signs of conflict had begun to appear as continued squabbles occurred between departments. Moreover, the informal social technology was moving towards more organic arrangements. Top management had begun to sense that the degree and kind of instability in markets and scientific technology facing the firm was not well served by its social technology.

The command–market model leads us to a similar conclusion. Top management seemed to lack information about the state of affairs, events, and problems at lower levels and at the boundaries of the organization. Enough information was not being transmitted upwards to top management. Messages travelling through the transmission channels bogged down.

As a result, operating rules issued by top management were often inappropriate, and their efforts to coordinate the complex flow of products through the functional departments to customers were ineffective. The use of the vertical channels, and attempts to follow higher-level plans, meant delays which cost market opportunities. The centralized style of management arrangements was inappropriate to the amount of information available at the upper strata of the organization and top management's ability to assimilate it.

Next, the cybernetic model hinges on the difference between the amount of information required for processing relative to the amount already processed by the organization. The amount required is a composite of output diversity, input resources utilized, and the difficulty of the level of performance required. For AC's mix of activities it seems clear that the level of both output diversity and input resources utilized was high. The highly technical nature of the product lines bears witness to this conclusion. A high degree of engineering competence was required for most products.

According to the cybernetic model, the amount of information required for processing was great, if not vast. Although we have no direct documentation of the amount of information already processed by the organization, several factors indicate that it was much lower than required. Coordination of the functional departments, particularly on new product development, was not adequate. A comprehensive plan did not exist. Sensing the need for more communication had begun to produce financial information relating to product line. All this suggests that a strategic organizational response was badly needed in order to close the gap between the amount of information required and the amount available.

From the perspective of the model, AC had experienced a remarkable change in its uncertainty-creating variables over the past decade. Products, services, and customers had become highly diverse. The number of departments, offices, and manufacturing centers had multiplied many times over. And the technical tolerance and quality levels for products had risen significantly. In consequence, the traditional means of handling information through rules, hierarchy, and budget targets could no longer keep up with the amount of information processing required to cope with the new uncertainty level.

Finally, using the historical–dialectical framework it seems clear that AC was at the end of a successful and prolonged direction phase but now faced an autonomy crisis of severe proportions. The company had grown in complexity and diversity. Lower-echelon employees felt restricted by a cumbersome, centralized hierarchy. They knew much more about the technology, markets, and recent scientific developments of the information technology market than did the bureaucrats at the top. The functional hierarchy, the booster engine for direction during the 1980s, was now sputtering. Within its historical dialectic, the forces of direction and leadership had severely repressed those of autonomy and creativity. A reversal seemed urgently needed if AC was to survive its autonomy crisis.

More Case Information

> At this critical juncture in its history, an event occurred that may well have saved AC from stagnation and even demise. The major owner, who had amassed a tidy fortune from AC's public offering of common shares and its stock option program, had joined an awareness cult based on the Kegon School of Buddhism that stressed the Buddahood of all sentient beings, the identity of nirvana and samsara, and the wisdom and compassion of the bodhisattvas. At forty-four years of age he wanted to devote the rest of his life to finding his inner spirituality and working for the green earth movement.
>
> At the same time, three former middle-level managers of Dionysus Corporation (DC), one of the world's largest computer companies, had cashed in their stock options, taken the company's early retirement payoff, pooled their personal assets, and bought a controlling interest in AC. They believed that with proper management skills and AC's superb technical expertise, they could get the company back on track and eventually even challenge some of the world's computer giants.
>
> The new owners moved swiftly to reorganize the company along lines similar to Dionysus. They instituted a new profit-centered organization featuring decentralized profit-responsible product groups. These groups were assisted by HQ staff departments such as R&D, engineering, software, sales, and finance. Several product departments were established within each product group. Each product group was headed by one manager in charge of marketing, engineering, and production units for his or her assigned products. Their main responsibility was to coordinate all activities of their product lines, including the HQ sales force, R&D, and engineering departments, which were not under their command. Their mandate was to assure that the products were profitable.
>
> Within the product group, the marketing unit worked on marketing strategy, pricing, contacting customers on special requests and factory problems, promotion, and new product development. The production unit was responsible for efficient manufacturing, meeting delivery dates, and

production costs. The engineering unit designed new products, devised new production processes, and worked on special customer requests. A production control manager looked after scheduling of work, supervised expediting, shipping and delivery, inventory, and purchasing. The product group managers, however, were the kingpins of the new organizational arrangements.

More Analysis

These changes correspond closely to the prescriptions of our theoretical models. They would, for example, prescribe a shift to organic management structures to suit the high rate of change in the external circumstances of their scientific knowledge, markets, and in all likelihood, their production technology. This seems to mirror the actual events. Powers and duties shifted from functional and vertical responsibility to product and lateral accountability. The product department managers became responsible for managing the lateral relationships involved in the design, marketing, production, and selling of the products assigned to them.

So they had to have considerable understanding of the various overall goals and tasks of the entire AC organization. Matters such as R&D and the sales force, however, were not under their jurisdictional authority. Duties and powers in these areas would be in a state of continual negotiation and flux. Top management was no longer in a position to know everything about the various product centers. The new management structure had shifted from a mechanistic to an organic social technology.

We also see that AC no longer attempted to coordinate its complex flow of products from the top of the hierarchy by issuing clear, firm commands based upon omniscient information and following up with punitive enforcement rules and sanctions. Rather, they reorganized around quasi-independent, miniature, free-market product groups—each with responsibility, if not the entire authority, for profitable survival in its own market sphere. A market structure replaced a command one.

In terms of our cybernetic model, the company previously seems to have experienced a gap between the amount of information required for processing and the amount available. The response followed two major strategies. First, AC created slack resources by some duplication of manufacturing resources and by investing in inside marketing personnel. Second, AC created self-contained units in the form of profit-responsible product departments. These two moves reduced the need for information processing.

At the same time, AC also increased its capacity to process information. The new product–profit responsibility required a horizontal accounting information system that provided information on financial performance for product groups. To effectively manage their new responsibilities, the product group managers needed to be involved in a significant amount of lateral communication, negotiation, and building of reliable relationships. They were required to manage the product flow from R&D through engineering, sales, manufacturing, delivery, customer service, and into the customers plant.

The company, then, employed all four coping responses: creation of new self-contained units, employment of organizational slack, development of vertical information systems, and creation of new lateral relationships. The new organizational design seems better suited to the amount of information processing needed.

Finally, viewing the situation from the perspective of our historical–dialectical model, we see that the new owners overcame an autonomy crisis by putting into effect a decentralized organization structure anticipating an era of growth through delegation. Much greater responsibility was assigned to the product group managers. The new top executives, accustomed to the DC system of restraint from interfering in operating decisions, were more than willing to follow the path of management by exception based on periodic reports from the product groups and staff technical offices. With the dialectic tension resolved in favour of autonomy, they could concentrate on long-range planning and searching for new acquisitions that could be lined up beside the other decentralized product groups.

These moves proved to be the right formula for surmounting the autonomy crisis. Delegation of product group responsibility provided heightened motivation and creativity at lower levels. The product group managers eagerly took the bull by the horns and, taking full advantage of their greater authority and expanded mandates, responded rapidly to customers requirements, penetrated larger markets, and developed new products. Another phase of growth and expansion was under way.

More Case History

Along with these changes came a new approach to management accounting and control system design and utilization. First, strategic planning was considered to be a live activity and was delegated to each of the product group managers. Each of these managers conducted a detailed assessment of their group's strengths and weaknesses as well as the risks and opportunities inherent in their selected product market spheres. They then prepared a qualitative report documenting the unit's competitive strategy for the next five-year period. The report included a very general balance sheet, income statement, and cash flow estimate for the next five years. This report was reviewed by a select committee of top-level executives. They reviewed it closely, especially its basic assumptions and either approved it or asked the product group manager to take part in a face-to-face meeting where the plan would be reviewed in detail. Usually these proved to be intensive and gruelling meetings and became known as hell sessions. The main focus was on opportunities that might be missed. The meeting ended when the plan was given final approval.

Each product group also prepared budgeted financial statements for the coming year. Upper management normally looked for an increase in the sales and profit targets of 40 to 60 percent because the computer and information technology industry was growing at this rate. Plans that came in below these levels were usually bumped up by top management to the required level. Actual performance compared to budget was reviewed monthly by upper management and each product group manager was given a formal face-to-face review every four months.

These new arrangements were received enthusiastically by the product group managers and their employees. Several managers developed their own internal operational control systems including profit–volume–cost information systems, including standards and actual outcomes to help them identify specific variances from a plan and to track down the reasons for them. Each department within the group participated in developing these data. These self-designed systems helped product group managers focus their attention and efforts on developing new products, seeking new customers, servicing old customers better, improving marketing efforts, or improving manufacturing processes. The field units were free to run their own show.

The relationship between bonuses and promotions to budget performance was deliberately left ambiguous because the industry could unexpectedly surge or go flat. The general feeling was that short-run profits were important. Furthermore, product group managers knew full well that the new owners' performance in relation to their corporate plan was being constantly monitored by stock market analysts in Los Angeles and San Diego. They also knew they were expected to take any steps necessary to meet the profit objective. During periods when they were below plan, there was considerable pressure to increase sales efforts, meet with R&D to develop new products, and to reduce asset levels.

The philosophy of control that pervaded Dionysus Corporation's global operations had been transferred to Apollo Computers. The main feature was complete decentralization of profit responsibility but with built-in tough targets. This way profit center managers were limited in the degrees of freedom available for building in organizational slack in operating budgets. And the pressure was kept on them for growth. The philosophy held that this was an excellent way for managers to develop business acumen. Those that did were readily identified and rewarded; those that did not were quickly relegated to technical jobs or dismissed.

Further Analysis

The characteristics of the new management accounting and control system follow closely the prescriptions for firms operating in relatively unstable circumstances. Much critical environmental information about customers, technology, and markets is known only by the product group managers at the firm's lower levels. The new management accounting and control system highlights the effectiveness of the flow of product through the organization to the customer. So lateral consultation and exchange of information, especially about products and product performance, ensued. An important part of these exchanges took place during the formulation of the profit budget and later when profit performance was reported against plan.

Turning to the command–market model, since AC operated under highly uncertain circumstances it was best served by decentralized managerial arrangements which featured a free-market economy. The reason for this is that knowledge of environmental conditions, market prices, and customer idiosyncrasies were in the hands of the product group managers at the tentacles of the organization, rather than with upper management. Product group managers were closer to customers so they could respond quickly and accurately to marketplace changes. Responsibility for the profit of a cluster of products was decentralized to a product group manager who had command over most of the resources and people involved in a cluster of product–market relationships. The sundry product-based, profit centers employed by AC were an attempt to create a miniature free-market economy.

The new operating rules for the product groups were general—maximize profits and long-run growth. These operating rules were backed up by strong enforcement rules. Top management met weekly to review the performance of each product group to pinpoint trouble spots. And product group managers were given a formal performance review every three months. The mandate called for profits *and* growth.

Actual profit and sales growth performance was constantly being evaluated against planned levels. These enforcement rules were designed to encourage product group managers to increase profits and sales as much as possible within their sphere of responsibility. Managers who did so would be recognized and appreciated. The new management accounting and control system, along with the new organizational decentralization and the revised operating and enforcement rules, conform closely to the theoretical prescriptions for organizations that operate in relatively uncertain environments.

From the perspective of the cybernetic model, the new organizational design helped considerably to reduce the need for information processing in two ways. First, it reduced substantially the amount of vertical information flowing to upper management. They now relied on short, standard, and general profit budget reports from the self-contained product groups to keep track of operations. Only the self-contained units processed the specific information they needed to manage their sphere of responsibility. Lateral information flows to and from other product groups were now minimized; they were limited to exchanging sales forecast information and other strictly business transactions. These new arrangements brought the amount of information processed in line with the amount required.

Finally, these new developments fit closely with prescriptions from the historical–dialectical model. Communication between upper management and the operating product groups was now infrequent. Upper management reviewed and approved the various strategic plans and restrained themselves to management by exception using monthly reviews of actual versus planned profits. The product groups, as witnessed by events in the AC profit center, responded very positively to their newfound autonomy. AC seems well poised to enjoy a prolonged period of steady growth and progress.

But there is no doubt that the new management accounting and control system creates a good deal of tension in the organization. The motivation to meet short-run profit and sales targets is strong, as is the motivation for long-run growth. It is up to the autonomous units to figure out the appropriate (and often difficult) trade-off between the short- and the long-run. And while participation in target setting is widespread it is clear that top management have the prerogative of having the last word in this regard. And for some, the ambiguous relation between rewards and short-run budget performance might be a worrisome aspect of the new management accounting and control information systems.

Yet some tension was required to make the new concept of product profit centers work. The regular review of performance provided the necessary motivational force; and participation in budget setting was widespread even though top management, with their broader view of the total scheme of things, was entitled to have the final say in setting budget levels. Delegation of profit responsibility without some way of providing motivation, without a means of monitoring results, and without a recourse to adjustment of inappropriate budgets, would seem to be both impractical and imprudent. Finally, research has indicated that formula-based reward systems are more appropriate to relatively certain operating conditions than to dynamic environments. And some ambiguity in rewards keeps people on their toes.[3] This macro-perspective does provide a broader picture and the models do seem capable of giving us new insights into problems of designing effective management accounting and control systems.

OTHER ILLUSTRATIONS

We close this chapter by returning briefly to the case studies in Chapter 2 examining each in light of the frameworks in Chapter 3. Wedgwood led the potbanks in both product and production technology and his competitive strategy was light-years ahead of the competition. With the exception of the period of worldwide economic recession, Wedgwood's environment was relatively stable. Not surprisingly, then, Wedgwood operated a highly mechanistic social technology. Overall knowledge of the firm was available only to him. Strong, clear commands were issued verbally during his inspection tours and were written down in the potter's instructions. His control system was simple but forceful. Thus Wedgwood seems to be a prime example of the successful employment of a mechanistic social technology that was well suited to its relatively stable environmental circumstances.

Wedgwood also seems to be archetypical of the command organization. Josiah Wedgwood, a remarkably low-cost and efficient information channel, had access to almost complete information about the sundry conditions prevalent throughout the firm. He operated with an extremely centralized and highly structured vertical hierarchy which he used to coordinate and control the flow of products through the organization. He wrote and disseminated specific and detailed operating rules for every aspect of the operation including comportment on the job. He followed these up with enforcement rules which detected and punished any violations of the operating rules. Wedgwood's centralized management system, a paradigm of the command organization, was logical, internally consistent, and complete.

Finally, from the perspective of our historical–dialectical model, Wedgwood is an outstanding illustration of a firm that successfully moved through the creative phase, overcame the leadership crisis, and settled into a lengthy and prosperous evolution. Josiah Wedgwood, even though immersed in mundane, petty administrative matters, such as the crisis in the countinghouse, foresaw the need for strong leadership and direction, and put in place sound management practices and formal systems of rules and hierarchy necessary for the direction phase. He successfully bureaucratized the company and came up roses.

The glass products division (GPD) of Empire Glass can also be slotted into the various frameworks. GPD seems to be clearly operating under stable environmental conditions in a mature market with a commodity-type product. Once the sales forecasts are analyzed, adjusted, and agreed, the plants are given their program for the coming year. These instructions become more explicit at each successive layer of the hierarchy. They are the specific and detailed operating rules which persuade the plants to execute their programs as per the detailed plan. These are the trademarks of successful mechanistic and command organizations.

In terms of its historical evaluation, the GPD seems to fit comfortably into the direction phase of the dialectic model. Cost accounting systems had been successfully introduced, including standard costs and comprehensive budgets. A formal hierarchy with clear-cut positions had been institutionalized. Order and routine reigned and fostered a climate of efficient growth. Large-volume vertical information systems along with self-contained units, including the glass plants and the various sales districts, seem more than capable of processing the requisite information. The mechanistic social technology was well suited to a long period of evolution through direction.

Finally, turning to Johnson & Johnson, we see an outstanding example of the market-based, decentralized organization preferred by most economists. Each of their 155 subsidiaries traded with the external markets as if they were independent, autonomous units. Any trading between subsidiaries was handled on the spot by the subsidiary managers on the basis of external market prices. The operating rules dictated that each subsidiary develop a prospector strategy paying particular attention to share of market and long-run profitability. The enforcement rules encouraged each subsidiary to increase these two measures as much as possible within their own product market spheres. The use of self-contained units and a highly interactive and intense vertical information system reinforced the market-based organizational design.

Johnson & Johnson also illustrates a highly successful delegation phase company. Each subsidiary had a great deal of autonomy and freedom to act as they knew best, albeit under the gaze of a small group of HQ watchdogs. The subsidiaries were given full responsibility for all aspects of their business—research, development, manufacturing, marketing, sales, information systems, etc.—and they took great pride in running their own businesses. They had every incentive to penetrate larger markets, respond quickly to customers' needs, and to develop new products. Upper management refrained from making business decisions for the subsidiaries operating on the basis of periodic reports and brief visits to the subsidiaries while keeping an eye out for acquisitions which could be lined up beside the other subsidiaries. Johnson & Johnson typifies the successful delegation phase organization.

CONCLUSION

There is a dialectic relationship between the abstract theories and the live case situations. The cases inform the frameworks but, at the same time, they act as a lens through which we can view specific issues of management accounting and control. Each time we shuttle from case to framework and vice versa, we have an opportunity to enhance our understanding of both. I believe it is important to recognize that the analyses offered here are my interpretation and mine alone. Your interpretation might well be quite different. Each framework necessarily directs attention to some aspects of organizational control systems and ignores other aspects. These other aspects also may loom large in effective system design. We now turn to another attribute of organizations that has equally important implications, strategy and strategic planning.

NOTES

1. The name of the company is disguised. The case data were collected as part of a research study reported in Daft and Macintosh (1984), Macintosh (1985), and Macintosh and Daft (1987).
2. The name of the company is disguised.
3. See Stedry (1960) and Simons (1987).

6
Strategy and Control

Most authorities on organizations agree that today strategy formulation is the single most important task of top management.[1] In fact, the strategic planning school of management, currently leading the pack of business school disciplines and subjects, proclaims that strategy and strategic planning are the keys to the governance of organizations big and small, private and public. Strategy deals with both mission and governance. Mission reflects the major aims of the organization and the tasks to be performed to reach them. Governance is the means by which the organization is controlled and regulated. While strategy and organization structure dictate the appropriate type of control, management accounting and control systems are used to enhance and influence the strategic planning process. The basic premise is that there are important links between environment, strategy, organization structure, and control and that a congruent matching of these variables is essential to performance as depicted in Figure 5.

This chapter explores that relationship. The purpose is to bring together some findings and speculations about how an organization's strategy influences the design and use of its management accounting and control systems.[2] To set the stage, we present a brief review of what is meant by the terms strategy and strategic planning. Then we outline a framework that includes four distinct types of strategies along with the appropriate organizational structure and management control systems most appropriate to each strategy. We sketch out another conceptual scheme linking strategy and control based on the idea of product life cycle. Finally, the chapter speculates on the strategy–control relationship from the perspective of ideology in a critical social theory sense.

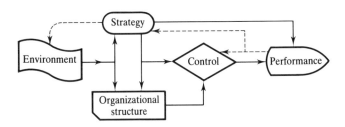

Figure 5 Environment, Strategy, and Control

STRATEGY AND STRATEGIC PLANNING

Chapter 3 looked at some ways of thinking about how an organization's environment and its history affect its management accounting and control systems. These frameworks imply that environmental and historical forces are out there beyond the organization's boundaries. This line of thought relies on a biological metaphor of survival, organisms must adapt to their environment or become extinct. We now want to revise this metaphor somewhat to argue that organizations enact their environment by way of strategic planning rather than simply adapting to it.

Enacting the Environment

Enactment, an important concept in the organizational behavioral literature, requires a little explaining.[3] In order to make sense out of the infinite set of variables and constantly changing events in the world in which we live, we must focus our attention on only a small number of variables. So we bracket off a manageable subset for closer attention as a basis for action. Our actions, however, produce an ecological change in the environment which appears to be in a continuous state of flux. In order to cope we again select a manageable subset, take action, produce more ecological change, and the process repeats. This ongoing process is called enactment.

An instance may help to make the point. Two friends on vacation tour the same city but experience it in quite different ways. One seeks out and tours the museums, art galleries, and book stores and purchases, at a handsome price, a painting by a new artist. The other heads for the beaches, cafes, and discotheques and eventually starts a brawl that brings the police on the scene. For the arts *aficionado*, the environment appears to be civilized, calm, and lively; for the rambunctious fun lover, it seems pretentious, boisterous, and definitely action packed. Both initially faced the same circumstances but each enacted the environment—selected a subset, took action, and changed it—in a quite different way.

Organizations enact their environment in a similar way. Top executives select part of the environment for close attention and action. They describe which parts of the environment present opportunities to exploit, which parts to keep an eye on for possibilities, and which parts to avoid as dangerous. These considerations become the basis for the organization's strategic plan. Organizations choose and act on their environment and as they do so, as with our two tourists, they inevitably effect some change in it. Such choices are a result of the organization's strategy and its strategic planning process.

The enactment concept brings to the surface the fact that a strategy is very much a subjective interpretist enterprise. That is to say, a strategy is a human creation, even though in most of the management literature it is treated as an objective reality that exists outside the minds of the managers who produce it. This reification error, the process whereby human creations, such as social norms or marriage, are treated as entities created by nature or God with an existence independent of human minds obfuscates the fact that strategy is a human creation.

The first two models in this chapter treat strategy as if it were an objective reality. Such reification permits us to analyze the strategy–control relationship from a convenient structural functionalist vantage point. The third framework, however, adopts the subjectivist position and shows how strategy works as an ideology that comes to be experienced as an objective reality by managers who feel compelled to implement its injunctions even if they go against the grain of their personal values and beliefs.

Strategy

A strategy is a master plan for how an organization intends to compete in its environment and what sort of structure, including coordination and control devices, is required to implement the plan. Strategy is concerned with fundamental, large-scale problems relating to how an organization defines itself including its relation to the environment. It is the road map for organizational effort.

The concept of strategy is not limited to business organizations. In fact, the idea was imported into business from the military where strategy means generalship and the art of war. A *stratagem* is an artifice or device for deceiving the enemy by deploying troops, ships, aircraft, and so on, in a way that imposes upon the enemy the place, the time, and the conditions for fighting that are preferred by the strategist. Business strategy still retains the spirit and jingoism of warfare, but strategy is also applied to nations, governments and government agencies, universities, and similar institutions. And for individuals, strategy can be valuable as a grand plan for getting on in life.

Strategic Planning

The strategic planning process usually begins with a detailed analysis of opportunities and threats in the environment. During this process, the economic climate, availability of critical factors of production, political atmosphere, and strengths and weaknesses of competitors are carefully assessed and predictions for the future attentively considered. Will the environment be stable, or will it be turbulent? Will the environment be friendly or hostile? Will successful operations bring large rewards or will it be slim pickings? How will competitors react and compete? These are the kinds of questions that must be answered in any thorough environmental analysis.

Such analysis is followed by carefully identifying the organization's current skills, strengths, and weaknesses. Distinctive competencies, those things it does better than most competitors, are identified and become the cornerstone of the strategic plan. This internal appraisal includes evaluation of the current product line, the state and condition of production technologies, the R&D abilities, and the capabilities of key employees. Each functional area, such as manufacturing, finance, engineering, and marketing, is carefully assessed for strengths and weaknesses. "Know thyself" is the rule.

The next phase in strategic planning involves matching distinctive competencies to opportunities uncovered during the environmental analysis stage. Success here calls for entrepreneurial skill, something we do not understand at all well. Hunch, judgement, cunning, intuition, experience, and optimism are difficult enough to define, perhaps impossible to teach or program. Luck, chance, and random events also seem to play an important role. More an art than a rational science, entrepreneurship is nevertheless the consummate business and political skill.

The final stage in the strategic planning process consists of drafting a blueprint for an organizational structure to get the strategy into action. The blueprint outlines the appropriate way of allocating authority and responsibility, putting in place integrative and coordinative mechanisms, and most importantly for our purpose, designating appropriate control systems for measuring and rewarding performance. The control system directs resources towards accomplishment of strategic results.

Strategies and their formulators come and go. But industrial and commercial enterprises have a life of their own, one which frequently lives longer than that of the current strategy, the incumbent executives, and the present administrative systems they use to run the company. To put it differently, enterprises have a unique history that transcends individual executives, strategies, and administrative systems. The trick seems to be for the enterprise to find a viable congruence among these three elements. A crucial part of this process is to work out the appropriate match between strategy and control systems.

This point is illustrated dramatically by Alfred Chandler in his classic and seminal book, *Strategy and Structure*. Chandler documented over many years the strategic and administrative responses to their shifting environmental conditions of four US giant industrial enterprises: du Pont, General Motors, Standard Oil of New Jersey, and Sears and Roebuck. Using highlights from du Pont, we illustrate this important point.[4]

STRATEGY AND CONTROL AT DU PONT

In 1902 the hundred-year-old venerable du Pont company, the nation's largest producer of black and smokeless gunpowder and dynamite, nearly vanished when the president, Eugene du Pont, died suddenly and the four remaining elder partners decided to sell the company to its largest rival. The fifth partner Alfred I. du Pont, visibly outraged, quickly arranged with his two young cousins Coleman and Pierre to buy out the other partners and run the company themselves.

The One-Man Show

Before the buyout du Pont existed as a loose federation of small firms and plants spread around the country. Henry du Pont, president from 1850 to 1889, cunningly purchased several of the larger powder plants during the post–Civil War depression. He also ran the Gunpowder Trade Association, a cartel which set production quotes and prices for the entire industry. Henry was little interested in administrative systems. He simply told each plant what to produce and at what price to sell their quota. The absence of administrative control is evidenced by Chandler's observation that he "carried on single-handedly from a one-room office overlooking the powder mills on the Brandywine, most of the business of his company and that of the Association. He wrote nearly all the business correspondence himself by longhand" (Chandler 1962, p. 55). It was the classical one-man show.

Centralization With Tight Controls

Upon taking over the reigns, Alfred and his cousins quickly moved to "consolidate and centralize" the company. They built a new central office in the center of Washington, where they installed a general research and development department and three new administrative line departments, one each for black powder, high explosives, and smokeless powder. Each line department was to set policy, coordinate, plan for, manufacture, form a nationwide marketing organization, and appraise the operations of its plants. The cousins also established staff offices, including treasurer, legal, real estate, purchasing, and development.

Pierre also came down hard on du Pont's tradition of nepotism by insisting that any family and relatives be well trained and competent before assuming a position of responsibility. He also clearly defined everyone's authority and channels of communication. The new owners aimed to transform du Pont from a laissez-faire collection of relatively small firms to a well-managed, centralized, integrated, and tightly controlled industrial empire. A vital part of the new arrangements proved to be the treasurer's department, headed up by Pierre himself. Chandler describes the set up as follows:

> The Treasurer's Department provided the central office with even more valuable and more regular information. Beside their routine financial activities—the handling and supervision of the myriad financial transactions involved in any great industrial enterprise—the financial executives concentrated on developing uniform statistics essential to determining overall costs, income, profits, and losses. The Department, at first divided into three major units—The Treasurer's Office and the Accounting and Auditing Divisions—came to have, by the time of World War I, additional units that administered credit and collections, salaries, and forecasts and analyses. The auditing unit gathered information on general external financial and economic conditions as well as on the company's internal performance, while the accounting office continued to develop cost data for production, sales, construction, research, and other activities. The creation of these statistical offices, like those in the sales and manufacturing departments, provided the executives administering the duPont properties with a steady flow of accurate information. Such data were not and, indeed, could not be assembled until Pierre and Coleman had created this centralized, functionally departmentalized operating structure (ibid., p. 60).

These arrangements served du Pont well. They allowed the executive committee to concentrate their efforts on long-range planning and overall appraisal while the operative committee met weekly to coordinate and watch over the operations of the field units.

During this period, du Pont continued to strengthen and hone its centralized administrative systems, particularly the development of statistical reports and procedures for using financial controls. Formal financial forecasting and capital expenditure systems were developed using sophisticated return on investment estimates for the growing number of new projects and proposals including expansion of existing operations and even new ones such as chemicals. Young Donaldson Brown, working under Treasurer Raskob, devised what was to become famous and widely copied by industry in general during the rest of the century. This was a financial control device which related rate of return on capital invested to capital turnover, sales volume, costs, and profit for overall operations as well as for almost each component throughout the enterprise. This control mechanism enabled head office executives to monitor and appraise in great detail the operations of each department by means of the common discourse of accounting.

These managerial innovations soon proved their worth many times over. World War I led to an explosive increase in du Pont's munitions business and in its new chemical department. The centralized organization with its detailed and sophisticated financial controls "proved admirably suited to meet the needs of the resulting phenomenal growth" (ibid., p. 66).

New Blood and Diversification

After the war the du Pont cousins, opting for a well-deserved and badly needed respite, turned the company reins over to a new set of top executives. At the time du Pont was recognized as one of the "best managed" enterprises in the world. The new top executives took stock of the situation and agreed that even more direction, coordination, and control from the center was required to cope with already large and ever expanding operations.

At the same time, paradoxically, they decided to vigorously pursue the strategy of diversification which had been started by the previous management. Before long many new products and markets, especially those related to its chemical operations were added. These included, for example, artificial leather, pyroxylin, dyes, rubber, paints, varnishes, vegetable oils, and a variety of organic chemical products. The company also expanded into new geographic areas domestically as well as setting up operations in Canada and Europe.

The Great Crisis

The ensuing success of the diversification strategy, however, was to lead to a key moment in du Pont's history, one which brought the company to the brink of bankruptcy. Chandler insightfully describes the impending crisis.

> By 1919, then, the du Pont Company was rapidly changing the nature of its business. Where before the war it had still concentrated on a single line of goods, by the first year of peace it was fabricating many different products whose manufacture was closely related to the making of nitrocellulose for smokeless powder but which sold in many very different markets and in some cases used new types of supplies and materials. This strategy of product diversification was a direct response to the threat of having unused resources.
>
> Yet, in carrying out a policy of diversification to assure the long-term use of existing resources, the same executives failed to see a relationship between strategy and structure. They realized clearly that structure was essential to combine and integrate these several resources into effective production, but they did not raise the question of whether a structure created to make and sell a single line of goods would be adequate to handle, several new and different products for new and different markets (ibid., pp. 90–91).

The problem, which went almost unnoticed at the time, proved to be that the rapid expansion into new products, industries, and geographic areas had vastly increased the demands placed on the centralized head office. Coordination requirements mushroomed. Close monitoring and performance appraisal of the operating departments greatly increased the amount of information flowing into head office. Overloaded with data and figures, top executives now had to make strategic decisions about products, equipment, and markets with which they were nearly totally unfamiliar. The centralized, functional organizational structure, with its now cumbersome and voluminous vertical information flows, was strained to bursting point.

The Solution: Found But Rebuffed

Noticing this strain and alarmed by the company's poor results during the sharp depression of 1920 and 1921, the executive committee formed an organization committee of younger key executives to study the problem. This committee called in outside experts, studied the organizational arrangements and administrative systems of eight leading industrial companies, and generally carried out their mandate in the typically methodical, rational and thorough du Pont manner. The committee's findings proved to be a watershed in du Pont's historical evolution. "Their work and the resulting report had an impact far beyond the company's immediate difficulties; for the first time a new form of management structure was proposed" (ibid., p. 94).

They had discovered, by dint of a lot of sharp pencil pushing, that du Pont's raw and semifinished products were quite profitable, based on an ideal return on investment of 15 per cent. Further analyses revealed, however, that du Pont's finished products for ultimate consumers were not profitable even though their chief competitors for the same products were reaping good profits. The problem, the committee concluded, must lie with the organizational structure and administrative systems. Under the current arrangements it was impossible to determine who was responsible for profits.

The committee concluded that a fundamental change in organization structure was urgently needed. Du Pont should organize around product responsibility, not functional responsibility. More specifically, they proposed to place responsibility for purchasing, manufacturing, marketing, and accounting for each of the company's main product lines under one executive with the executive committee having only general supervision over these product department general managers.

Yet, in spite of a very strong recommendation, the proposal was initially rejected by the executive committee. Its members argued that only closer attention to specific problems by the central office executives would solve the problem. "The answer was not reorganization but better information and knowledge" (ibid., p. 99).

Another special committee of three older executives was struck to study the whole question of organization. Their findings not only mirrored those of the organization committee, but also went further to suggest that the entire company be regrouped into product profit departments with the head office staff departments establishing and maintaining general policies and procedures to guide the line departments. Once again, however, the proposal was rejected. President Irénée du Pont, stalwart protector of the status quo and the old guard, argued against tossing out the proven organizational and administrative arrangements which had worked so well until recently. He sent the report back for more study.

In Through The Back Door

An unofficial committee without portfolio, however, brought the new ideas in through the back door. They developed a plan outlining how du Pont could make 10 percent return on investment (ROI) on the paint and varnish business, presently making a loss, by regrouping it into a single unit. The executive committee, impressed by the plan, gave it the go-ahead.

Soon after, another group of executives presented a report that ostensibly recommended improvements in financial statistics but in substance calling for broadening product profit responsibility to include all the company's products. The executive committee almost immediately accepted these proposals and by 1921 du Pont "was beginning to move toward a de facto structure based on product divisions rather than functional departments" (ibid., p. 103). The inevitable move to the delegation phase was underway.

The Reorganization Ordained

A financial crisis in the first six months of 1921 heated up the debate. Old guard members of the executive committee, dyed-in-the-wool centralists, argued for the appointment of a strong, single, dictator-president with absolute jurisdiction and full authority to take whatever steps he felt necessary to remedy problems anywhere in the company. In opposition to this idea the younger executives, now more than ever convinced about the need for a decentralized structure, pressed their case hard, and persevered in their struggle until finally the old guard capitulated. They had won the day.

The new organization featured five product (or industry) departments, eight staff departments, a treasurer's department, and an entirely distinct executive committee with no direct operating duties but with responsibility for overall planning, appraisal, and coordination. The eight staff departments were given no line authority whatsoever. Rather, they were to act as consultants to the line personnel and to provide services to the company as a whole. The treasurer's department, however, continued to have line authority for overall management accounting and control practices including the type of controls required and the timetabling of the various reports, while the general managers of line departments looked after the details of the cost accounting methods and systems for their own components.

The new arrangements went into effect in September 1921. The results were felt almost immediately. Red-ink operations soon turned into black-ink operations. Du Pont never again, not even during the great depression of the 1930s, encountered as severe a crisis. While over the years new product departments have been added and a few more staff offices opened, the organization model and administrative systems have remained fundamentally the same.

The du Pont history, in sum, is a vivid example of the importance of the need for a strategy–structure–control systems match. At the turn of the century, the three cousins converted the company from a laissez-faire federation run at the top by a one-man show, into a highly centralized, tightly controlled, functional organization, marketing a related group of products. A key feature of the new organization was its sophisticated, highly detailed, financial control system run by the active, energetic and professional personnel in the treasurer's office.

The company's successful diversification strategy, however, resulted in overloading the central office with information and decision making. It pushed the centralized structure, with its voluminous vertical flow of information, to the breaking point. A new decentralized structure, eventually put in place in spite of fierce resistance from old-guard executives, realigned du Pont into a set of product departments responsible for manufacturing, marketing, and purchasing. The head office top executives, with no direct operating decisions, looked after overall planning and general appraisal of the operating components. Staff offices at headquarters provided services to the line operations and to head office. The new arrangements, which brought strategy, structure, and controls into alignment, served du Pont well for the rest of the century.

Corporate Versus Business Strategy

The du Pont history provides an impressive illustration of the potential power of strategic planning as well as the importance of the strategy–control system interrelationship.[5] It is necessary, however, to distinguish between global strategy and business strategy. Large, complex, widely diversified multinational companies, which dominate the scene today, have a global or corporate strategy to define the pattern of the firm's diversified activities. Acquisitions, divestitures, joint ventures, relations with governments of nation states, and major capital market transactions are grist for the mill of global strategy.

Business unit strategy, in contrast, deals with the specifics of how particular strategic business units (SBUs) within the firm will compete including policies for matters involving marketing, production, engineering, purchasing, R&D, and so on. It also outlines what kind of organizational structure and administrative systems are appropriate to implement the SBU strategy. In small companies, global and business strategies are usually one and the same.

STRATEGY, STRUCTURE, AND CONTROL

During the past couple of decades scholars in general management and business policy have developed, following structural functionalist assumptions, typologies which are valuable for thinking about the link between strategy and management control. The first typology identifies four types of organizations on the basis of their competitive strategies and outlines the best fit for each type in terms of organizational structure and management control systems.[6] The basic notion is that each type has a unique configuration of strategy, organizational structure, and management control system. The key ideas are summarized in Table 2.

Before describing each of these types of organizations a few general caveats are in order. It is important to recognize that they are presented only as archetypes; some organizations will not fit perfectly into one of the four moulds. Particular firms may closely resemble one archetype at one time, and another archetype at another time. A small prospector, for example, that pioneers a new technology may metamorphose into a defender organization as other firms catch up on or copy the technology. And, finally, large Fortune 500 type companies are made up of a portfolio of businesses including all four types. Nevertheless, the typology can be very valuable for conceptualizing the strategy–management control system relationship.

Table 2 Typology of Strategy, Structure, and Control

Type of Organization	Key Characteristics		
	Strategy	**Structure**	**Control System**
Defender	Aggressively maintain a prominent position in a carefully chosen, narrow product-market domain	Traditional centralized functional organization	Efficiency focus with close detailed, and tight controls
Prospector	Create turbulence by continually bringing new products to the market	Organic management arrangements and product group organization	Effectiveness focus accenting innovation, entrepreneurial effort and self-evaluation at lower levels
Analyzer	Highly selective in its stable sphere and rapidly copies successful innovations in its dynamic domain	Dual core organization: centralized functional for its stable domain and organic for its dynamic sphere	Tight controls and an efficiency focus for the stable sphere, and looser controls with an effectiveness focus for the dynamic domain
Reactor	A well-defined but obsolete strategy or a "running-blind" strategy	Politics and careerism dominate over any logical arrangement of authority and responsibility	Treated as merely a bookkeeping system

Defender Organization: Efficiency Controls

The defender organization pursues a strategy featuring a well-defined, stable, and frequently mature product market domain. It aims its products at a limited segment within the total potential market, often the healthiest and most lucrative. Within this narrow catchment, it tries to offer customers a full range of products. Success for the defender depends on aggressively maintaining a prominent position in a carefully chosen market.

By carving out a narrow niche in the market, defenders can concentrate on internal processes, procedures, and problems. There is little need to scour the environment for new markets nor to diversify into unrelated products. The focus of energy is on finding new ways to reduce production and distribution costs, cut marketing expenses, and improve product quality. Japanese auto companies in the 1970s and early 1980s typify this approach. By choosing to compete in the subcompact and compact segment of the market, aggressively reducing costs, continuing to improve their products, and pricing moderately they discouraged new competitors from attempting to enter their carefully selected market niche. Operating from this solid base in recent years, they have carefully and gradually established beachheads in the midsize and luxury segments of the auto market. "Lean and mean" is the maxim.

The traditional functional organization structure is well suited to the defender strategy. Specialists with similar skills are grouped within functional units such as manufacturing, engineering, sales, R&D, and finance. Each specialist unit becomes extremely adept at its particular task. Work flows sequentially from one unit to another. Coordination between the functional units is routine, having been initiated during the planning phase. Finally, a small group of key top managers acts as watchdogs over operations.

Defenders also have distinctive administrative features. The simple strategy allows the organization to develop formal operating procedures specifying clearly how each employee is to carry out specific duties. Production plans and coordination needs can be meticulously mapped out well in advance. Planning tends to be concentrated over a few key areas of needs. Output and cost objectives are set and translated into budgets and specific operational goals. Only then is production set in motion.

During production, control is centralized, formal, detailed, and far-reaching. A small group of top management personnel, usually dominated by finance and production, keeps a close eye out for any deviations from plans. They do this by means of long-looped, vertical control systems which reach from the top of the organization deep into the lowest levels. These controls usually focus on efficiency indices and comparisons with previous years. Formal progress reports and explanations for budget variances flow upwards from the lowest levels to the top of the hierarchy; formal standard operating procedures, job descriptions, and specific directives flow down. Formula-based bonus schemes are explicitly connected to budget performance. For defenders, with a very clear understanding of what they want to do and how to do it, tight controls fit like a glove.

In sum, defenders concentrate on narrowly defined, stable, product market domains which they aggressively defend by means of low costs, high quality, and reasonable pricing. This strategy is implemented within a functional organizational design not unlike the mechanistic organization described in Chapter 3. It features rigorous planning, long-looped vertical information flows, formalized standard procedures, and close monitoring by upper management of progress on plans. Not surprisingly, in successful defenders, business proceeds smoothly

Strategy and Control

in a predictable manner. The glass products division of Empire Glass described in Chapter 2 is an archetype of the defender firm.

The Case of Glass Products[7]

> The glass products division (GPD) of Empire Glass is an archetypical defender. GPD competes with a full line of glass products ranging from small orders of fancy specialty jars, to middle-sized orders of jam and honey jars, to long-runs of standard size beer and soft drink bottles and aggressively maintains a prominent position in this narrow, stable, and mature market domain.
>
> GPD does not invest much time in strategic planning as its strategy is clear-cut, well understood by all, and changes little from year to year. Nor does it spend much at all on product R&D. Instead, managers at all levels focus their energy on finding ways to reduce the costs of production, distribution, and marketing and on improving product quality. GPD vigorously defends its market niche with continuous cost reduction and product improvement. Potential competitors think twice before taking them on and usually decide to look elsewhere for opportunities.
>
> GPD also typifies the defender in its organizational design. HQ and divisional executive groups are small and financial officers play a dominant role. The glass division is organized along clear-cut functional lines. Coordination needs are taken care of in advance by means of an elaborate, careful forecasting and planning process. These plans are translated into budgets and specific operational goals which are discussed and debated at all levels until agreed upon. From then on the field units, sales districts and plants, merely execute the program.
>
> The field units are monitored closely by both divisional and HQ offices. Detailed profit planning and control reports, which cover nearly every aspect of operations, convey performance information from the lowest echelons to the top. Variances from the plan are analyzed in detail and explained in writing. And the bonus scheme is spelled out explicitly and linked to budget performance for the year. Tight controls and close monitoring are the order of the day. GPD is not a place that tolerates deviations from plan during execution.
>
> For GPD strategy, structure, and management control formed a complementary and coherent package. Not surprisingly, it not only held but improved its market share in a highly competitive mature industry. As well, it recorded small but steady growth in profitability at a time when some competitors went under. GPD is the prototype of the successful defender.

Prospector Organization: Effectiveness Controls

Organizations following a prospector strategy are almost the antithesis of defenders. While defenders focus on a narrow, well-defined product market niche, prospectors compete in a broad product market domain. Their distinctive competence lies in their ability to find and capitalize on new product opportunities before the competition. They systematically add new products, shed old ones, retrench in some markets, jump into avant-garde product lines, and buy new technology. It is almost as if they continually try to get ahead of themselves.

More accurately, prospectors deliberately create turbulence within their industries. With the environment in constant flux, prospectors bring their distinctive competencies—developing and marketing new products—into play, keeping the competition off balance. In fact, their entire product line may undergo transformation within three or four years. The idea is to try to keep the product market domain in a state of perpetual change and so manoeuvre the competitive turf in its favour. While change is anathema to defenders, it is opportunity knocking for prospectors.

Prospecting calls for a unique style of management. It is not like a game of chess, where management carefully plans programs and closely monitors their faithful execution by skilled specialists down the line. It is more like a game of chance where they must roll the dice against large odds, deploying limited resources over numerous ventures and hoping for the best. So planning, such as it is, takes the form of just-in-case scenarios with only sketchy and tentative plans for contingency moves for both successful and losing ventures. And monitoring performance involves knowing when to cut the losers short and go with the winners. It is not a strategy for the cautious conservative or the weak-hearted.

Firms employing the prospector strategy also feature special organizational arrangements. Characteristically, these firms localize entrepreneurial and engineering activities by grouping them into temporary project teams and task forces where they can be most effective. Talented professionals are similarly shuttled in and out of projects as needed. The project groups are free to explore new markets and develop avant-garde products as they see fit. The necessary project coordination and integration is decentralized to coordinators at the project level rather than handled through the vertical hierarchy.

It is not surprising, then, that the product group organization fits prospectors well. All the talent necessary to research, engineer, manufacture, and market a set of related products is grouped into one organizational component. In order to give each component a free hand, formalization is kept to a minimum. Rigid planning and tight control are avoided in favour of loose, organic arrangements. Flexible, lateral relations are given priority over vertical hierarchies to give the product groups room to make important decisions.

Prospectors' management accounting and control systems are designed to support these arrangements. They promote effectiveness rather than efficiency, with an accent on entrepreneurial effort and market results. Nevertheless, access by the product or project groups to financial performance information is essential in order to respond rapidly to market signals. In consequence, horizontal feedback loops prevail. While efficiency measures such as cost cutting are available to upper management, primary control is located within the product group, which actively assesses its own market performance. With competent, self-motivated professionals in charge, these arrangements work well.

Careful research supports these ideas. One study, for example, uncovered evidence that for high-performing prospecting firms (using return on investment as the criteria) management control systems featured output effectiveness goals, forecast data in control reports, frequent reporting, tight budget goals, and, surprisingly, an emphasis on cost control (Simons 1987). The study also found that the use of tailor-made controls by prospectors was negatively correlated with performance. For defenders, however, performance was positively correlated with budget-formula-based bonus schemes, but negatively associated with informal control information, interperiod budget performance, and output effectiveness goals. The study also confirmed that prospectors perform better in rapidly changing conditions while defenders do better in stable environments.

The prospector strategy places unique demands on the organization. Sometimes prospecting pays off handsomely and sales suddenly take off. But this puts heavy strains on employees, who must constantly deal with new and unfamiliar technology and face the frustration of never approaching optimal efficiency. At other times, market expectations may fail to materialize, causing morale to sag and with it the confidence and bravado essential to prospecting. Environmental turbulence induced by the prospector can cause it to spread resources and talents too thinly; then experts turn into dilettantes, merely trifling with new products and markets instead of mastering them. Volatility can mean opportunity, but also uncertainty, risk, and frustration.

Corning Glass: An Archetypical Prospector[8]

Corning Glass (CG) seems almost the epitome of the prospector firm in terms of its strategy, organizational structure, and management control systems. It invents, develops, and produces a wide range of domestic cooking and glassware as well as specializing in related high-tech products, such as refractory materials, electronic devices, and medical instruments, which it markets around the globe. CG aims to be on the cutting edge of developments in these fields and can boast of its well-earned reputation as a world leader in speciality glass and related inorganic materials products.

This strategy paid off handsomely over the years. Its reputation, growth, and profitability stems from its remarkable stream of inventions with superior technical qualities, including breakthroughs in heat resistance, mechanical strength, chemical stability, and light transparency. The market acceptance of these products resulted in an enviable record in profits growth.

Many of CG's products are developed for original equipment manufacturers (OEMs). Upon receipt of what seems a viable OEM request, the scientists in CG's technical staffs division (TSD) get to work inventing the necessary technology. From there, projects are passed on to the manufacturing and engineering division (M&E). M&E, staffed by a pool of CG's best engineers, develops the applied technology working closely with the field units that will eventually produce and market the new product. CG strives to be first in the marketplace with these high-growth-potential products. This lead, backed up with patents, manufacturing know-how, and large investments in production facilities, permits CG to capitalize on its strong market position and the product's high growth potential.

This prospector strategy is accomplished by means of an organization structure featuring the functional TSD and M&E units alongside a matrix of product divisions and area managers. The product divisions are responsible for strategy formulation, resource allocation, and technological development for their assigned products. The 130 area managers are assigned responsibility for the day-to-day operations and assets in their geographic territory. The product division is broken into seven product line divisions, twenty business groups and sixty businesses. Both the product and the geographic units share responsibility for marketing, capital investments, and acquisitions. Ambiguity and overlapping responsibilities, typical of the matrix organization, are accepted by all as going with the territory.

CG has an array of management control systems to reinforce its prospecting strategy. One important control is its strategic planning system. Each product and geographic unit prepare a business strategy plan once a year. This document features a breakdown of products on a grid including on one axis four categories of competitive strength (ranging from weak to strong) and on the other axis four categories of market life cycle (ranging from embryonic to aging). It also includes estimates of five key financial statistics (with sales as the top line and return on assets as the bottom line for the current year, the next year, and the fifth year out); a qualitative description of the unit's competitive strategy, its goals and how they would be measured; the major resources required in the next few years to implement the strategy; an analysis of the unit's competitive position, threats, risks, and opportunities; and a statement of any significant changes in the strategy since the previous year. The strategic plan is prepared annually, reviewed by the worldwide business managers and the relevant geographic area managers, and then submitted to corporate management for approval.

Large investments are controlled with a traditional capital budgetary system. Business and unit managers submit a resource request document to the corporate office outlining the resources needed for new projects and for expanding existing ones for each capital project as it arises. The request includes cash flow and rate of return estimates for the project and indicates how the request supports the current business strategy of the unit. In addition to money requirements the proposal also includes a request for M&E engineers. The M&E pool of talent is considered to be a much more scarce resource than is money capital.

Requests are reviewed and prioritized at the levels of worldwide business manager and geographic area manager before submission to the corporate resource allocation committee, which consists of the financial vice president, the corporate controller, the director of strategic planning (who reports to the financial vice president and the directors of M&E and TSD). The resource allocation committee (known as an exclusive club for accountants) recommends priorities and funding levels, including cash and M&E talent for development projects.

The business development committee, also at the corporate level, complements the resource allocation committee. Its role is to identify opportunities for new markets and products that dovetail with CG's special technological competencies and to stimulate managers to exploit new technologies and markets. Thus, it acts as a counterbalance to the natural tendency of the accountant's club to stifle creativity, initiative, and enthusiasm of the unit managers necessary for the prospecting strategy.

The business strategies and resource requests are the basis for the annual operating budget and the annual capital budget. All units down to plant level prepare the traditional operating revenue and expense budget including three measures of profit: gross margin, operating margin, and contributed margin (after assigned corporate expenses). They also prepare and submit return on assets (RAO) and other standard financial ratios, as well as a capital expenditure budget for the coming year.

Variance from budgeted operating margin, however, is considered the most important statistic. Fifty percent of a manager's salary increase is directly tied to the manager's operating margin variance. The individual performance factor (IPF), the other half weight, includes the manager's ability to meet the budgeted operating margin. Bonuses for corporate executives and divisional managers, however, are based on corporate profit only. CG's compensation system also includes the typical stock option participation plan to help soften the emphasis on operating margin.

In sum, CG has a well-defined strategy of seeking leadership in its preselected market domains. Managers at all levels are motivated to strive for cutting-edge developments in their particular area of responsibility. This strategy is implemented by an organization design featuring the centralization of the crucial scientific and engineering functions which feed products to the matrix line organization. An array of management controls and key committees constantly remind managers of the importance of prospecting. These controls are not free from the normal politics and careerism that go along with highly motivated and ambitious managers but they do provide a great deal of force and motivation for institutionalizing the prospecting spirit throughout the company.

Summary

The prospector's strategy involves competing in a broad product market domain which it deliberately keeps in a state of flux by constantly introducing new products and developing new markets. These moves enable prospectors to keep a step ahead of the competition. Formal planning, vertical hierarchies, and tight controls give way to self-directed, often temporary, product groups. These product groups are run by specialists who rely on performance information with short horizontal feedback loops and self-evaluation. Sometimes the prospector may be riding high on a wave of innovation, at other times limping along in the doldrums. Defenders prepare meticulously, take careful aim, and deliberately fire. Prospectors fire first then take aim.

Analyzer Organization: Dual Control Systems

The next type of organization, the analyzer, is a hybrid of the defender and prospector. In one domain, like the defender, the analyzer competes with a stable core of products for a set of traditional customers. It does not, however, try to dominate the market for these products. Rather, it takes a piece of the action by developing clearly identified market niches. These are usually in the most lucrative corners of the market where it relies on its marketing expertise and strength in channels of distribution. It is a player in the market, but not the predominant one.

The analyzer also operates as a quasi-prospector in its other domain. However, instead of developing new products by investing heavily in R&D, the analyzer vigilantly monitors the environment for new products that competitors successfully introduce into the marketplace. When such a product is spotted, the analyzer's applied engineers spring into action and rapidly develop a good-value "me too" version. Then the product is marketed, posthaste, through its well-developed distribution channels. Market surveillance, product engineering, and distribution channels are all key skills in this dynamic domain. The analyzer never leads the market, but will always be close behind those who do.

The analyzer goes for the best of both worlds. It tries to skim off the cream in its stable market domain where it captures a small but lucrative niche rather than holding a commanding position. Simultaneously, for its other domain it aims to minimize the risks of new product opportunities by never being first in the water but by rapidly copying products successfully introduced into the market by prospectors.

The analyzer strategy calls for a unique kind of organizational design, one that can handle both the stable and the dynamic domains. The stable domain is best served by high levels of routinization, formalization, and mechanization. The emphasis here is on efficiency and strong functional departments such as engineering, manufacturing, and marketing. By contrast, the dynamic domain is best served by an organic, flexible organization design where the stress is on the effectiveness of getting new products into standardized production and through the distribution channels. The dual domains demand a dual organization.

The analyzer often resolves the tensions between its two domains by relying on a matrix organization. The matrix features the combined presence of functional groups housing similar technical specialists, and product groups which are partially self-contained. This permits the functional specialists to hold sway in the stable core with the product engineering groups playing second fiddle. While in the dynamic domain, the influence swings in favor of the engineers with integration needs handled by special product managers working in the crossroads of the matrix. Their job is to coordinate the tricky but very necessary integration of the production-oriented functional domain and the market-oriented development and engineering one. Even so, the analyzer can never be either fully efficient nor fully effective.

The dual marketing strategy and the matrix organization places unique demands on the analyzer's management accounting and control systems. The stable domain requires tight controls to monitor performance against standards and promote the cost-efficient production of standard products. The dynamic core calls for loose control systems which focus the effectiveness of adapting new products to existing technological skills. Controls must balance flexibility and stability.

The Case of Brothers Ltd.

> Brothers Ltd., the Japanese consumer goods firm, is typical of the successful analyzer.[9] Brothers competes in Japan and around the world in a variety of household products, including sewing machines, vacuum cleaners, typewriters, calculators, and personal computers. Its products are never the top of the line; but they work well and are reasonably priced. The vice president of marketing outlined the finer points of Brothers' strategy this way:
>
> We monitor our markets and competitors very closely, keeping a special eye on the prominent

firms in each product area. When a product is nicely in the growth stage of its product life cycle, we like to enter the market with a pretty good copy of the best product on the market, one that we know we can manufacture efficiently. Because of our strength in distribution channels and dealership networks, we are able to quickly capture 5 to 10 percent of the world market. You would be surprised at how much this amounts to in total sales.

In order to do this successfully we must watch the movements of the prominent firms as closely as a crane watches frogs in a pond. We know instantly when a new product hits the market and we begin at once to sketch out the design for a similar product. If the innovator's product starts to take off, we get down to serious business, finish the detailed design, get it into production, and push it out through our distribution channels.

We have a large group of dedicated engineers at headquarters who are very good at imitating products—without infringing on any patents—and who find this work very challenging and exciting. Our great strengths are in copy design and distribution channels. We feel it is better to be second with a good imitation, than to be first in unchartered waters. So far, the result has been very lucrative "frog catching."

Reactor Organization: Ineffectual Controls

The final organization in the typology is the reactor. A key characteristic of a reactor is inconsistency in the way it responds to change in its environment. For some companies this is due to a mismatch between strategy and technology. For others it is because upper management clings to clearly articulated strategy and structure when they are no longer viable. More frequently, it is simply a case of running blind rather than following an explicit strategy. In this case new products and technologies are introduced at random without the benefit of strategic planning, or systematic planning and control systems. In any case, the common factor is a misalignment of environment, strategy, and organizational systems.[10]

In a way, reactors are a residual type of organization. Such firms have fallen out of, or have been unable to shift into, one of the more stable patterns of defenders, prospectors, or analyzers. They seem to be continually playing catch-up, so they chronically respond in an inappropriate way. They may, for example, move to a prospector strategy when a defender strategy is called for. Poor performance is frequently followed by a loss of confidence, highly tentative initiatives, and backpedaling on positions. Reactors become overcautious and shy away from acting firmly and aggressively in the face of environmental change and internal chaos. Luck may bring success to some reactors, but in the main they seldom prosper.

In reactor firms, politics and careerism frequently feature prominently. Decisions are made in secret and communications are often distorted. Management accounting and control systems tend to be ignored and financial managers are treated as bookkeepers. In such a situation, the disciplinary effects of planning and control systems, no matter how technically sound in design, are not much help. Not surprisingly, reactors make stressful workplaces.

Universal National: A Case Example[11]

> Universal National (UN), a large US aircraft and missile company with operations spread across the USA as well as abroad, provides a clear-cut example of the reactor organization. Its aircraft division (AD) had won a large and potentially lucrative cost plus incentive contract from the Department of Defense for troop carrier aircraft. The division had not been profitable for the past couple of years and so came under considerable pressure from UN's president to become

profitable. As a result, it was vigorously pursuing a formal profit improvement program. The new contract was seen as the chance to restore AD's position as a major profit contributor within UN.

One vital piece of high technology for the carrier, a sophisticated automatic range finder (SARF), was to be subcontracted out to some sophisticated, specialty electronic firm. Henry Hall, the director of purchasing identified three possible sources of supply including Wade, an aircraft electronic subsidiary of UN. He was reluctant, however, to do business with Wade because severe problems of quality, delivery, and changes in specifications on a previous contract had led to acrimony and bad feelings between the two divisions. He decided, after consulting other AD executives not to ask Wade to bid on the SARF. AD managers felt deep down that Wade should not even be in the aircraft electronics business and that Wade saw AD as a captive customer.

Hall had all but agreed to accept the bid by Bolster Inc., an outside supplier, when UN President Joe Sullivan called a meeting of AD's division manager and Wade's marketing director. Wade's marketing director was extremely upset at not even being asked to bid and reported that Wade had recently put a great deal of money into R&D on SARFs, claiming that, with this new technology and Wade's reputation for technical proficiency and low cost, it was now the leader in the field. He also explained that Wade needed this business to carry it over until its traditional radio receiver market recovered. He also stated that Wade's good reputation in the marketplace had suffered badly as the word spread that AD had not even asked Wade to bid, and that Wade was all but out of the SARF business. The SARF contract, he argued, would either make or break Wade. At this point, UN's president ordered AP to let Wade bid on the order.

Hall phoned Bolster's marketing manager, telling him to put the contract on hold. The contract had been sent to Bolster in final form but had not been executed. In the next several weeks, Hall tried without success to get a final bid out of Wade, who claimed to need time to evaluate and price the highly technical blueprints. In the meantime, Bolster requested AD to pay a substantial amount of the anticipatory costs, stemming from short delivery dates, they had incurred on the project with Hall's knowledge and agreement before final execution of the contract. At this point Hall's ulcer acted up and he spent one week in hospital and another week in bed at home. When Wade's bid finally came in it was nearly 40 percent higher than Bolster's with delivery one month later than required. Meanwhile, AD's divisional manager had received an angry phone call from the Air Force's contracting officer questioning the wisdom of subcontracting with Wade at such a high price.

The case study ends there. Decisions were made in secret, communications got distorted, and political manoeuvring ruled. So instead of managing on the basis of messages from the management accounting and control system, politics played a central role. Clearly, UN's executives face an unpalatable and thankless situation. No miracle cure looms up. At best, Sullivan can mandate that Wade be given the contract and that for internal management accounting measurement purposes the transaction be recorded on Wade's books at their bid price but on AD's books at the Bolster bid price with the difference charged to some headquarters special expense account. This way Wade is not strategically finished in the SARF business and AD's profit improvement drive is not undermined. Though not very satisfactory, it is, perhaps, the best that can be done in the circumstances.

The case, then, provides a dramatic illustration of those situations that occur all too frequently in reactor organizations. The lack of any clear-cut strategy makes it difficult to put into effect management accounting and control systems which support the strategy. It is not surprising that UN has not mapped out any explicit and logical procedures regarding how divisions should deal with each other nor has it put in place any explicit rules and policies for transfer pricing on interdivisional transactions. So instead of decisions being made by the appropriate managers involved at the divisional level in a rational way, they came to be dominated by internal power and politics, inevitably leading to the point where the president had to interfere to save the day. As stated earlier, reactor firms are not pleasant places in which to work.

Some Research Findings

The framework presented above is one of the newest to emerge in the area of behavioral management accounting. These ideas have led to research designed to explain the interrelationship between strategy and management control systems. The central idea is that an organization's competitive strategy shapes and guides the key characteristics of its management accounting and control systems. Recent research, however, has yielded the tantalizing idea that the relationship is a two-way street. Management controls are influenced by strategy but they can also have a profound influence on strategy.

One important study uncovered evidence that use of different parts of management control systems seems to vary with the type of strategy being pursued (Simons 1990, 1991, 1992). Today most organizations of any size have in place and use a similar set of management controls—operating budgets, long-range plans, strategic planning systems, standard operating rules and policies, performance evaluation systems, and nowadays activity-based costing systems. The research indicated that some controls were treated in a programmed way, handled in a mechanical and cursory fashion without a great deal of attention.[12] Other parts were handled interactively as the input data for debate, dialogue, and learning. The mix of programmed and interactively used controls differed between defenders and prospectors.

For example, in one successful defender organization, the controls included specific value-chain cost-reduction programs for nearly all its products. Value-chain analysis involves tracing vertical cost-effectiveness of products from a suppliers' production via the firm's manufacturing to the ultimate customers. Top management closely monitored information reports regarding progress on these programs, scrutinizing them closely for clues about possible improvements. A great deal of learning took place and management made several changes in the production process along the chain as well as in the products themselves.

The controls had affected the organization's competitive strategy in an interactive way. In contrast, other parts of the management control system, such as the strategic planning and the operating budgets, were handled in a programmed and routine manner. In fact, the weekly review of actual and target cost reports for all its products was treated on a management-by-exception basis and completed in less than twenty minutes.

In one successful prospector firm, the mix of programmed and interactively used controls was different. The strategic business plan, the two-year forecast, and the operating budget, which was updated three times during the year, were the raw material for almost continual discussion and debate about products and markets by managers at all levels. As a result, competitive strategies were constantly revised and changed. In contrast, detailed cost and efficiency reports, also part of the regular management control system, were not an agenda item for top management. Instead, they were dealt with by staff offices and treated as programmable.

In sum, both the defender and the prospector had the usual assortment of management accounting and control systems in place. Yet each treated different parts of the controls as either programmed or interactive depending on which were most relevant to the strategy. The interactive use of controls for both organizations led to dialogue and learning, which in turn led to changes in their competitive strategies. The point to underline is that controls can feed back into and change strategies. This means that the relationship between strategy and controls is not simply static and one-way; it is dynamic and two-way.

The idea that management accounting and control systems are used differently according to strategy and are also used interactively in a way that gets top executives to focus their attention on matters of strategic importance opens up an exciting new terrain for looking at the roles these systems play in organizations.[13] They can be thought of as strategic accounting and information systems. They go beyond the traditional job of providing financial information after the fact to monitor operational performance towards predetermined plans. And they can be explicitly designed and used to focus on strategic variables. One way to do this is to think about the strategy management accounting and control systems link from a product life cycle perspective. We now turn to a framework based on this idea.

PRODUCT LIFE CYCLE, STRATEGY, AND CONTROLS

The key notion in the life cycle idea is that products pass through four major stages in their journey from initial appearance on the scene to their extinction: introduction, growth, maturity, and decline. The central argument is that different generic strategies are compatible with each stage. It follows, importantly for our purposes, that key characteristics of management accounting and control systems should also differ at each stage, as shown in Figure 6.[14]

During the *introduction* stage of the product life cycle, the strategic focus is on research and development. The firm is looking for a new product or process that will eventually yield a competitive advantage. At this stage, controls should concentrate on identifying opportunities for such breakthroughs. Measuring performance, decision making, and problem solving, the traditional functions of management accounting and control systems, must be set aside in place of assessing the prospects that appear promising. During the latter part of the introduction stage and the early part of the growth stage, a *build* strategy is appropriate. A build strategy is achieved by expanding the product line, more R&D, investment in production facilities, vertical integration, joint ventures, and acquisition of technology. The strategic mandate at this stage calls for investing capital in projects that build strength. The

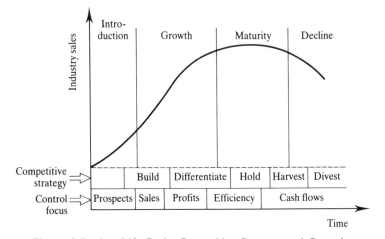

Figure 6 Product Life Cycle, Competitive Strategy, and Control

prime concern of the management accounting and control system should be directed to getting advance signals from the market in terms of its response to the products being developed. Early market reaction is the bottom line.

During the *growth* stage, the focus of assessment and control shifts to profit performance and market share. Strategies begin to feature product differentiation, which must be supported by capital injections. As a result, especially in the late growth stage, measuring return on investment becomes a priority. So also does developing elaborate methods of calculating interest charges for different types of capital needs, such as cash, receivables, inventory, facilities and equipment. Control systems featuring residual income (volume of profit after an interest charge on invested capital) help fine-tune strategic choices.

As the life cycle approaches *maturity*, the focus of the management accounting and control system shifts to efficiency measurements and cash flow budgets. Management must tread a narrow path between replacing old production equipment with newer models and on getting caught with overcapacity as competitors follow suit in an attempt to reap the benefits of economies of scale available from the most recent technologies.

The clever response to this flux in the market is to rationalize product lines with an eye to concentrating on high profit products while phasing out unprofitable items. Astute scaling-down requires pinpoint accuracy in costing. Sophisticated product costing becomes a key competitive skill and management and control systems must introduce new accounting techniques such as activity-based costing and cost-driver identification. Costing information becomes the key ingredient for strategic choices about product pricing and product line pruning.

Also during the mature stage of the life cycle, the control side of management accounting systems increases in importance. Budgets and standards must be tightened. Detailed analyses and responses to variances from standards become mandatory. And close coordination across engineering, manufacturing, marketing, and other functions is essential. General indicators of profit and growth appropriate during the previous stage give way to detailed analysis and pinpoint scrutiny of operations. Tight control and strict discipline are the keys to survival.

Finally, at the end of the maturity stage, the right strategy calls for walking a precarious tightrope between *harvesting* and *divesting*. Harvesting means minimizing cash outflows and maximizing cash inflows while keeping a close eye on the proper time to bail out. Management accounting and control systems here take on a new function. They must provide estimates of breakup costs such as labor settlements, site cleanups, loss on liquidation of finished goods, and raw materials inventories as well as disposal prices for equipment, land, patents, and trademarks. There may also be unanticipated costs of experts, including lawyers, accountants, and appraisers.

Limitations

Do recognize that use of the product life cycle notion to look at the strategy–control relationship is more like a sensitizing device than a solid theory and be aware of the potential problems of this approach. For one thing, the life-cycle curve, indicating progression from one stage to another, is never as smooth as the figure suggests. Further, it is never easy at the time to decide where particular products lie on the curve. Sometimes products skip entire stages. Moreover, many firms follow a strategy of concentrating mainly on products in one or other of these stages.

Some very successful companies, for example, concentrate on innovating in high-technology products at the R&D stage. When the technology is developed and the product potential spotted, the firm then sells or licences it to a company with marketing skills, especially one that concentrates on products during the growth stage. Other firms concentrate on products at the mature stage of the life cycle, operating with a very lean organization. Still others, like most multinational conglomerates, try to have a portfolio of products spanning the entire life cycle. General Electric, for example, competes in electrical products ranging from kettles and light bulbs to highly sophisticated, secret, Star Wars products for the Department of Defense. There are lots of possibilities.

Nevertheless, the framework can provide helpful clues about the appropriate strategic focus for management accounting and control systems at each stage. Undoubtedly, strategic management accounting systems will be the subject of a lot more research attention in the future. Even now, it is much more than a tantalizing idea; the practical applications can be decisive to success or failure. And for us it provides a way to develop some insights into strategic management accounting systems.

STRATEGY AS IDEOLOGICAL CONTROL

This chapter closes with some speculations on the strategy–control relationship by looking at it from the radical humanist paradigm of Chapter 2. Ideology is one of the most important and controversial concepts to emerge from developments in critical social theory, and from the radical humanist vantage point it seems to be vitally concerned with how individuals within any social system are controlled. Yet it has been by and large ignored in the conventional management accounting literature.[15] Its central argument is that a strategic plan is an organization's ideology and as such it acts directly as a powerful mechanism of control.

An ideology from the social theory perspective is a socially constructed worldview about what activities the participants in a social order should pursue, how they should be pursued, and why they are important. So individuals in a social order, such as a firm, are seen to have more than just a bundle of randomly collected beliefs, attitudes, career goals, professional aspirations, and aesthetic endeavors regarding their working life. Rather, they have access to some sort of coherent social knowledge whereby each piece fits into a coherent package in which the parts are related, albeit in a complex way.

This package of ideas is called an ideology. It has a characteristic structure, deals with the central issues of human life, is shared widely by the various participants, and has a deep influence on their behavior. It is a major factor in determining the patterns of social action and their routine reproduction. Thus we can conceive of a strategy as an ideology for a particular organization. When participants come under the sway of the precepts for proper behavior incorporated into the ideology, they are *ideologically controlled*. In this way, strategy acts as a very powerful direct control on the individuals in the organization. An example of the way strategy can act as ideological control may be helpful at this stage.[16]

The Case of Tandem Computers

> Tandem Computers, following a simple strategy, grew from sales in the millions to a place on Fortune's second 500 listing in less than a decade. Tandem's strategy was to build computers that were guaranteed *never* to fail or to ever garble data. These products were targeted at very

precisely identified market niches, such as on-line systems for car reservations, travel booking arrangements, and bank transactions recording. The system consisted of two computers that worked in tandem, splitting the work load evenly. If one had a problem, work simply shifted automatically to the other computer, thus safeguarding the data and not interrupting the work flow. In short, Tandem would make foolproof systems for very special applications.

Mr. Treybig, the founder and president, made sure every employee clearly understood Tandem's strategy. He held meetings with employees where he explained the strategy and reviewed the five-year financial plans. He ensured that employees understood how one little mistake or oversight could affect profits in a big way, and so leave less money for R&D and, importantly, employee bonuses. He constantly preached his five cardinal points for running a company: all people are good; all employees, managers, and shareholders are equal; every employee must understand the essence of the business; every employee must benefit from the company's success; and an environment must be created where these rules are applied. This ideology was reinforced by giving every employee stock options.

New employees went through a mandatory two-day orientation where they studied and discussed the company book, Understanding Our Philosophy. Company newsletters and a glossy magazine repeatedly reminded employees of the company's business strategy and the need for loyalty, hard work, and respect for coworkers. At Tandem, the corporate strategy, acting in an ideological way, proved to be a powerful mechanism of control.

Ideological control, however, does not necessarily work in a way that enhances the lives of the individuals under its sway. In fact, it can have important negative consequences for them. More specifically, it can delude them about their real values and interests, their true position in the social order, and their notions about how the social order works. When this happens, ideology produces a delusion—a false consciousness—and a fettered existence whereby individuals are misguided, partly by themselves, as to their situation and life goals. The history of ITT provides a vivid example of strategy working in this reprehensible way.[17]

The Case of ITT

Sosthenes Behn founded ITT in 1920 as a tiny telephone business in Puerto Rico. Behn saw a great opportunity in the telephone—the newest form of instant communications—for expansion in Latin America and eventually around the globe. Early on he got a lucky break by acquiring Western Electric from ATT when the US Anticombines Agency ordered ATT to divest its network of telephone equipment manufacturing companies including its large UK subsidiary. From this new safe base in the US and the UK, ITT expanded to equip, install, and operate telephone systems in Central and South America as well as throughout Europe. Behn's mission and opportunity, as he put it, was to bring the telephone systems of the rest of the world up to US standards and to make a tidy profit along the way.

But Behn's plans to rival ATT by manufacturing and operating telephone systems around the globe constantly came up against a major hurdle. National governments were bent on controlling their own telephone and telegraph systems in the interests of national security. In consequence, Behn formulated his notorious chameleon-like strategy: do business in each country by enthusiastically supporting the local regime regardless of its political beliefs, aims, and methods, hire nationals as managers to run ITT's local subsidiaries and always retain a controlling interest. In pursuing this strategy, however, Behn "...gradually wove a web of corruption and compromise which left idealism in ruins, and his company with deep kinks in its character" (Sampson 1974, p. 23). Subsequent chief executives at ITT, including Harold Geneen in the 1960s and 1970s, seemed content to carry on the chameleon strategy.

Since its inception in 1923 and through the 1960s and 1970s, ITT executives engaged in a litany of corruption and unsavory and illegal escapades around the globe. Of the many well-documented illustrations, a few stand out. For example, ITT colluded with the highest-ranking

Nazi officials before and during World War II. Its factories in Europe and its communications networks and spy operations in Latin and South American pro-Nazi countries contributed mightily to the Axis war effort just as they had done for Franco's dictatorship in the Spanish Civil War. At the same time, its plants and facilities in the USA and the UK helped the Allies in their battle against the Germans. Even so, ITT did not lose its German and Italian operations after the war. They successfully played it both ways.

ITT continued with its nefarious activities in the postwar era. During the Cold War, ITT engaged in considerable espionage and sabotage in Eastern Europe, which eventually led to the execution by local governments of some of its executives and life sentences in jail for others. ITT also collaborated with CIA officials in a plot to prevent the election of communist President Salvador Allende of Chile, and later on abetted in his subsequent removal from office. Allende was later assassinated under mysterious circumstances. And ITT has been accused of unsavory lobbying in Washington as well as giving huge donations to very high-ranking US elected officials and bureaucrats.

These are but a few examples of the scandalous activities carried out under the umbrella of ITT's strategy of operating as a sovereign global conglomerate transcending the control and aims of nation states, including even the most powerful. The mandate for its executives has been to do whatever is necessary for business and to maintain operations in the local host country. The consequences, early on and years later, were behind-the-scenes activities so reprehensible that "...ITT's record of mendacity and doubletalk... implies a deep irresponsibility: and the prospect of its being further involved in communication must be viewed with dread" (Sampson 1974, p. 307).

The Tandem and ITT cases illustrate how a strategy acting in an ideological manner can have a powerful direct effect on its managers, motivating their behavior as much or more even than a firm's management accounting and control systems. But in a sense, ideological control acts behind the backs of the participants. The strategy is taken for granted and participants' actions become second nature. This can be positive, as in the case of Tandem, or it can be reprehensible, as in ITT, where participants were deluded about their real personal interests and went against the grain of society and their own deeply held values.

SUMMARY AND CONCLUSION

This chapter described two frameworks for thinking about the strategy–control relationship. The first identified four distinct strategies—defender, prospector, analyzer, and reactor— and outlined the key characteristics of management accounting and control systems for each situation. The second framework sketched out a scheme based on the product life cycle notion and speculated on the appropriate focus of the controls for each stage. These frameworks provide a focus for management accountants and controllers to go beyond the conventional ideas about cost accounting and control. They also enable them to develop strategic management accounting systems.

The single most important idea in this chapter is that there are important links between an organization's strategy and its management accounting and control systems. A strategy determines how an organization enacts its environment, how it intends to arrange its organizational structures, and what sort of controls it needs in order to successfully implement its strategy. The relationship between strategy and control, however, is a two-way street. While strategy places unique demands on the important characteristics of an organization's management accounting and control systems, they can be used interactively to help managers realign strategy. Moreover, strategy acting in an ideological way can be a powerful control in its own right over the actions and behavior of an organization's participants, even to the extent of pushing them into unsavory actions. Either way, strategy acts as a powerful means of control over organizational participants.

NOTES

1. See, for instance, Daft (1992).
2. See Dent (1990) for an overview of the literature on the strategy–control systems relationship and Dermer (1990), for some interesting ideas on the significance of accounting to the pre-enactment stage of strategy.
3. Weick (1979) in his classic book, *The Social Psychology of Organizing*, brought the enactment process to the attention of scholars in organization and management theory.
4. The following summary is based on Chandler's (1962) detailed, historical research and documentation of the du Pont company.
5. For another careful and detailed analysis of the strategy–management accounting relationship in a UK conglomerate in the 1980's, see Roberts (1990).
6. This typology is based on the work of Miles and Snow (1978). Simons (1987, 1990, 1991, 1992) has pioneered the research in behavioral accounting, linking Miles and Snow's (1978) work to issues in management control.
7. See Anthony, Dearden, and Govindarajan (1992) for a detailed description of the Empire Glass case.
8. This synopsis, based on the case study by R.F. Vancil, was reported in Anthony, Dearden, and Govindarajan (1992).
9. From the case study Brothers Ltd., Queen's University (1974) by N.B. Macintosh.
10. Gordon and Miller (1976) describe in some detail the accounting issues in running-blind firms.
11. See Anthony, Dearden, and Bedford (1989) for a detailed description of this case study.
12. Simons (1992) also uses the term interactive to describe this style of using management control systems, "when top managers use that system to personally and regularly involve themselves in the decisions of subordinates."
13. In addition to Simons's work cited above, Shank (1989) offers a highly readable and insightful article on this topic as do Shank and Govindarajan (1989) who provide several case illustrations of the evolution from cost accounting to strategic accounting in various firms.
14. See the following for seminal articles along these lines: Govindarajan (1984, 1986, 1988); Govindarajan and Gupta (1985), Govindarajan and Fisher (1990), and Gupta and Govindarajan (1984a, 1984b, 1986).
15. Notable exceptions include Tinker and Neimark (1987) who provide a careful analysis of the ideological control of women in the work force at General Motors between 1916 and 1976. Also see Tinker (1980) for a theoretical exegesis of the ideology-control problematic. And Macintosh (1990) offers an ideology-based critique of IBM's annual reports in terms of the control of women in the information technology workplace.
16. See Daft (1986), pp. 491–493 for a more detailed exegesis of Tandem's ideology.
17. Sampson's (1974) blockbuster book, *The Sovereign State of ITT*, provides a carefully documented and startling exposé of these activities over the years. The case details are based on Sampson's book.

7
Technology and Interdependency

The frameworks presented in previous chapters focused on the entire firm or large parts of it, such as a subsidiary or a strategic business unit. In this chapter we shift our attention to lower levels in the organization—departments, offices, and plants—and work with some concepts which treat technology as an important factor influencing the shape and use of management accounting and control systems.

The realization that technology is a significant factor in shaping administrative systems came nearly two decades ago when organizational theorists discovered "...patterned variations in problems posed by technologies and environments result in systematic differences in organizational action."[1] Following this lead, accounting researchers began to uncover important links between technology and management accounting and control systems.[2]

By the 1980s technology had been accepted as one of several important contextual variables in models of management accounting and control systems.[3] Studies of budgeting focusing on participation and based on the human relations model of organizations also began to include technology as an important moderating variable in participation and performance.[4] And recently, technology has resurfaced as an important variable in debates about the relevance of traditional management accounting techniques in automated and computer-aided design and manufacturing processes.[5]

The logic is as follows. Most of the actual work getting products and services through the organization and out the door is done at the lower strata. Here techniques and knowledge (i.e., technology) are applied to some sort of *raw material*, such as metal, ore, symbols, or people in order to make some change in it. The type and method of work should be major factors determining the kind of administrative arrangements employed to direct the efforts of those doing the work. This is, of course, a *materialistic* conception which assumes that the way we produce our material life determines the nature of our social relations, particularly the way members of any social order are controlled. In this chapter we look at two structural functionalist frameworks based on the assumption that the nature of the tasks done in a work unit is a major property structuring (organizing) the design and utilization of management accounting and control systems.

A TECHNOLOGY FRAMEWORK OF CONTROL SYSTEMS

Over the years, intensive investigation of the effectiveness of management accounting systems points to a central factor, still often overlooked, that places a critical constraint on the design and functioning of these systems, *organizational technology*. There is a strong relationship between the technology of a work unit (or the entire organization in the case of a small company) and the characteristics of management accounting and control systems managers require to perform effectively. Mismatches between the characteristics and work unit technology account for a large percentage of management accounting and control difficulties.

Technology Defined

Organizational technology pertains to the nature of work activities. It is "the actions that an individual performs on an object, with or without the aid of tools or mechanical devices, in order to make some change in that object" (Perrow 1967, p. 198). This definition stresses the conversion process which changes inputs into outputs. Inputs may be any sort of raw material—people, ideas, orders, paperwork, steel castings—upon which organizational skill and knowledge are brought to bear. This definition implies that organizations are not characterized by a single technology. The conversion process used in the personnel department differs from the marketing department, which in turn differs from the production line and the R&D group.

Two aspects of departmental technologies are especialy important to management accounting and control system requirements: the variety involved in the conversion process, and how well the conversion process is understood. Task variety is the frequency of unexpected and novel events that occur in the conversion process. Variety can be considered high when individuals encounter a large number of novel situations, with frequent problems. Variety is low when there are few exceptions, as when day-to-day job requirements are considered repetitious. Variety in a work unit can range from the simple repetition of a single act, such as on an assembly line or sorting mail in a post office, to work that is a series of unrelated problems or projects.

The other dimension of technology concerns how individuals respond to problems that arise in the course of their work. When the conversion process is well understood, participants typically follow an objective, computational procedure to resolve the problem. Solutions may involve the use of stored procedures such as instructions, manuals, programs, and standards, or conventional technical knowledge such as textbook, handbook, or computer data base. When the conversion process is well understood, there are known ways of responding to problems.

When tasks are well understood, often the task knowledge is stored in analyzable programs. Analyzable programs span a wide range. Some are simple, as in the case of procedures for operating a drill press or checking a book out of a library. Others are complex but still understandable in their deep structure. The knowledge about how to operate an oil refinery or handle difficult financial reporting problems is complex but is contained in analyzable programs such as blueprints and generally accepted accounting principles. Cause–effect relationships are understood, so the knowledge base represents correct ways of responding to problems that arise.

In contrast, some work is not well understood. The cause–effect relationships characterizing the conversion process are unclear, so when problems arise it is difficult to find the correct solution. There is no store of techniques and information to tell a person exactly what to do. Thus a different procedure is called for. Participants may have to spend time thinking about what to do. They may actively search beyond the available procedures. Or they may rely upon accumulated experience, intuition, and judgment. Personal experience may be difficult to articulate, but it is a source of knowledge in which individuals have confidence and trust. The final solution to a problem is not the result of a computational procedure, nor is there certainty that the chosen solution is the correct one.

When the two dimensions of technology are combined they form the basis for four major categories of technology: programmable, craft, technical–professional, and research. These four categories are placed on a grid as in Figure 7.

Programmable Technology

Programmable technologies are characterized by little task variety and the use of objective, computational procedures when contingencies do arise. The work is typically routine. Examples are the repetitious work on an assembly line, the audit verification function in the professional accounting firm, and the filing activity in government departments that handle birth and death certificates. Here work programs can be planned and efficiency, orderliness, and low risk are the order of the day.

Craft Technology

Craft techniques are characterized by a fairly predictable stream of activity, but the conversion process is not well understood. There is no store of procedures and techniques to apply to problems that arise. Tasks require extensive training and experience on the job, because decision makers respond to problems on the basis of wisdom, intuition, and experience. Examples include master chefs, professional athletes, jazz bands, manufacturers of fine glassware, and money market managers. Many managerial activities also fit in the craft category. And teaching management accounting may be a craft.

Professional Technology

Professional technologies tend to be fairly complex since there is substantial variety in the tasks performed. Very seldom do professionals face an identical problem in their work tasks. As the expert bridge player says, "No two hands are ever the same." Complexity also stems from the nature of the stocks of knowledge professionals rely on to get the work done. These consist of a repertoire of analyzable programs which, although standardized, are often lengthy, intricate, complicated and difficult to fathom, especially for the novice.

TASK VARIETY	TASK KNOWLEDGE	
	Well Understood (analyzable programs)	Not Well Understood (unanalyzable programs)
Low (few exceptions)	Routine technology Close control	Craft technology Results control
High (many exceptions)	Professional technology Comprehensive control	Research technology Prospects control

Figure 7 Technology and Control Style

Importantly, the professional can get a lot of pleasure from skillfully applying the well-designed techniques to a variety of problems which are difficult but understandable in their deep structure.[6] Believe it or not, professional accountants get a lot of joy out of much of their technical work.

Engineers, for example, can refer to books and technical manuals to discover the correct formulae to use in calculating tolerances and stress loads. Tax experts search through statutes, interpretation bulletins, and judicial rulings to complete tax returns. Professional accountants look to generally accepted accounting principles, accounting board pronouncements, and Securities and Exchange Commission (SEC) requirements when preparing financial statements for clients. In these examples the responses are complex, but professionals are trained to understand them. There is a set of analyzable programs available regarding the response; yet it is hard to anticipate the exact kinds of problems which may arise.

So tasks in professional technologies tend to be more difficult and time-consuming than for routine or craft technologies. A management accountant, for example, might be called upon to design and computerize a cost accounting system for a new refinery. A great deal of technical knowledge is available through advanced accounting books and well-documented case studies. Generally accepted ways of accumulating and allocating costs have already been developed, so most accountants tend to design similar systems. Even so, such projects take months to complete; and extensive professional training and experience is necessary to complete the project.

Research Technology

Research technologies in cell 3 (Figure 7) are almost the opposite of routine ones. Task variety is great and the correct solution to a problem is not usually identifiable through an established store of analyzable knowledge and procedures. Uncertainty prevails due to the wide variety of problems encountered in the tasks and the inability to predict the best way to obtain the desired output. Most discretionary cost centers fall into this category. Purpose, mission, and objectives are difficult to articulate in any concrete way. Neither the length of time required to complete the work nor its eventual form can be confidently predicted. The work of experts on accounting standard setting boards who develop new generally accepted accounting principles (GAAPs) for the profession is like that.

In consequence, research technologies require a great deal of analytical effort. Typically it is difficult to identify a single correct solution because several acceptable options can be found. This is frequently the case in research components in large industrial enterprises and government organizations because problems are often unexpected and a great deal of energy must be devoted to finding the most acceptable solution. Other examples include one-of-a-kind machine tool manufacturing, government policy analysis groups, long-range planning units, and financial statement qualifications by professional accountants. Many of these situations contain nonuniform raw materials which are poorly understood. Experimentation, hunch, and discretion prevail.

Control Systems, Styles, and Technologies

Generally considered crucial among the dimensions of management accounting and control systems are the detail of the information, the frequency of reporting, the difficulty of any target levels in the controls, the pattern of participation in setting the target levels, the degree of association of rewards with the measurements, and the use of the controls for planning and coordination. A systematic examination of these characteristics suggests four distinct styles (or patterns) of control: close control, results control, comprehensive control, and prospects control. Each of these styles seems appropriate for one of the four technologies described above.

Close Control for Programmable Technologies

The characteristics of programmable technologies—low variety and well understood tasks—call for the *close* (but not punitive) control system style. Close control is effected in several ways. Management accounting reports (such as a budget, a standard cost report, and statistical summaries of department inputs, outputs, and activities) are detailed and issued frequently. Target levels in the reports are influenced more by upper management than by employees in the department. But targets tend to be moderately easy to achieve. These reports are important for measuring and monitoring performance, as well as for pay and promotion decisions, but they are less important for coordination and planning. And standard operating procedures are well developed and documented. This profile suggests a close and supportive control style, one which seems well suited to programmable technologies. The rationale for this fit is as follows.

The best way to perform the routine and repetitive tasks is readily determined and this knowledge can be committed to policies and procedures as a permanent guide for job behavior. As more know-how is incorporated into policies and procedures, the number of controls and the extent of their influence increases. The need for steady, reliable, and orderly output explains the high frequency of reporting and the increase in upper management influence over target setting.

Moreover, since the repetitive tasks are well understood, performance expectations are well established and widely agreed upon. Consequently, target levels are based on past experience and medium difficulty, and employees in the department have little influence in setting them. Nevertheless, the measuring and monitoring of performance is critical to ensure maximum results from routine work. For the same reason, the coupling of rewards to controls increases. Since much of the work can be programmed in advance, the role of controls in planning and coordination is less important.

Controls, however, are not normally used in a coercive manner to pressure the department for more output. The potential danger of creating a coercive system is defused because the target levels that are set are reasonably easy to meet. A *kind* but *firm* disciplinary *process* is the apt control system rule.

In sum, the need to guarantee reliable, orderly, and efficient output in programmable technologies is well served by detailed and frequent reporting, upper management influence over target setting, and rewards coupled to measurement. Repetitive, well-understood tasks lead to reasonable expectations of performance that are widely accepted because the targets are only moderately difficult. Since much of the work can be programmed in advance, the planning and coordination aspects of the controls take on less importance. Controls usually work well in routine situations and satisfaction with them is higher than in more complex technologies. A close and supportive style of control is well suited for programmable technologies.

Results Controls for Craft Technologies

Controls for craft technologies should be simple. Task variety is low but task knowledge is hard to put into standard programs. The optimum relationship between inputs (such as materials, labor, and capital) and results is not known. So close monitoring of work patterns and preparation of detailed reports about costs and resource utilization are not very helpful. However, since the desired result can be determined in advance, the appropriate controls are basic effectiveness tests. The focus is on *results*.

The trading vessels and official pirate ships of Elizabethan England were prototypes of this bottom-line style of control. Venture capitalists would sponsor a captain and take a share of the merchandise or loot when the ship returned a year or so later. The test of success was the value of the booty. For a high fashion boutique, reports about markups, profits, and cash balances provide the suitable controls. For the investment fund, the appropriate test is the annual percentage increase in the fund's value, which can be compared readily with other funds. For an advertising agency, gross billing to clients is the appropriate test. In professional sports, the team's won–lost record and gate receipts are sufficient. For a couture, it is the buyers' orders after the fashion show that matter. And for the orchestra, the number of curtain calls tells the story. These simple tests assess effectiveness.

Management accounting and control reports in craft technologies tend to be general rather than detailed and they are issued less frequently than is the case in routine technologies. The controls are used more for measuring performance and determining rewards than for planning and coordinating. Targets and standards in the reports are only moderately difficult to achieve and department managers have a good deal of influence in setting them. Yet the controls have a good deal of motivational effect since they are used in making pay and promotion decisions. Moreover, controls tend not to be administered in a heavy-handed way; they simply indicate whether or not the desired outcome was achieved.

Comprehensive Controls for Professional Technologies

The appropriate style of control for professional technologies is almost the opposite of that for craft technologies. Here the variety of tasks encountered and the number of exceptional cases is high, but well-established procedures for handling the work are usually in place. Management accounting and control reports in professional technologies have several distinguishing features.

For one thing, they tend to be much more detailed than is the case for the other technologies. This is probably due to the fact that there is greater variety in the tasks undertaken. In addition, tasks are complex but well understood and they can be broken down into many parts, so more detailed information is available. Control reports also tend to be issued more frequently. When there are a lot of complex details to keep track of, it seems likely that frequent reporting on events and outcomes is important for effective control.

The participation pattern in setting targets in the controls is also noteworthy. Department managers and employees in professional work tend to have more influence than for other technologies in setting control system targets, but upper management influence is also strong. This widespread participation in target setting is probably due to the fact that power and knowledge in professional technologies are widely distributed. Nevertheless, target levels incorporated into the controls are more difficult to achieve and often less accurate than is usual. The reason for this may be that it is easy in preparing the budget to underestimate the magnitude of complex tasks ahead.

Planning and coordination functions of control reports become increasingly important in professional technologies. Although increased variety leads to more complexity, tasks are well understood so planning and coordination can greatly ease the managing of such complexity. Consequently, the information contained in control reports can be extremely helpful for fulfilling these functions.

Overall, developing appropriate management controls for professional technologies seems to be more problematic than for others. The wide variety of complex tasks makes for difficult administration even though tasks are analyzable in their deep structure. Attempts, therefore, might be made to develop *comprehensive* and *elaborate* controls which turn out to be of some help but do not provide ready-made answers. These circumstances may account for the fact that departmental managers in professional technologies tend to be less satisfied with controls than those in other technologies.

Prospects-Oriented Controls for Research Technologies

Research technologies feature a great deal of variety in the tasks required to get the job done but the tasks themselves are not well understood at the outset when the work is begun. Not surprisingly, then, the appropriate pattern of controls for research technologies is strikingly different than for programmable ones. Controls are less detailed and reporting is less frequent. They are used more for planning and coordination than for measuring performance and allocating rewards. Less emphasis is placed on meeting target levels in controls than in routine technologies. But the influence of budgets on activities increases slightly. Standard operating policies and procedures are not much in evidence since they are ill suited to the uncertain and ambiguous environment.

Further, money and effort cannot be traced to outcomes, and it is difficult to know when, if at all, the benefits will appear. In consequence formal controls and reports lose their value for measuring performance and for reward schemes. So it is not surprising that managers are less satisfied with controls than they are in programmable technologies. Budgets, however, can be valuable for inducing managers to coordinate with other departments and to speculate about future prospects. Control reports can also be useful for coordination and planning.

The pattern of target setting for budgets and other control reports in research technologies is unique. Employees in the department have much more influence in determining target levels than do either the department managers or upper management, maybe because employees are often highly trained and know most about the specifics of job requirements. Target levels, however, tend to be very ambitious and difficult to achieve. Perhaps employees are overly optimistic and underestimate the difficulties of achieving the goals they set for themselves. Or it may be that employees are less cautious than they otherwise might be, since control reports are used very little for measuring performance. Nevertheless, since research tasks are frequently prone to overrun estimates of time and spending, controls can bring large estimation errors to light. Although they have a part to play, controls are less useful here than in other technologies and overall satisfaction with them is low.

In sum, a *prospects-oriented* style of management control is well suited to research technologies. Budgets and statistical reports include only general information and are issued less frequently than for routine technologies. Controls are most useful for planning and coordination rather than for monitoring and measuring performance. Employees in the department have a lot of influence in setting the targets in control reports but these standards tend to be set at an ambitious level. Overall satisfaction with controls is lower than for other technologies.

Implications of Technology Modelling

The technology model of management accounting and control systems contains important lessons for executives designing and administering them. Such lessons are needed since mismatches between technology type and control system style are common. A striking example involves the control system installed by Secretary of Defense Robert McNamara to help manage the war in the Vietnam jungle.

McNamara insisted on receiving detailed computer reports daily regarding battle manoeuvres, tactical deployment of troops, ships, and planes, as well as damage reports and kill counts. His detailed and sophisticated control system had worked well at the Ford Motor company but was ill suited to the technology of conducting jungle warfare with its variety, surprises, and unanalyzable programs. In consequence, officers on the spot had to make up fictitious kill and damage numbers for their reports to indicate progress. When the discrepancy between reported and actual performance in Vietnam came to light, McNamara's star rapidly faded. This was a contributing factor in his resignation as well, some argue, in President Johnson's shocking decision not to run for reelection. McNamara wanted a *close* management control system when a *comprehensive* system suited the professional technology of an extended jungle war.

A second example involves the R&D department of a multinational aluminum company. The department was the subject of a budget system that provided an intensive line-by-line analysis of actual spending compared to budgeted amounts for energy research project. Over a period of years, as the detail and intensity of the control system increased, the satisfaction and productivity within R&D actually decreased. Every month, the R&D vice president spent several days explaining and bickering with the budget officer over differences between actual and budget. Finally, the R&D vice president resigned.

A technology-based analysis indicates that a close control system was used, whereas a prospects-oriented one was called for. Research is not a well-understood process. Precise, detailed reports, along with close scrutiny and analysis of each line item, are quite inappropriate for the ambiguity of the R&D activities. Minor deviations from the budget are the rule rather than the exception. Prospects-oriented control, which focuses on plans and future accomplishment and features infrequent and general reporting, would be more suitable. This would shift the concern of the R&D department to future problems and to keeping the research output consistent with the overall company strategic plan, and away from minor past deviations from budget.

Summary

The matrix of technology types and control system styles proposed above is based on the idea that systematic variations in the structure of departmental technology call for patterned differences in the characteristics of managing accounting and control systems. Using two aspects of technology, task variety and task knowledge, a fourfold typology was developed. In programmable technologies, where analyzable programs are available for the predictable stream of raw materials, a close but supportive control style is called for. In contrast, for research technologies, characterized by high variety and unanalyzable programs, management accounting and control systems should focus on prospects. In the case of craft technologies, bottom-line, results-oriented controls are appropriate. Finally, professional technologies seem to pose the most problems for control systems due to the complexity caused by the variety in tasks and the complicated stocks of knowledge. These technologies call for comprehensive reports which are issued frequently. By analyzing the technology of the work unit, the appropriate style of control can be identified and costly misfits avoided.

TECHNICAL RATIONALITY, INTERDEPENDENCY, AND CONTROL

Our second framework of technology and control is based on the idea that the design and focus of management accounting and control systems are related to the overall pattern of the work flow through organizations.[7] The pattern of work flow (or technical rationality) is seen to influence the type of interdependence among departments within the organization. Interdependency refers to the extent to which departments depend on each other and exchange information and resources to accomplish their respective tasks.[8] The framework identifies three patterns of technical rationality interdependency and relates each to a distinct means of control. The main ideas of the framework are outlined in Table 3.

Serial Interdependence: Control Through Standardization

Serial interdependence is a result of *long-linked* technical rationality. In this pattern of work flow, a value-added system contains components organized along technical specialized lines. Each component performs its part of the larger task and passes the job on to the next one. This is epitomized by the factory assembly line, but is also the basis for the organization of clerical work, military training, and public sector organizations such as the post office. Specialized exchange and value-adding arrangements can also be interorganizational as is the case in the production, distribution, and marketing of foodstuffs such as grains, vegetables, fruits and wine. Long-linked technologies approach instrumental perfection in cases where they repeatedly produce a single, standard product at a consistent rate. The pattern of work flow is shown in Figure 8.

The automobile assembly plant is the archetype. The serial technology required clearly dictates the requirements for the machinery, plant layout, raw materials, and workers. The experience gained by a repetitive production process provides a learning curve effect. Imperfections in the technology are eliminated, preventive maintenance is scheduled, and workers' motions are studied in order to minimize energy losses and errors. Once the optimal constant

Table 3 Technical Rationality, Interdependence, and Control System

Technical Rationality	Interdependency	Control System Focus
serial	long-linked	planning and measurement
mediating	pooled	standardization
intensive	reciprocal	feedback from the object

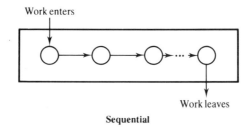

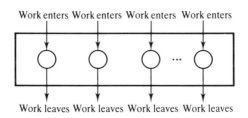

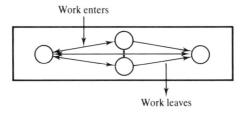

Figure 8 Sequential, Pooled, and Reciprocal Interdependence

rate of production is settled and the flow of work adjusted accordingly, the proportions of resources in each component can be standardized so that each works to its full capacity. No components go underutilized. In this dream realm of the management scientist, instrumental perfection can be closely approximated and economies of scale approach optimal levels.

In long-linked technologies the components are serially interdependent. Each component in the chain is highly dependent on all the others and each can only perform its part of the task after work has been successfully completed in the previous component. Serial interdependence puts great demands on the organization for coordination and close control. Disruptions and breakdowns in the work flow must be minimized. This is accomplished by means of detailed and rigorous planning, scheduling, and measurement systems.

Budgets and statistical reports are important in long-linked technologies. Component managers find budgets and statistical reports helpful for coordinating with other components and valuable for monitoring performance. Upper management has a lot of influence in setting targets in these controls. The targets tend to be difficult to attain with a lot of emphasis on meeting them. Component managers keep targets in mind in going about their daily administrative duties and report that they have an important impact on component activities. Standard operating procedures and policies, however, are not so important since each component in the chain is highly specialized and it would be difficult to develop standard rules suitable for them all. Instead, coordination and control is achieved through planning with follow-up feedback reporting on whether or not plans and quotas are met.

Pooled Interdependence: Control by Standardization

Pooled interdependence predominates in the case of organizations operating mainly a *mediating* form of technical rationality. These organizations provide a facility for mediating between sets of customers or clients who only occasionally want to transact with each other. Banks are a prime example as they link depositors who want safekeeping and who also want to put their savings to work, with lenders who want to borrow money. Mediating organizations act as the go-between for large numbers of customers and clients who only wish to deal with each other in small doses.

Operating components in mediating organizations share a pool of common resources but, unlike those in the long-linked organization, have few dealings with each other. Insurance companies, for example, provide for the pooling of common risks, such as life and home insurance, for a large number of clients. The branch offices are spread across the nation and only occasionally engage in transactions with each other. They normally get on with the job of selling insurance and collecting premiums without regard to actions in other branches. Yet each makes a discrete contribution to and is supported by the whole organization.

Such organizations are common today. Telephone companies maintain ground lines, switching facilities, and satellite equipment to connect up callers and receivers (including computers) around the globe. Credit card services link consumers needing credit with retailers who want to be certain of getting paid within a short time and who want to minimize record keeping. And post offices provide a pool of facilities capable of linking nearly everyone in the world who wishes to send or receive correspondence. Rather than having each operating component geared to the demands of the next one as in serial interdependence, the key is to have components extensively distributed in space, with each operating in a *standardized* way.

So standardizing each operating component is vital. For the bank, no matter how diverse the depositors or loaners, each transaction must be handled in a uniform way and be recorded using uniform accounting and record-keeping procedures. Each borrower must be handled according to standardized ways of categorizing the risk involved and determining the appropriate interest rate. For the telephone company, equipment and operating procedures must be standard across the network. Such organizations can offer their service or facility at a price far below what it would cost the parties to make the transaction by themselves. They can also handle vast numbers of transactions and are capable of achieving vast economies of scale.

The key to harvesting these economies lies in the ability of the organization to assure that each operating component does business in exactly the same way as all the others. The same services and techniques must be available across time and space. And work must be carried out uniformly and unfailingly. Thus, the critical control problem is how to effect standardization.

While perfect standardization is not always possible, it can be approximated through the bureaucratic rules, standard operating procedures, and routinization. Standardization also means that the sundry components can carry out their transactions without being dependent on any of the other components. So interdependency and coordination needs between components are limited. Not surprisingly, these organizations invest heavily in developing precise, detailed, standard operating procedures as well as in intensive on-the-job, apprentice-like training of personnel about how to apply these procedures.

Budgets and statistical reports about operating details are also important in these organizations, but less so than for those featuring long-linked technology. Target levels in these controls tend not to be too difficult to achieve even though the component managers have little influence in setting them. Budgets and statistical reports are standard for all components and so facilitate comparisons between components, identify less efficient components, and pinpoint trouble spots. Even so, these controls are less important for measuring and monitoring component performance than is the case for long-linked organizations.

Thus efficiency and conformity *across* the entire organization are much more important than efficiency *within* the individual components. Impersonal rules, clear-cut job descriptions, and detailed standard operating procedures, hallmarks of the traditional bureaucracy, are most beneficial for achieving the requisite standardization.

Reciprocal Interdependence: Control by Mutual Adjustment

The third kind of technical rationality is *intensive*. Departments embedded within intensive technologies feature *reciprocal* interdependency. Typically, departments work jointly with the other codependent units on the same raw material, customer, client, or project. Treatments and services are customized on the basis of feedback from the object. Based on that information, the work is coordinated through mutual adjustment by the respective departments involved. Reciprocal interdependence places a heavy demand on management to coordinate and integrate effort.

For example, in a psychiatric unit for children with learning disabilities, both the children and their parents required diagnosis and testing by psychiatrists, psychologists, educators, social workers, and occupational therapists (Williams, Macintosh, and Moore 1990). The unit layout featured a large kitchen in the centre of the building where coffee was available all day. The professionals frequently dropped by the kitchen several times during working hours, to grab a cup of coffee or a snack and to exchange information about the clients. These kitchen chats were a highly valuable and efficient complement to the formal meetings. The process became known as the "kitchen control patrol."

The US Desert Storm operation against Iraq in 1991 provides a more striking illustration. Initially a multiplicity of highly specialized capacities were brought to bear: satellite readings, weather reports, bombing raids, pinpoint missile attacks with TV cameras, spy networks, radar, advance ground observation and sorties, even sophisticated social and psychological intelligence units. As the war progressed, the treatment changed to long-range missile attacks, stealth bombing, mass missions, sea bombardments, and mock invasions. Finally, the operation shifted to a lightning ground encirclement, ground warfare, and mopping-up exercises including prisoner internment. The highly interdependent sundry branches of the military included fighting components from a host of nations, as well as the United Nations. Coordinating and controlling the sundry units required the expertise of highly sophisticated communications and intelligence centres with the most up-to-date information technology. Control happened as a result of feedback about the effectiveness of the many initiatives.

As these illustrations show, intensive technologies features the customized application of nonstandard techniques applied to the raw material. But neither the selection of techniques, nor their particular combination, nor the order of their application can be determined in advance. Instead, these matters are decided by diagnosing the specific, most urgent problems with the raw material on the basis of feedback from it. Success depends on getting the right mix of specialty expertise and on maintaining an appropriate flow of information. Interactions between specialists feature intensive dialogue, question and answer exchange, and lots of debate, notably without the help of any intermediary or coordinating unit. Intensive interdependence requires real-time, intensive information flows among the various components.

Management accounting and control systems are not very well suited to these needs. Information in plans such as budgets and feedback reports after the fact are not as useful as for the other two types of interdependencies. In fact, research has indicated that budgets, statistical reports, standard operating rules, and procedures are relied on much less than in either the long-linked or the mediating technologies.[9] Instead, the requisite coordination and control is achieved through personal interaction, frequent communication, and mutual adjustment by the various experts and administrators concerned.

Thus, traditional control systems lose their primacy and effectiveness under the uncertainty and rapid adaptation called for by reciprocal coordination. The need for quick response and sophisticated innovation, with different departments joining forces around specific problems means that the organization cannot rely on rules, standardization, and all the regular bureaucratic mechanisms including management accounting and control reports. Expert knowledge, mutual adjustment, building community spirit, and dedication to the organization replace formal controls.

Nevertheless, accounting systems such as budgets are still valuable. They are used for planning the overall resource requirements for the budgeting period. Similarly, statistical reports (such as the number of clients or projects started, completed and in progress) are important to rationalize activities and efforts to superiors and important outsiders such as funding agencies. Controls are less useful for the traditional measurement and control functions.

Thus, the basic tenet of management accounting that managers should not be held accountable for costs and other aspects of performance over which they have no control, does not fit very well in the case of intensive technical rationality. Such accountability is widely shared and the work of any one department depends greatly on the ongoing work in other interdependent departments. Formal accounting and control systems may be beneficial for longer-term planning, but they are not very helpful for either coordination or measuring the effectiveness of the highly interdependent groups.

CONCLUSION

This chapter described two structural functionalist frameworks of management accounting and control based on technology or the nature of the work done by departments in organizations. The basic theme was that patterned variations posed by technology result in systematic differences in characteristics of management accounting and control systems. The first framework described a fourfold typology of control based on two aspects of technology, task variety and task knowledge. It proposed that a close control style is well suited to programmable technologies while a prospect style is more appropriate for research technologies. Similarly, the framework prescribed results control for craft work and comprehensive control for professional technologies.

The second framework outlined control system characteristics for three different kinds of technical rationality. In the case of serial interdependence and long-linked technical rationality, plans and performance feedback information are crucial for control. For the mediating technology, with its pooled interdependence, traditional management accounting and control reports are not quite so helpful; standardization is the key. While for reciprocal interdependence and intensive interdependence, feedback from the object and mutual adjustment by the various players is the appropriate means of control. The effectiveness of management accounting and control reports and other bureaucratic techniques such as standardization and routinization varies considerably with the nature of the interdependencies involved.

While these frameworks provide a systematic way for analyzing the fundamental nature of an organizational component and then relate this to the appropriate characteristics of management accounting and control systems, we are not advocating a technological imperative to design problems. Rather, we argue that technology, along with environment, history, and strategy as discussed in earlier chapters, is an important consideration in designing and using management accounting and control systems. In the next chapter we set aside, for the most part, the structural functionalist paradigm to look at the way managers subjectively interpret management accounting and control systems, especially in a proactive way.

NOTES

1. Thompson (1967) p. 2. See also Woodward (1965), Perrow (1970), and Khandwalla (1972).
2. Bruns and Waterhouse (1975), Hayes (1977), Daft and Macintosh (1978, 1981), Macintosh (1981), Waterhouse and Tiessen (1978).
3. See particularly Otley (1980), Otley and Berry (1980), Birnberg, Turopolec, and Young (1983), Kim (1988), Duncan and Moores (1989), Abernethy and Stoelwinder (1991).
4. See especially Hirst (1981, 1983), Brownell and Hirst (1986), Mia (1989), Dunk (1990), Brownell and Dunk (1991).
5. See, for instance, McNair and Mosconi (1989), Dunk (1992).
6. Mintzberg (1979) especially Chapter 19, "The Professional Bureaucracy," and Macintosh (1985), Chapter 9.
7. These are Thompson's (1967) definitions of technical rationality and interdependencies.
8. The following framework is based on the research reported in Macintosh and Daft (1987) and Williams, Macintosh, and Moore (1990). Other behavior accounting studies incorporating interdependency include: Watson and Baumler (1975), Hayes (1977), Ginzberg (1980), Otley (1980), Kilmann (1983), Emmanuel and Otley (1985), Merchant (1985), Chenhall and Morris (1986, 1991).
9. See Hayes (1977), Macintosh and Daft (1987), and Williams, Macintosh, and Moore (1990).

8
Uncertainty, Scorekeeping, and Control

This chapter continues the flow of thought from the previous chapter to look at two frameworks that link scorekeeping and control needs to the nature of the uncertainty involved in the work done by the organization or organizational components. It tries to provide a solid understanding of the precise way in which a few different patterns of uncertainty influence how organizations handle these important functions. Whereas the previous chapter focused mainly on formal controls, such as budgets and statistical reports, this one expands the terrain to include informal, unofficial, and nonaccounting-based scorekeeping and control systems. The basic premise is that the cutting edge of uncertainty affects the actions organizations take to cope with these requirements.

More specifically, the chapter talks about different ways of keeping score—efficiency, instrumental, and social tests—as well as different types of control—bureaucratic, charismatic, market, collegial, and tradition. It shows how patterned variations in uncertainty result in systematic differences in these practices. It will be helpful at the outset to make a distinction, at least in the abstract, between *closed-rational* systems and *open-natural* ones. The frameworks include a valuable mixture of the objectivist and subjectivist positions. For the closed-rational part of the organization, the objectivist position is adopted and scorekeeping control systems are treated as concrete social facts existing independently of social actors. While for the closed-natural part of the organization, the subjectivist position is assumed whereby these systems are treated as social phenomena existing subjectively and intersubjectively in the minds of the managers and employees.

CLOSED-RATIONAL VERSUS OPEN-NATURAL SYSTEMS

Under conditions where certainty is pretty much the order of the day, organizations are amenable to the logic of closed-rational systems. By this we mean a system where the number of variables and the relationships between them can be comprehended and controlled in a predictable manner. That is to say, they can be effectively treated as rational.

In order to determine whether or not a component is amenable to the logic of rationality we might ask a few questions. Are the component's goals and mission known? Are the necessary resources to reach these ends on hand or available? Are the component's tasks well understood? And is the output absorbed pretty much automatically? If the answer to most of these questions is yes, then it is likely that organizations will treat such components as closed-rational systems and try to protect (or buffer) them as much as possible from environmental disturbances. In these circumstances, the rules of rationality can reign.

In fact, we are surrounded by such systems. Routinely they deliver newspapers, carry conversations over long distances, transport people back and forth to work, stock supermarkets, keep track of money, educate children, and even entertain us on picture tubes. These daily miracles are achieved through marvellously efficient action systems within organizations which are buffered from environmental disturbances that might upset their routines. Following the rules and norms of rationality, they perform with such precision that we are outraged if the morning paper is late or if the supermarket runs out of our favorite brand of catsup.

These circumstances do not pose large problems for the selection of organizational scorekeeping and control systems. Operations proceed smoothly according to the rules of rationality. This is the case, above all, in the technical core of large industrial enterprises (e.g., a factory, post office, or a mining operation) which are buffered and shielded from environmental shocks by such means as inventory reserves, production scheduling, and idle capacity. It is also true of clerical units, where large numbers of similar transactions are processed day after day as was the case for the finance company branches in Chapter 5. Predictable environments are congenial to organizational effort. Specialized units with repetitive tasks are established and operated according to plans and predetermined yardsticks.

For example, a mass-production factory operates under relatively certain conditions. Scorekeeping is straightforward and focuses on whether or not optimal results have been accomplished. Management accounting and control systems, backed up by inducements geared to ensure conformity to plans, provide accurate and timely information for performance assessment. They play a critical role in the relentless drive for efficiency.

Another example is the case of a national supermarket chain that recently restructured its retail arm. Under the new arrangements nearly all management decisions are taken centrally. Purchasing, advertising, payment of bills and employees' wages, personnel policies and records, banking, and special promotions functions are handled at the corporate and district offices. Even store layouts and shelf-stocking methods are dictated by a central blueprint. Thusly the individual stores are buffered from almost all environmental uncertainties.

The premise that components should behave rationally is reasonable enough when organizational goals are known, the necessary resources are available, tasks are understood (explicitly or intuitively), and output is absorbed automatically. Organizations go out of their way to operate their basic core technologies as closed systems by buffering them from environmental turbulence and then treating them as rational systems.

In contrast, when the answers are mainly no, an open-natural system logic is called for. A natural system is not closed off from the forces and changes in the larger environment. Nor are the number of variables and their interrelationship fixed. Uncertainty is the order of the day and the system must adapt to environmental perturbations in order to survive. It must, as the rule of nature dictates, change and evolve.

These circumstances mean the organization is not amenable to rule by rationality. It is open to the uncertainties and vagaries of the environment. Many parts of an organization operate under such conditions. A marketing department, for example, has no way of knowing whether or not it has optimized operations. The same applies to accounting and computing departments; and for R&D units, optimizing resources is not even the goal. Survival overrides efficiency.

Turbulent environments, then, pose major challenges. Organizational structures and management control systems, so well-suited to closed-rational logic, are inappropriate. Nevertheless, while it may appear at first that developing suitable scorekeeping and control systems is a random trial-and-error process, evidence is mounting that organizations faced with similar sources of uncertainty respond in similar ways. The major thesis is that systematic variations in uncertainty tend to lead to patterned, and therefore predictable, variations in scorekeeping and control needs and practices.

Following this line of reasoning, we can move on to describe two insightful frameworks which treat the kind of uncertainty in the component's ends and means as the driving force behind scorekeeping and control needs. We start with a typology of scorekeeping and assessment based on the above ideas.

A TYPOLOGY OF ASSESSMENT AND SCOREKEEPING

Our first framework deals with the appropriate type of assessment and scorekeeping for different kinds of uncertainty.[1] Uncertainty is defined as the interaction of two antecedent conditions: whether task instrumentality is rational or nonrational and whether purpose is crystal clear or ambiguous.

Task instrumentality refers to the available means for task accomplishment. It is a continuum; at one end the ways of doing the work to get the desired output are believed to be known, either by calculation or through experience. Well-understood actions produce highly predictable results. In the machining department of a tractor parts factory, the correct sequencing of work through the machines is well understood, the precise technical tolerances are predetermined, and the exact quantities of raw materials, direct labor, and supervision are known with a high degree of precision. When task knowledge is complete, the effect of instrumental action can be traced to known results.

But sometimes the effects of instrumental action cannot be predicted with any degree of certainty. At this end of the continuum, outcomes may be the result, not only of actions taken inside the department, but also of events and actions taken outside the department. Further, some consequences of actions may be known, others generally agreed upon but unproven, some suspected, some occurring in the vaguely far-distant future, and still others go entirely unnoticed. Task instrumentality is problematic.

In a marketing department, for example, sales results cannot be related to specific instrumental action within the department. The effect of pricing decisions, choice of distribution channels, selection of advertising media and message, and bonus schemes for salespeople cannot be traced directly to specific sales transactions. External actions of competitors, governments, and banks also have an important but undetermined influence over the outcome.

The other continuum concerns beliefs about the organization's ends, mission, and purpose. At one extreme, ends are rational, that is to say, they seem to be factual and objectively determined, one-dimensional, and with clear-cut preferences over other possible ends. Profit, for example, is preferred to market share. Health is preferred to illness, wealth is fancied over poverty, efficiency is deemed better than malingering, and world peace is clearly preferable to global holocaust. Also the direction for improvement is obvious: from market share to profit, from poverty to wealth, from inefficiency to efficiency, and from hostilities to peace. When there is general agreement on one clear-cut end, goals can be treated as unambiguous.

But often ends and missions are nonrational; they are not always represented by a single unambiguous criterion. Sometimes they involve a choice between two or more dimensions. It is not merely a matter of health over illness, but a choice between health and wealth. Shades of health and shades of wealth may be involved in the choice. Ends are value-oriented and subjective. Much of the time we are dealing simultaneously with some degree of health and some amount of wealth.

To further complicate the issue, we may also feel ambivalent if the choice lies between two roughly acceptable alternatives, such as long-run cash flows and short-run profits, or even between two equally repulsive alternatives, such as illness today versus poverty next year. As well, ends usually change with time; one day we prefer wealth but the next we favor health. When ends and mission are nonrational the choices are agonizing; we are hard-pressed to decide. When there is no general agreement about which end(s) are preferrred, goals can be treated as ambiguous.

Efficiency Tests

These two continuums, particularly their extreme values, can be combined to derive different assessment and scorekeeping situations, as depicted in Figure 9. When instrumentality is believed to be complete and goals clear and unambiguous, the optimum economic relationship between inputs and outputs can be derived. In such circumstances, efficiency is the appropriate test. Efficiency is achieved when the maximum amount of output results from a given level of input resources.

Was the result produced with the least cost? asks a scorekeeping question for a closed-rational system. Alternatively, efficiency may be the minimum input resources for a given level of output. The scorekeeping question asks, Was the given amount of input used in a way to achieve the greatest result? Either way, a known scientific relationship exists between resources consumed and outputs produced. Perfection is the goal and the efficiency test measures relative perfection.

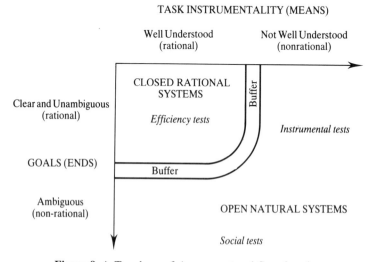

Figure 9 A Typology of Assessment and Scorekeeping

Instrumental Tests

But when uncertainty enters the scene—nonrational means and/or ends—efficiency scorekeeping quickly loses its utility. One source of uncertainty is poorly understood task instrumentality. The relationship of inputs to outputs cannot be known with any degree of certainty and it is not possible to determine whether the desired result has been achieved efficiently. There is no way of knowing if the best instrumental action was taken because the possibility always remains that a better course of action exists.

We do know, however, whether or not the instrumental action achieved the desired state. So, the scorekeeping question should center on effectiveness; Did the action result in the desired goal? The robust efficiency test gives way to the more appropriate *instrumental test*.

Instrumental tests, of course, are widely used. In professional sports, for example, won-lost columns, and which teams reach the playoffs, are common ways of judging performance. Managers and coaches are well aware that their continued employment hinges on these simple instrumental tests. In the public domain, even though objectives for desired outputs such as levels of employment, inflation, and interest rates are less clear, voters and political commentators alike set arbitrary levels of satisfactory attainment. These are straight forward bottom-line assessments.

Most large industrial and commercial organizations aim for instrumental scorekeeping. They carve themselves into smaller components, each of which is relatively self-contained in terms of resources, products, and markets. Performance is deemed satisfactory if predetermined desired profit targets are met. But optimum profit levels are indeterminate, since the possibility always remains that some other combination of resources, products, and markets would yield more profit. And we do not know whether or not factors outside the system, such as a spurt in the economy, influenced the result. We do know, however, whether or not the *desired* level of profit has been achieved. Instrumental tests are never perfect, but they can be appropriate.

Social Tests

Our third assessment situation occurs when both ends and means are nonrational. Here uncertainty reaches its maximum and assessment takes on a new dimension. Both efficiency and instrumental tests are unsatisfactory, so organizations must retreat to an even less satisfactory but more appropriate means of assessment, the *social test*. The idea of a social test may be foreign, even repugnant, to many accounting and control systems professionals, but the basic idea is to judge accomplishment and fitness by the collective opinions and beliefs of one or more relevant groups, not on the basis of efficiency or instrumentality.

A human resource department, for example, provides an organization with services such as hiring, firing, pay and promotion schemes, safety programs, and union negotiations. The collective opinions and beliefs of its client departments regarding these services are good indicators of its fitness for future action. Another good source of assessment comes from personnel associations that periodically award prizes for outstanding achievement in the field of human resources. In these examples, assessment leaves the realm of economic fact and enters the domain of social opinions, beliefs, and values.

If we pause and think about it, social tests are not so uncommon. In fact many of them are a regular and official part of organizational life. Scientists in R&D departments present papers at conferences where their research undergoes close scrutiny and criticism by colleagues from other organizations. Students elect the teacher of the year. Panels of journalists decide best article awards. Small groups of critics make artists, playwrights, and writers famous, or even infamous. A panel of expert accountants from industry, government, and universities judge the annual reports of corporations. Research funds are granted by groups of distinguished academics, who judge the fitness of applicants to conduct the research work. And for many, the ultimate social test is the Nobel Prize, which is awarded for outstanding contributions to science and the humanities. In each case the social tests are anchored not in organizational rationality, but in the collective and subjective opinions and beliefs of relevant social groups.

Such tests, of course, are not without problems. It is well to recall that when Giuseppe Verdi applied to the Academy of Music in Rome, he was turned away for lack of ability; the French impressionist painters were shunned by the Academy of Art; and several universities rejected Albert Einstein's application for enrolment due to weakness of background and lack of promise. Still, in the absence of efficiency and instrumental tests, social ones, even though capricious and precarious, provide at least some information about fitness for future action.

Social tests, in fact, play a critical role in organizations. More often than not an organizational component, even if deemed to be a separate and autonomous subunit, is in reality highly dependent on several other components. A typical example is a parts and assembly plant treated as an autonomous profit center within an integrated home appliance company. Transfer price systems and methods for allocation of joint costs are used to develop efficiency (costs per unit) and instrumental (profitability) tests for the plant.

Efficiency and instrumental tests, however, lose some of their bite and credibility since output is highly dependent on the performance of other interdependent components, such as engineering, purchasing, marketing, and sales. In such a case, social tests, although informal and even invisible, come to the fore. The expectations, beliefs, and opinions of the other managers in the network are the crucial test of performance and fitness for future action. Does the component fill its quotas? Does it deliver as promised? Does it follow the rules? Are its members good team players? The confidence expressed in a component by the other coordinate interdependent units is an important and relevant assessment test.

In addition to scorekeeping for individual departments, organizations also attempt to assess the overall performance of the entire organization. In stable environments, historical improvement, particularly growth, is taken as evidence of both current fitness and past performance. In dynamic environments, historical improvement is of little consequence and organizations turn to comparisons with similar organizations.

Business firms, for example, try to convince those they depend on, like investors, bankers, shareholders, suppliers, customers, and employees, that they compare favorably with competitors. To do this they point to increasing market share, amounts spent on R&D, the number of new products brought on the market, and concern for the environment. Historical improvement and favorable comparison with similar organizations are the vehicles for convincing relevant social groups of organizational fitness.

Assessment and scorekeeping, however, become even more difficult when there are several important external groups to satisfy. Organizations, realizing they cannot compare favorably on all criteria, try to hold some constant, and show improvement in other more crucial areas. A business firm needing a bank loan will attempt to score well on the balance sheet, especially working capital and liquidity ratios, by investing in inventories and paying suppliers quickly. Whereas firms seeking new equity capital will attempt to score well on the income statement, particularly in the earnings per share category.

Universities must also respond to many external groups. They try to convince government funding agencies that their operating budgets are kept at a minimum consistent with some quality standard, while simultaneously demonstrating to alumni and prospective students that the already high quality of teaching is getting even higher. They also publicize the more rapid than average improvement in the quality of their students and the quantity of research done by their faculty. And they try to convince accrediting agencies and research funding committees that faculty scholarly output is better than ever. The trick is to hold some aspects of performance constant and show improvement in those areas important to critical external assessment.

Coping with the need to satisfy simultaneously many different elements is no easy matter, but the problem becomes all the more difficult when dependency on various groups fluctuates. Here the organization must adjust the relative weightings of the multiple and varying criteria. Although the scorekeeping task is difficult when critical elements change and weightings shift, keeping alert for clues about the shifting weights enhances flexibility and helps organizations shift to a different but more viable competitive stance consistent with changing demands.

The US automobile industry in the 1970s provides a concrete example. For many years the industry was motivated by shareholders' requirements for growth in earnings per share, customers' wants for bigger and more powerful cars, and managers' needs for large annual profit-based bonus and stock option schemes. The Big Three US automobile companies concentrated on the larger, accessory-loaded automobiles which produced the highest profits. In the interim, however, important environmental elements shifted the industry's concerns away from traditional performance criteria. Safety, pollution control, quality of working life, energy conservation, and eventually jobs for auto workers emerged as dominant criteria. The Big Three firms, spurred on by declining profits, the success of Japanese competitors, and government regulations, slowly but surely shifted their assessment criteria to stress factors such as safety tests, fuel consumption, improvement in working conditions, and the number of jobs at stake.

Summary

When objectives and missions are clear and unambiguous and the means for getting work done are well understood, organizations can assess the performance of their components according to past efficiency. If, however, the means are not well understood, assessment must be based on instrumental tests dictated by goals and mission. But if components are internally autonomous, assessment must shift to extrinsic measures. And when ends and missions are nonrational, organizations must turn to the opinion of relevant social groups. Finally, under conditions of multiple criteria, organizations must be sensitive to, and quick to change the relative weightings of the various criteria as the importance of relevant social groups also shifts with the fluctuating environment.

That accountants should gather, store, and report information about social opinions and beliefs remains a novel and perhaps uncomfortable idea. The idea, however, is creeping into the accounting domain. Socially and environmentally concerned groups of citizens are forming mutual investment funds to purchase securities of only those firms with a good record in these areas. Their opinions count, even in the pocketbook. And assessment and scorekeeping in the public sector at the federal, state, and municipal levels have been working on this front for many years. It is a good time for management accounting and control systems managers to take the idea seriously.

A TYPOLOGY OF ORGANIZATIONAL CONTROL STYLE

Our second framework, also based on the cutting edge of uncertainty, is concerned with the way organizations control their managers and employees. It deals with the way they keep their members in step with each other and keep them working towards overall organizational purpose. The ideas underlying this typology stem from Max Weber's classic and seminal work on the sociology of organizations (Weber 1947). The framework, outlined in Figure 10, identifies five generic types of control—bureaucratic, charismatic, market, collegial, and tradition—and indicates the circumstances under which each can be used to advantage. It also brings into consideration the significance of authority and power, aspects of organizational life which most conventional treatments of management accounting and control systems tend to ignore or push off center stage.

Bureaucratic Controls

The rise of the rational–legal bureaucracy (or hierarchy) to its dominant position in society today, is one of the most remarkable events of the last century and a half. Bureaucracies, such as churches and military organizations, have existed for centuries. Today, however, the bureaucracy is ubiquitous throughout the world in both public, private, and mixed organizations. Most of us spend all our working day, and much of our nonworking day, in bureaucracies of one kind or another.

The most outstanding feature of the bureaucracy is that authority is vested, not in persons, but in offices (bureaus). Thus, at least in principle, officials who are appointed to the offices

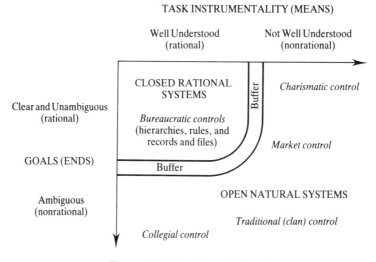

Figure 10 A Typology of Control

act without prejudice or personal feelings. Governance is not by personality, hereditary rights, or tradition, as was formerly the case, but through offices. Bureaucracies rely on three major mechanisms to control their members: observational hierarchies, rules and procedures, and written records.

Observational Hierarchies

Bureaucracies feature a hierarchy of supervisory positions each of which endows incumbents with the authority to oversee the work of others. Direct observation involves having superiors watch and guide the actions of subordinates. Supervisors, for example, watch the workers on an assembly line closely, knowing that if the work is done exactly as prescribed, the expected quantity and quality of product will automatically appear. Personal observation and the issuing of commands assumes that superiors have a relatively complete understanding of the task instrumentality involved in the work. It also assumes that ends and values are rational and clear. When these conditions obtain, observation and command are appropriate.

Department managers in a retail store, for example, keep a watchful eye on the way sales personnel dress, approach customers, ring up sales, and wrap parcels. Apprenticeships for trades, internships for medical students, and training camps for aspiring athletes involve intensive training periods where neophytes work under the close surveillance of certified and respected experts. Supervisors observe the actions of individuals as they go about their activities and issue appropriate commands when corrective action is judged to be in order.

The acceptance by subordinates of this command hierarchy is an essential ingredient in hierarchical control. They exchange their talents for organizational rewards and must willingly give up their autonomy and submit to authority of superiors up the line to command and guide their actions. Such acceptance is made easier by the fact that observation and command is not necessarily hard-hearted. In fact, it is more often than not exercised within a considerate, warm, and flexible relationship. Moreover, supervisors are also under the watchful eyes of superiors further up the line and they are obliged to follow the rules prescribed for their particular office.

Another key feature of bureaucracies is that they attempt to define as clearly as possible each office in terms of its jurisdictional boundaries: section, department, plant, and so on. Each component is specified in terms of its functional specialization and its functional relationship to all the other bureaus. And each office is assigned a place in the chain of command. Bureaus are specialized, serialized, and ranked in terms of hierarchical authority.

Rules and Procedures

Bureaucracies also use formal rules to control their members. The rules outline the rights and duties of each office and position as well as, to the extent possible, the proper procedures for getting work accomplished. They provide the impersonal codes for conduct at work as well as the criteria, based on technical competence, for promotion and pay. Rules and regulations are conducive to the development of impersonal relationships between superiors and subordinates and among members of the rank and file. They depersonalize the work environment and mitigate the effects of personality, whims, and nepotism. Importantly, rules are binding on *both* subordinates and superiors.

Records and Files

Bureaucracies also invest heavily in keeping complete records and files. Such records include both standards and actual outcomes for inputs, outputs, and resource utilization. Management accounting and control systems are some of the most important types of records and files used by bureaucracies. Organizations make full use of accounting records and data files. Cost accounting systems, variance analysis reports, operating budgets, and profit–cost–volume charts abound in the bureaucracy.

Records and reports have one distinctive advantage over hierarchies and rules. Upper management do not have to be involved in direct observation of either the specific actions and the rule-following behavior of subordinates. Nor do they have to be knowledgeable about the instrument action used to achieve the results. They can merely rely on records and files, such as accounting reports, to know whether or not results conformed to expectations. They can control at a distance.

Consider some practical examples. For a chemical refinery, the actual yield of chemicals and consumption of feedstock are measured and compared with predetermined standards for quantities. Direct surveillance and issuance of instructions is not necessary. Upper management can afford to ignore the behavior of employees as long as outputs meet required levels. For a television show, network officials can rely on the ratings without knowing how the producer went about putting together a successful production. Teachers can use examination results to control students without knowing anything about the students' study or classroom habits. Formal, impersonal information is used instead of direct personal observation.

Accounting reports are an important part of these controls. They must, however, be used judiciously. In some circumstances they are a mixed blessing. At higher levels of management, for example, they are far from ideal. High levels of interdependencies obscure individual contributions; complexity is high, specific technical expertise about top-level jobs is thin, and role ambiguity is common. Yet ambitious upper-level managers often go overboard to provide evidence of their contributions. They seek out and even develop accounting-based measures.

A budget, for example, is welcomed since it appears to reduce role ambiguity, thus giving managers a sense of knowing where they are going and how they are doing, and, most importantly, providing them with objective proof of good performance. The temptation, however, is to treat accounting controls as unambiguous evidence of performance even when they contain only vague information.

But observational hierarchies, rules, records, and files, the distinguishing features of bureaucratic control, are not cost-free. Accountants, information system specialists, analysts, personnel people, lawyers, and a host of other administrative support staff are paid to design and administer the bureaucracy. They are, in effect, the costs of bureaucratic arrangements.[2] In spite of the transaction costs, bureaucratic controls can be effective, especially when task instrumentality is well understood and ends are unambiguous.

This, of course, is not always the case. Sometimes accounting records do not contain reasonable performance standards. Standards are often only approximate representations of desired behavior and output levels. So they lose their effectiveness when they do not contain reasonable performance information. Also, standards are prone to idiosyncratic interpretation by managers and employees alike. They become even more problematic when tasks are unique, when task knowledge is not well understood, or when interdependencies on the job are high. When standards no longer provide either the necessary motivational force or the direction necessary for goal congruence, accounting controls lose their effectiveness.

Summary

Bureaucratic control is ubiquitous in today's world. It works by means of observational command hierarchies, rules, records, and files. These controls function reasonably well for those parts of the organization that can be treated as closed-rational systems since ends and values are unambiguous and the knowledge regarding means of getting work accomplished is well understood. But they do not work as well for open-natural systems where these conditions do not prevail. Here, organizations look to other forms of control.

Bureaucratic controls, so well suited to the certainty of closed-rational systems, tend to lose their potency when means are not well understood and when ends are ambiguous. Under these conditions, other types of control must be brought into play to supplement bureaucratic controls. These include charismatic, market, tradition, and collegial controls.

Charismatic Control

Charismatic control is almost the antithesis of the bureaucratic mode. It works best in situations where there is a clear-cut mission or calling but where the means of accomplishing it are uncertain. Certainty and routine are anathema to charismatic control. If bureaucratic control feeds on constancy, regularity, and order, then charismatic control thrives on radical change and promises of emancipation from the ordinary. And if bureaucratic control conjures up a vision of faceless bureaucrats poring over records and files and looking up rules and procedures, charismatic domination brings forth a vision of a dynamic, heroic leader with personal magnetism, the gift of grace, and the capacity to perform superhuman feats.

Charismatic control works particularly well when purpose and the leader's claim to authority is in conflict with the claims of a well-established, fully institutionalized order. It requires some sort of revolutionary mission aimed at overturning the status quo. The leader's charismatic qualities, however, must be proven from time to time in order to be recognized as genuine by followers. In order to maintain support the leader must also mete out gifts and share any booty and glory with followers, especially outstanding ones. Exceptional feats, revolutionary missions, and the sharing of spoils are the basis of the charismatic leader's legitimacy.

Charismatic control, importantly, means that the leader does not have to watch subordinates to see that they follow orders and carry out their ad hoc assignments. The leader simply tells them what to do (and sometimes how to do it) knowing it will be done. Duties are never taken as orders but accepted as obligations, almost in the nature of holy missions to be carried out at any cost. Followers are disciples, not subordinates.

Thus the charismatic leader does not have to be concerned with gaining the consent of followers nor with trying to understand their will. And, unlike the bureaucratic control situation, charismatic leaders frequently have no established legal rights over subordinates. Likewise, there are no rules and regulations spelling out subordinates rights in terms of their protection from superiors. Instead, enthusiasm for the cause, hero worship, and personal loyalty to the leader guarantee compliance.

There are many examples in recent history. Early on in their reigns, Hitler and Mussolini, initially revolutionary and social democratic, were archetypes of the charismatic leader.

Both exuded charisma and gained the overwhelming support of the citizenry when their countries faced economic crises and general malaise of enormous proportion. Similarly, Churchill and de Gaulle, also relying on charisma, rallied their nations against the Axis powers during World War II. Field Marshall Rommel, nicknamed "Desert Fox," was a prototype of the charismatic leader. During World War II, outnumbered and with inferior firepower, he kept the Allied forces on the run despite overwhelming odds.

Formal accounting systems are little used in the case of charismatic control. Pure charisma is foreign to economic considerations (Weber 1947, p. 302). Economic resources (often in the form of booty) are required, but only as a means to achieve the mission or calling. Accounting is concerned simply with keeping track of the assets on hand and relating them to the mission. Charismatic leaders, however, often maintain their own private information network (including informants and spies) to keep tabs on followers.

While charismatic control works well during revolutionary periods, it can rapidly lose its robustness if the mission remains unaccomplished over a long time or if it runs out of steam. It also loses its effectiveness if the mission is successful, when it usually gives way to bureaucratic arrangements and a more stabilized order. Large business organizations seldom have success with charismatic control. Instead, they pursue clear-cut goals, such as profit maximization and rely more on market controls.

Market Controls

Market controls are based on the idea that victory in the marketplace is a key indicator of successful effort and that the information capacity of the market is a powerful vehicle for mediating among individuals. This is particularly so for spot markets where the goods or services desired are available and are exchanged and paid for immediately. Commitments and obligations are short-term and prices are the only information needed to regulate behavior. They enable individuals to pursue their goals in a wonderfully efficient manner.

The idea that organizations might use market controls may come as a surprise since this is not the main business of markets. Yet if circumstances are favorable it is possible to use them to advantage. The invisible organization of the market, as discussed in Chapter 3, is thought to capture and condense information about the sundry actors' wants and productive capabilities into market prices and so provides the disciplinary glove of the invisible hand. There seems to be no need for more complicated arrangements.

A practical example of the parts division of a major company may help to highlight the differences between bureaucratic and market controls.[3] In the warehouse, 150 overseers and supervisors oversaw the work of nearly fourteen hundred pickers and packers. The overseers gathered information about the work flow from two sources. First, they watched the workers to ascertain who was doing a good job and who was not. They queried workers about the way in which they performed their tasks. And, if appropriate, they gave orders to follow proper procedures. The other source of information was records of daily output for each worker. The overseers used these reports to confirm their personal observations; they worked within the limits of normal rank and organizational authority, along with the informal limits bestowed on them by the workers. Surveillance, output records, and the organizational hierarchy ensured an effective and efficient flow of work.

In the purchasing department, by contrast, only a few supervisors and a dozen or so clerks purchased hundreds of thousands of items each year from thousands of suppliers. The purchasing agents received bids for each order from a handful of suppliers and accepted the lowest bid quoted on the condition that the supplier had a reputation for reliability and honesty. Instead of undertaking costly surveillance and monitoring of the efficiency, quality control mechanisms, and delivery systems of each potential supplier, the company utilized market prices and competition to promote efficiency and quality on the part of the suppliers.

Within the purchasing department itself, market controls were also at work. As long as the purchasing agents were getting competitive bids for each order and sampling delivered products for quality, there was little need for supervisors to watch over the employees or for accountants to prepare output records. The manager needed only to spot-check that the purchasing agents were accepting the lowest bids. Market control proved effective and efficient for controlling both suppliers and purchasing agents.

The profit-centering concept is a more familiar example of market control. In fact, it is almost ubiquitous in today's world of huge, widely diversified, global business enterprises. These firms simply carve up their operations, as far as possible, into self-contained businesses (subsidiaries, divisions, product centers, strategic business units, etc.), which are granted considerable autonomy and discretion in making decisions relating to products, markets, production, marketing, engineering and R&D. Then upper management monitor progress and track performance by means of simple measurements of profit and share of market.

Profit measurement, it is important to recognize, is a market control in the sense that it reflects the prices received from customers for products and services as well as the prices paid for costs and expenses expended to earn those revenues. So the profit earned by the component for the accounting period is an indication of how well it competed in the marketplace, while share of market indicates the possibilities for future profits. Market controls, however, are usually complemented by some form of bureaucratic control by head office staff experts over their counterparts in the business units.

Market controls also have their limitations. While the normal objective is to maximize profits, it is usually difficult, if not impossible, to determine the maximum amount. So organizations settle for a targeted amount, which becomes the standard for judging market performance. Comparisons are also made with previous periods and other comparable companies. As long as the prices used are truly competitive and targets realistic, profitability measurement is a reliable indication of market performance.

As with the purchasing department example, upper management do not have to monitor each and every employee in the profit center as they perform their tasks. Nor do they have to gather detailed information on daily output. They simply periodically review the component's profitability and share of market. As long as the manager of the component behaves like a self-interested entrepreneur, competing in an assigned marketplace, the invisible hand of the market will reach in to provide the necessary motivation and discipline.

While market controls work well when profit is the clear-cut goal, they lose their efficacy when goals and missions are nonrational. Then organizations look to other means of motivation and discipline, such as tradition and collegial controls.

Control by Tradition

The next mode of control in our typology is control by dint of tradition. Until the advent of the Enlightenment project it was the major means of nonphysically violent domination in Western society when the legitimation of social codes for authority and domination stemmed from the collective wisdom of ancestors and the revered customs of antiquity.

The inherited right of clan chiefs, the divine right of kings, the authority given to fathers to discipline their wives and children, and the supremacy of the church over souls were ideas handed down from ancestors to posterity. Embalmed in the orthodox of custom and belief, these codes become imperatives for all members of the social order.

While traditions were normally beyond reproach and questioning and could not be challenged on the basis of rational knowledge, they could be changed on the basis that new or altered codes were traditions that had only recently been rediscovered. Generally, however, the received order was beyond reproval and rebuke.

An extreme form of domination by tradition is known as "clan control."[4] Membership in traditional clans was highly exclusive, sometimes limited only to persons of blood or marriage. Members had to undergo a long period of indoctrination and subtle value training into the ways of the clan before they knew how to behave. Clans had their own jargon and special meanings which were normally not understood by outsiders. Turnover was rare, even if the only way out was death; and this is still true for some Sicilian clans. Clans are insular and members have intense feelings of pride and loyalty towards them.

Clans stored blueprints outlining proper social behavior and their own means of communicating it to their members. Sagas and myths told of heroic deeds, almost superhuman abilities, and great achievements of ancient and legendary figures. Legends depicted dire events which befell the errant. Rituals and ceremonies reinforced the implicit but well-understood codes of correct conduct. Once these codes were absorbed, members knew precisely how to behave without being told or watched.

Relations of authority also took on special characteristics in clans. Leaders were neither seen as superiors nor considered as officials. A leader was simply the chief, the person designated by tradition as the ultimate authority. Similarly, the rank and file were neither considered subordinates nor subjects; they were simply members. Their obedience and loyalty was due not to the chief personally, but to the traditional authority granted to the chief. Nevertheless, members' obligations for obedience were essentially unlimited.

This does not mean the chief's powers and discretion were unlimited. On the contrary, the extent of the chief's authority, as well as the content of any commands, were also inscribed in traditional codes. These usually left the chief some space for personal decision and prerogatives, such as sprinkling grace and gifts on favored subjects. Thus, while chiefs had some freedom for arbitrary imposition of their own will, they had to keep within the bounds of traditional limitations accorded to them. The chief that violated these undermined the very source of his or her traditional status. While opposition to the chief could be legitimated by claiming he or she had violated or failed to observe tradition, it could not be sustained against the traditional system itself. The codes of conduct, etched in tradition, contained powerful obligations for both chief and subjects.

In Scotland, the Highland clans of the Middle Ages provide a dramatic example of clan control. Membership was by birth or marriage only. The selection of the chief was based on the hereditary rights of the chief's family. Clansfolk had a great love of their place of origin and a deep knowledge of the clan's genealogical roots. Songs, poems, and bagpipe laments encapsulated the tales of great feats, deeds, and battles in bygone years. Each clan had its own tartan, coat of arms, heraldry, and motto. The English invaders, as part of their efforts to eliminate the clans in the eighteenth century, outlawed the wearing of clan tartans. The annual gathering of the clan featured contests of strength and skill, dancing, bagpipe music, feasting, and the raising of the clan standard. Although they frequently fought internally about who should be chief, clansfolk happily followed the chief into battle against other clans or foreign invaders, even in the face of overwhelming odds and certain death.

Clan controls are extremely powerful. Members hold the belief, legitimated by tradition, that their interests are best served by complete immersion of every member in the interests of the whole. A strong sense of solidarity prevails and commitment towards global goals runs high. The goal incongruence problem is minimized, or even dissolves. The needs of the clan swamp those of its individual members.

Clan controls are subtle and illusive; it takes a long time to learn the codes for proper behavior. Members share a profound agreement about what constitutes proper behavior and new members are not able to function effectively until these codes are absorbed. In the US Senate, for example, it takes newly elected senators several years to discover and assimilate its traditions. Performance evaluation is a continuous process of subtle signals from long-serving members. Yet once taken on board by the individual, clan controls are more powerful than either bureaucratic controls or market controls.

A striking example of control by tradition comes from the exploits of the Japanese kamikaze pilots who shocked and stunned the world with their certain-death attacks on US warships during World War II.[5] The name kamikaze was taken from the legendary "divine wind" which miraculously destroyed Kublai Khan's huge invasion fleet in A.D. 1281 and saved Japan from certain defeat and repressive colonization.

The force consisted of a special squadron formed spontaneously in the autumn of 1944 from pilots and officers of the Japanese First Air Fleet in a last-ditch attempt to slow down the American invasion forces. Its mission was to destroy as many enemy aircraft carriers and other warships as possible. Each pilot attempted to crash his plane, loaded with a 250 kilogram bomb, into a warship. The resources of the First Air Fleet were severely depleted and the chances of scoring a hit by a kamikaze attack were very high compared with conventional bombing. Each pilot fervently believed his individual interests were served best by complete personal immersion in the needs of Japan and the emperor. It was implicitly understood that each flier would sacrifice himself and his plane by crashing into an enemy warship. National ruin without resistance was eternal ignominy. These beliefs were shared intensely by each pilot.

Rituals, ceremonies, and slogans played an important role in sustaining the organic solidarity of the force. Talks and memos from officers included slogans such as: "to the divine glory of his majesty," "win the Holy War," "save the divine nation," and "we are the imperial forces of heaven." On the evening before their fatal mission, chosen pilots meditated and then wrote a philosophical and cheerful letter to loved ones at home. Fellow pilots, not lucky enough to be chosen for that particular mission, sang the kamikaze song; and before take-off the pilots performed the Hachimaki ceremony of wrapping the traditional white cloth, with a red circle on front symbolizing the rising sun, around each other's helmets, all the while chanting patriotic slogans.

The history of kamikaze force is a rich example of the awesome potential of control by tradition. All the necessary ingredients were present: sagas, ritual, ceremony, agreement on correct behavior, dedication to global purpose, and exclusive membership. Surveillance and output measures were out of the question. Not only were they unnecessary, they would have been seen as disgusting. These controls were more powerful than even the human desire to live.

The kamikaze force, of course, is an extreme example. Yet if we stop to think about it, control by tradition is widespread. They are used to some extent in nearly all our institutions: families, schools, universities, fraternities, clubs, athletic teams, corporations, public accounting firms, professional associations, and governments. They are obviously a powerful means of motivating individuals toward global goals. Although according to the model they come to the fore when ends and values are unambiguous and task instrumentality is not well understood, they are also used frequently along with other types of control.

Collegial Control

Collegial control is closely related to traditional control. It has its roots in the idea of *collegiums* (colleges) where one particular group (e.g., the professors) enjoyed special privileges regarding authority and domination. Entrance to a collegium was usually by way of election with each incumbent having a vote (sometimes in the form of a veto or a blackball) or selection by an elite committee of colleagues. Members of this elite group, the colleagues, were usually experts in their field of specialty, persons of high social status, or individuals privileged through education. Each colleague had a say in most important matters.

Collegial controls are much more common than is generally recognized. It is the dominant means of control in many of our most important institutions, including universities, churches, fraternal orders, and international organizations such as the United Nations and NATO. This lack of recognition may be due to its undemocratic veneer in that an elite coterie holds sway. Its strength, however, is that the best, brightest, and most capable individuals have the upper hand in important decisions.

The distinguishing feature of collegial control is that administrators are the subjects of control by the collegium. This differs from bureaucratic control where the incumbent officeholder has the upper hand and from clan control where the chief is the ultimate authority. In contrast, the collegium has a monopoly on both creating the rules which govern the actions of the administrators and in determining the means of checking to see that they adhere to them. In many cases the collegium also elects the chief administrator, or leaves it to a committee of colleagues, or the position is rotated among colleagues. Either way, the chief administrator (dean, Pope, secretary-general, etc.) and other officials owe their obedience and loyalty to the collegium. The final authority rests with the colleagues.

These arrangements frequently result in considerable tension between administrative officers and the collegium. The colleagues antipathy towards the chief administrator stems from a deep suspicion in general about the intentions and motives of demagogues, charismatic leaders, dictators, and any type of strength. Colleagues resist monocracy of any kind. They also have strong apprehensions about the dilettantism of administrators in regard to important technical and social issues. The typical university president is often a distinguished scholar in a narrow field of specialty, but is almost totally ignorant of the technical, political, and social issues in most other academic areas.

Collegial control, on the one hand, can make decision making tedious and time-consuming. It also leads to apparent inconsistencies in policy, since every important decision is made ad hoc. And it tends to cloud individual responsibility. On the other hand, collegial control promotes objectivity and integrity in decision making. It also limits the power of any one individual to usurp power. And it champions the need to reconcile different points of view and divergent interests of the sundry experts. Collegial control ensures that debate and compromise are possible.

Not surprisingly, administrators and the collegium coexist in a state of dialectic struggle. In this contest, budgets and accounting systems play a vital role. Administrators attempt to gain control of the financial strings and control reports. This is readily accomplished if the colleagues see accounting as merely bookkeeping; administrators use the budget process to garner command over resource allocation and exploit the financial reporting system as a way of controlling debate and discussion. The administrator's plea, "That would be nice, but we simply can't afford this initiative," masks the fact that other programs and projects are automatically funded without debate. Accounting systems are used to problematize some matters, while simply taking others for granted.

In sum, collegial control can be used to advantage when means are well understood but ends are ambiguous. Although they are widespread in many of today's important institutions, they are generally not well recognized or appreciated. Collegial control is rule by those persons selected according to merit due to their outstanding abilities and accomplishments. Thus, it gives the impression of being undemocratic and elitist. Unless colleagues are on their guard, appointed administrators will frequently usurp the prerogatives of the collegium and use budgets and financial controls as a way to allocate resources according to their own priorities, not necessarily those of the collegium. When this happens, the benefits of collegial control, such as debate over important issues and necessary compromises over values and ends, can vanish.

Summary

This section outlined a typology of control styles that links Weber's general ideas with systematic variations in the uncertainty involved in the work and tasks of organizations or their components. Bureaucratic controls, consisting of observational hierarchies, rules, procedures, records, and files, are well suited to circumstances featuring rational ends and means. When ends are unambiguous but means are not well understood, market controls, which rely on the information content of prices, and charismatic controls, provided by a dynamic heroic leader, can be used to advantage. Finally, under circumstances where the means for task accomplishment are well understood, collegial and tradition serve as suitable controls.

Management accounting and control systems have a role to play in each of these types. They are a major part of the records and files needed for bureaucratic controls. They provide the necessary information about prices required for market control. In the case of charismatic and traditional control, they are less important but do provide information regarding the use of resources in terms of missions as well as indicating how resources are shared. Finally, in collegial control they become an important resource in the struggle between the collegium and the administrators.

These ideas are not wholly theoretical. In fact, they can have very important practical applications. We close this chapter by demonstrating their relevance for analyzing two case histories. The first describes the introduction of a new management accounting and control system in a health center, and the second involves a recommendation to use profitability measurement as the major means of control within a full-service advertising firm.

A COMMUNITY HEALTH CENTER

Case Information

The center was established under the auspices of a leading hospital as an autonomous unit in a separate location.[6] Its purpose was to provide a full range of preventative as well as therapeutic health care to the residents of a nearby slum district. The primary objective of the center was to be a prevention-oriented family-centered source of health care, available to all community residents. The administrators also hoped that in the long run the center would become financially self-sufficient.

The mission departments included pediatrics, internal medicine, community mental health, nursing, dental health, social services, and nutrition. Each department was staffed by high-calibre physicians and practitioners who held joint appointments at the parent hospital and who incurred a substantial loss in earnings by working at the center. The quality of their work was controlled by careful screening and selection of physicians, a continuing peer review by department heads, and random reviews of medical records by a review committee. The center prided itself on its competent, dedicated, and altruistic medical staff.

The center also had a good financial accounting system, as well as line-item budget controls. Feelings that costs were above average for this kind of establishment led the administrative director to have a consultant install a new management accounting and control system. The source data for the new system came from detailed forms which reported the actual time spent with a patient, the total minutes available, and the salary rates for each practitioner. These data were used to calculate the cost per minute for each physician and to create a series of reports, including total costs for each department, average cost for each practitioner per visit and by type of visit, and average cost per encounter for each department. The reports included standard and average cost data and highlighted whether or not each physician had spent more or less time than average for each encounter, and whether or not direct patient care time had changed from the previous period.

The new control system soon became an integral part of the center's administrative activities. It was used to assess the monthly performance of each physician. The director met individually with each physician, and used the new control system as a focal point for reviewing their allocation of time. Also, the data were distributed to the department heads and discussed at executive and departmental meetings. A management accounting discourse became an integral part of the center's activities.

Reactions to the new system were mixed. The administrators were pleased with it and believed it had greatly increased cost-consciousness behavior on the part of the staff. Evidence of this was the hiring of low-salary people to relieve the physicians from routine tasks. The department heads allowed that the new system made them more aware of time constraints and costs, but believed that it ignored important long-term effects of spending time out in the community. They also complained that the new system had brought about a philosophical change by the administration to increase the volume of direct patient care and a decreased emphasis on the quality of care and on family preventive medicine.

Analysis

Let us use some of the ideas in this chapter to analyze and highlight some critical problems with the new control system. For one thing, there are two quite different tasks involved, direct therapeutic medical care (TMC) and preventive-oriented, family-centered health care (PFC). TMC objectives are unambiguous, to effect a successful cure; cause–effect knowledge is complete for most patients (a broken leg) but incomplete for others (cancer). The new control system does a good job of measuring the efficiency of TMC activity. It is likely, however, that effectiveness is more important than efficiency. Yet the new control system paid it no heed. The motivation of the new control system is strong; and its message is efficient TMC.

PFC is quite a different matter. Objectives and ends are more ambiguous. Also the instrumental knowledge of how to install an effective preventive-family health care system into the slum is not well understood. Clearly, the new management control system with its efficiency orientation is not only unsuitable but also counterproductive to PFC activities. Routine, programmable tasks have a natural tendency to drive out nonroutine, nonprogrammable tasks. When this tendency is strongly reinforced by the control system, with the physicians consequently spending more time on TMC than previously, the results could prove highly detrimental for PFC activities.

The ideas in this chapter also provide clues about the selection of more appropriate assessment and control. In terms of our first framework, when efficiency and instrumental tests are of only limited usefulness, we turn to social ones. There are three relevant social groups involved in the center: the slum community, the funding body, and the physicians themselves. Periodic surveys should be taken of the families in the community regarding the center. A key aspect of PFC is the acceptance of the center by this social group. Such social tests, not the efficiency ones, are appropriate for assessment and scorekeeping.

Turning to our second model, we can see that a bureaucratic system of records and files like the new management accounting system is unsuited for controlling the physicians. Market controls are ruled out due to the inability and lack of inclination of the slum community to pay market prices for health care. But statistics should also be collected about increases in the number of families registered, the number of vaccinations, dental checkups, fluoride treatments, and the like, to demonstrate to funding agencies that historical improvement has taken place.

However, it seems likely that a clan atmosphere could be readily fostered. Weekly meetings of all practitioners to discuss PFC problems and progress, formation of specific family-designated teams including nurses, circulation of research material on health care progress in poverty-stricken communities, and visits to the center by dedicated minority leaders and politicians could all lead to a sense of a working in a very special, dedicated, and loyal group of medical professionals. The physicians have willingly given up a higher income to dedicate themselves, at least for a time, to solving the health problems of the slum. Clan control, in fact, is remarkably well suited to the type of work necessary to achieve the major aim of the center.

As well, our typology points towards the potential for collegial control. We already see evidence of the dialectic tension between the practitioners and the administrators. At the end of the case, it appears the administrators have gained the upper hand by invoking messages about efficiency in the new management control system. In order to restore power and control to the practitioners, it may be necessary for the most senior and respected practitioners to form a collegium to challenge and overturn rule by the administrators.

To summarize, we have a classical case of the skilful design of a technically sound control system which is, unfortunately, unsuited to the prevailing circumstances. The detailed cost accounting system was thoughtlessly designed and used without regard for the purposes of the organization or the nature of its work. Efficiency of TMC became the focal point for performance measurement. As a result, the new management control system started to have what could eventually be a profound effect. The practitioners have already shifted their efforts to TMC to the neglect of the center's fundamental purpose. A clan atmosphere and collegial control could effectively get the center back on the right track. While our frameworks do not provide a definitive answer, they point the way towards a means of redressing the balance.

KNOX WALTER AND THOMAS

Our second case involves the introduction of profitability and efficiency management controls into Knox, Walter and Thomas (KWT), the Canadian subsidiary (a disguised name) of Walter Thomas & Associates, a UK-based advertising agency with operations in thirty-seven countries. Agencies like KWT service advertising campaigns for brands marketed by existing clients. They also take on new clients after consideration of their long-run profit potential and possible conflicts of interest with their established clients. Agencies earn revenues through either a flat 15 percent commission on the gross billings by the media organization (television, radio, newspapers, etc.) where the media is placed and from cost-plus contracts and retainer fees. For cost-plus assignments the agency bills the client for the full cost of the work plus an agreed profit margin; in the case of a retainer, the agency receives a guaranteed fee for the job as well as reimbursement of expenses.

Case Information

KWT was a full-service agency and most of its work involved creating and implementing advertising campaigns for major corporations as well as for the federal Progressive Conservative party. It also took on cost-plus and retainer fee engagements. KWT had worked with most of its twenty major clients (involving over a hundred brands) for more than twenty-five years and with some for over fifty years and took on a major new client every year or so. In order to prevent advertising programs from stagnation, KWT followed a policy of rotating clients among the account supervisors every couple of years or so. As well, it was not uncommon for clients to hire KWT account supervisors as in-house product executives.

KWT followed the normal organizational structure (see Figure 11). Five directors reported to

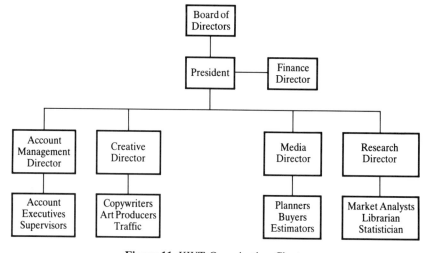

Figure 11 KWT Organization Chart

the president, John Knox, who reported to the board of directors. The account management group, the hub of the agency, worked with clients to develop the advertising approach and oversee the production process through the agency. Account executives and supervisors performed a dual role: consulting with clients on marketing strategies and plans and coordinating overall agency involvement with clients.

The creative group developed the communication concepts for the campaigns, sold the concepts to clients, and executed the creative work. This involved planning out the words and artwork and producing the printing advertisements and visual communication in keeping with the advertising objectives and strategy developed by the account people. The media group then developed the placement plan deciding where, for how long, and which media to use. They also got involved in presenting media plans to clients and were responsible for buying media. The research group provided basic market information to assist in the development of advertising objectives and strategies and also conducted market research studies for clients including pretesting, posttesting, awareness and recall testing, and opinion measurement.

KWT took pride in its team approach to servicing clients. The president, under advisement from the directors, formed teams from the four groups to work together on specific clients' advertising campaigns. Generally, a team consisted of an account executive, an account supervisor, an art director, a copywriter, a media buyer, and a media planner. Often the finance director sat in on team meetings.

Analysis

A full-service advertising agency seems to be very much in the nature of an open-natural system rather than a closed-rational one. Client's whims, customer's tastes, competitor's innovations, and the ups and downs of the general economy make for a highly uncertain and often turbulent environment. Agencies are very much open to the uncertainties and vagaries of the environment. This is particularly so for the technical core—the creative, media, and research groups—where there is little chance for them to know whether or not their efforts have optimized input–output relations. Survival, not efficiency, is the primary goal.

Analysis of the two antecedent conditions in Figures 9 and 10 helps define more specifically the nature of this uncertainty. The goals of an advertising agency are relatively clearcut and unambiguous. The mission is to sell the clients' products. But creating advertising that works is anything but a rational, well-understood task. Successful advertising is very much a hit-and-miss creative process.

As David Ogilvy, the creator of the now classic Hathaway Shirt eye-patch ad and the recognized father of the image school of advertising, put it, "I've been in this trade for more than thirty years and written as much advertising as anyone alive. In those thirty years I had only nine big ideas. It's not many, is it? But it's more than most people."[7] While *Advertising Age* summed it up this way: "Advertising is like electricity. We know a great deal about it and its uses, but we are not very successful in defining it or delimiting it."[8] Moreover, clients' sales of products are also affected by a large number of factors such as sales-force effort and pricing decisions in addition to those factors mentioned above.

More Case Information

The finance and administration group maintained the accounting records, handled treasury matters, and took care of the usual sundry office management duties such as personnel records and computer support for the other groups. They also developed a budget every six months for presentation to the board of directors. The budget included anticipated revenues, line-item expenses (salaries, rent, entertainment, supplies, etc.) and overall net income. In addition, the finance director kept a subsidiary client ledger in which expenses and revenues were recorded for each client and each client account.

KWT's salaries and wages amounted to 63 percent of gross revenues while other expenses (travel, entertainment, rough-copy costs, general research work, pretesting, occupancy, employee benefits, telephone, stationery, etc.) ran about 22 percent, leaving 15 percent profit before taxes. All employees, with the exception of those in the finance and administration group, filed weekly time sheets indicating the hours spent on each client account. Nearly 90 percent of the payroll total could be traced directly to client accounts with the rest posted to the indirect expenses account. About one-third of the other expenses also were charged directly to specific client accounts. The remainder and nonchargeable payroll costs were allocated to client accounts on the basis of direct salary payroll.

This data base was used to put together a profit and loss statement for each client account showing: gross billings; commissions and fees; direct payroll by each group (account management, creative, media, and research); and the allocated amount of direct expenses, unbillable costs and indirect expenses. Each month, the finance group also produced client P&L statements, which included the above items for the current and previous year to date and for the current and previous month. These statements were bound in a package each month but were made available only to the president and the board of directors. While the other directors knew they existed, they were not routinely given copies, although they could, if they wished, come to the finance director's office and have a look at the numbers for specific accounts.

Recently, one of KWT's large and long-standing clients appointed a new chief executive whose mandate was to improve the company's sagging earnings per share. Historically, the company had budgeted 12 percent of total sales to advertising but was now considering cutting this back to 8 or 10 percent. In discussing advertising with KWT's president she stated, "I know half the money we spend on advertising is wasted but we never know which half!" And her new controller had initiated a detailed analysis of the company's advertising expenses. The controller reported that a sophisticated sensitivity analysis produced no statistically significant correlations of changes in sales for the company's forty-five different consumer brands with changes in advertising dollars. John Knox responded with the traditional comeback, "advertising should be seen as an investment, not as an expense".

Nevertheless, Knox was concerned that the problem might be widespread. So he asked the finance director to undertake a study of the agency's management accounting and control systems and to make recommendations for any changes that might help KWT maintain or improve its overall profit picture. The director commissioned a consultant from a leading professional accounting firm to assist him in this charge. Six weeks later, the director informed Knox that, on the advice of his wife whose tolerance for Montreal winters had fallen in recent years, he had decided to take early retirement and move permanently to his condominium on Long Boat Key near Sarasota in Florida. The director recommended that Knox appoint the consultant, who had been chief controller of Ford Canada's huge Oakville automobile plant, as his replacement. When Knox approached the consultant, he immediately accepted the position.

The new finance director, comparing the controls in the agency with those he was familiar with at Ford, became concerned with the lavish expense accounts submitted by account executives and employees in the other groups. As a result, he designed an elaborate budgetary control system which collected expenses by every conceivable category, right down to individual taxi rides. He also introduced a new computer system including a network of on-line terminals. All the employees were on hand to witness the installation of the network and give a resounding cheer as the president pushed a button which put the system on line.

The new director also initiated a series of monthly meetings during which he met with key employees in each group to review the profit statements with them. He soon came to the conclusion that account supervisors and executives should be held responsible for the profitability of the client accounts they handled. He also concluded that each of the directors of the four operating groups should become responsible for the profit performance of their particular group. The calculation of group profit involved allocating revenues and expenses among the operating groups on the basis of the amount of direct labor dollars each group charged directly to client accounts, as well as a cost-plus-profit-margin-based transfer price for services provided by the research group to the other groups.

A year after the expense contract system was installed the director of finance learned that the group executives and managers no longer bothered even to look at the reports. The numbers, they claimed, did not seem to help them very much so the reports fell into disuse even though they were not unhappy with them. He also noticed that their interest had waned in the monthly meetings to review profit performance. While a few executives tried to make sense of the information and to understand the intricacies of cost and revenue allocations and transfer price calculations, most seemed bored and indifferent. Behind his back, the executives referred to the new expense and profit reports as so much "smoke and mirrors bean-counting." Shortly thereafter he left the agency, blaming the executives and managers for the failure of the new controls.

Analysis

The two frameworks in this chapter indicate clearly that the finance director was a long way off base. He had treated the advertising agency as a closed-rational system instead of an open-natural one. Instead of efficiency tests, like the detailed expense reports and profit breakdowns, the model in Figure 9 indicates that instrumental assessment and scorekeeping tests are much better suited to the work of an advertising agency. And, in fact, gross billings are a simple, yet highly relevant, instrumental test and are readily available. By focusing on expenses, which are recovered from clients in any case, instead of tracking gross billings, the new scorekeeping system was out of sync with the key factors involved in advertising work. Similarly, the model in Figure 10 indicates that market controls are better suited to the nature of work in advertising agencies than are bureaucratic controls such as accounting reports. Moreover, market controls are readily available; executives can simply track the sales volume of clients' products.

Furthermore, the new accounting-based controls precipitated a number of dysfunctional actions. For instance, the account management, creative, and media groups cut back on services from the research group. So the research director, in an effort to show a profit, raised the transfer prices. This backfired as the other groups cut back even further on research services. In addition, intergroup squabbles broke out as each group tried to pinpoint the blame for low profitability on other groups while taking the lion's share of credit for the high-profit accounts. Furthermore, executives found themselves spending a lot of wasted time trying to make sense of the accounting reports and going to budget meetings instead of servicing clients.

When the finance director left, Knox hired a newly qualified certified general accountant as treasurer, *not* as finance director. She quickly dropped the comprehensive expense and profit reporting in favor of a system that simply tracked gross billings and direct costs for each account. The executives and managers reverted to spending money in their own way to increase gross billings. The control system once again matched the nature of the agency's environment.

SUMMARY

The two frameworks presented in this chapter provide alternative ways of thinking about the problems of designing appropriate management accounting and control systems. The approach is based on the need to go beyond, although not to exclude, closed-rational systems, and the need to understand how patterned variations in uncertainty call for systematic differences in management accounting and control systems. The rational approach is highly satisfactory for stable and predictable parts of the organization, particularly the technical core. Here efficiency tests and bureaucratic controls are a valuable and integral part of the drive to optimize the use of resources.

But efficiency tests and bureaucratic controls lose their power in the face of changing and uncertain environments. The rational approach, so well suited to a predictable closed-system, should give way to an open-natural perspective which focuses on the mechanisms whereby organizations react, adapt, and survive in the face of an incessantly changing environment. Historical efficiency tests are replaced by criteria of instrumental effectiveness and ultimately to the opinions and beliefs of relevant social groups. Similarly bureaucratic controls give way to market, charismatic, tradition, and collegial controls.

The other basic element of this new perspective is that it rejects either the quest for universal truths, contained in generally accepted management accounting principles, or the preoccupation of detailed case studies with their exhaustive analysis of the uniqueness of the scorekeeping and control systems in a particular organization. Instead, a middle ground is advocated where the patterned variations in the uncertainty are linked with systematic differences in organizational action. Different kinds of uncertainty lead to different patterns of assessment and control.

By following these premises and looking for patterned variations, we were able to capture the full range of assessment and controls mechanisms used in organizations: efficiency, instrumental, social, bureaucratic, charismatic, market, collegial, and tradition. These ideas, which put accounting and control systems into the wider and more realistic organizational context, can provide a richer understanding of the issues and problems of management accounting and control than the traditional textbook approach, which focuses on technical problems such as better cost allocations and more accurate profit measurements.

NOTES

1. This framework was developed by Thompson (1967), one of the great pioneers of organizational theory.
2. Williamson (1973) who developed an elegant theory of organization calls these "transaction costs."
3. Ouchi (1977) describes the two departments in detail.
4. Ouchi (1977, 1979) popularized the notion of clan control. See also Macintosh (1985) and Ezzamel and Hart (1987).
5. See Inoguchi, Nakajima, and Pineau (1958) for a detailed description of the kamikaze pilots.
6. See the Hyatt Hill Health Centre case study in Anthony and Dearden (1981), p. 720.
7. Quoted in Wright et al. (1984), p. 137.
8. "World of Advertising", *Advertising Age*, November 15, 1983, p. 10.

9
Interpretivist Models

This chapter leaves behind the objective realist presupposition of the structuralist position to look at several frameworks based on interpretivist premises. The former implicitly assumes that the meaning in a management accounting and control system preexists both its capture by the accountant and its semantic content for the manager. This implies that accounting systems objectively reflect or mirror some already-there reality, and so anybody who uses them can, if they try, discover that reality in the reports and take actions accordingly. We now take a radical turn away from these assumptions to concentrate on several frameworks which adopt the subjectivist presuppositions introduced in Chapter 1.

From this alternative vantage point, the meaning in any management accounting and control system is seen to be the product of the subjective experience and intersubjectively of the accountants and managers involved. The accountants and users construct a story about that reality using as materials the principles and theories that accountants and managers are trained in and skilled at putting into practice. This does not mean, however, that a world doesn't exist until the accountant or manager thinks it up. Rather it recognizes that accountants and managers subjectively construct the "facts" about an organization's reality even though they may then come to reify them, as if they existed independently of their subjectivity. It also gives full recognition to the subjective processes which shape and control meaning and to the fact that the realities of organizational life are *socially constructed*. Simply put, socially constructed means that the social structures and arrangements (e.g., who shall rule, how to speak, what is deemed to be ethical) are not given by nature but made by people, therefore they are potentially mutable.

In constructing meaning we must rely on a common medium such as language in order to communicate our subjectivity to others and to ourselves. One such instrument especially applicable to organizations is accounting; as the saying goes "accounting is the language of business." From the subjective interpretivist perspective, accounting is an important part of the material out of which reality is constructed.

A simple illustration may be helpful at this point. In accounting for an oil refinery the accountant has a lot of choices. It can be accounted for as a cost center, as a revenue center, as a profit and return-on-investment center, even as a discretionary cost center. There are lots of possibilities. But the responsibility center is not, say, a profit center until the managers and accountants construct it as such. It does not preexist as a profit center. Importantly, however, the social relationships of the center will likely vary considerably depending on which of these accounting meanings is attributed to the refinery. The "reality" is socially constructed.

Another example is the case of the health center discussed in the last chapter. It was originally set up to provide adequate preventive as well as therapeutic care, become a family-centered source of health care for the local low-income community, and to act as an experiment to determine the impact of a community health center on the community and the parent hospital. The aim was to bring family-oriented and preventive health care services to a community that traditionally was resistant to and suspicious of outsiders including social workers, psychiatrists, and to a lesser extent medical doctors. The new management accounting system, however, defined the center as a financial performance and efficiency center. The new financial discourse reconstructed the social reality of the center in a fundamental way.

In this chapter we look at three valuable frameworks which adopt this subjective interpretivist perspective. The first of these outlines the *ideal* way that management accounting and control systems should be used in organizations and then contrasts these with the way they are *actually* used. The second examines the redundant information processing phenomenon and describes how managers utilize management accounting and control systems for strategic and symbolic purposes. The third framework takes the wraps off the notion of objective accountability to reveal how accounting systems are used subjectively by organizational participants in the invisible war of self-interest in the workplace. Each shows us how accounting and control systems, rather than providing an objective neutral mirror on reality, play an important role in the way reality is subjectively constructed by organizational participants, especially accountants and managers.

UNCERTAINTY AND THE USE OF ACCOUNTING AND CONTROL SYSTEMS

Our first framework is anchored on two familiar dimensions of work: the extent of knowledge available about the task conversion process and the degree of certainty in objectives.[1] The interaction of these two dimensions produces four types of problem situations; it also dictates the ideal way to use accounting and control systems in each situation and reveals how managers actually use them in quite a different manner.

Ideal Uses

The ideal uses are depicted in Figure 12. In the first situation the task conversion knowledge is believed to be complete and objectives are seen as certain. Under these conditions tasks can be programmed so that predetermined rules, formulae, and algorithms can be applied to the work and decisions can be made by computation. Accounting and control systems often provide answers on the spot. Examples are standard cost systems, economic order quantity inventory systems, credit inquiry systems, and linear programming models for transportation problems. For these situations, accounting systems can provide accurate, timely, and unequivocal answers.

Sometimes uncertainty arises; there may be disagreement over which objective is primary; managers may be ambivalent about the major choices amongst multiple objectives; or objectives may simply be unstated. Whatever the reason, uncertainty over objectives, spurred on by individual self-interest and rapidly changing environments, brings with it conflict over principles and perspectives.

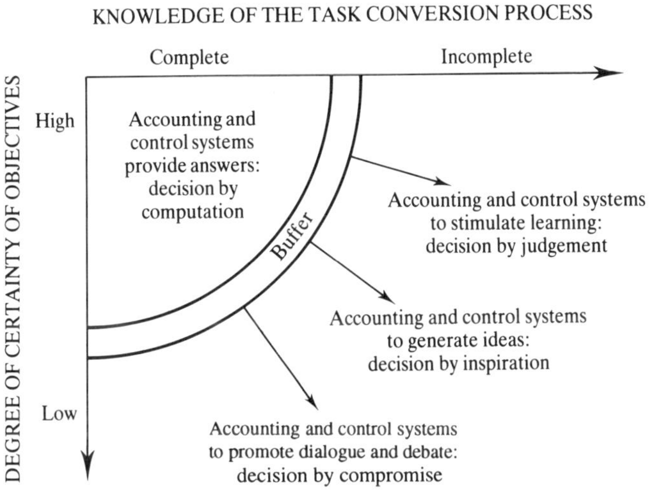

KNOWLEDGE OF THE TASK CONVERSION PROCESS

```
                          Complete          Incomplete

DEGREE OF CERTAINTY OF OBJECTIVES

   High    Accounting and
           control systems
           provide answers:
           decision by         Accounting and control systems
           computation         to stimulate learning:
                    Buffer     decision by judgement

                               Accounting and control systems
                               to generate ideas:
                               decision by inspiration

   Low     Accounting and control systems
           to promote dialogue and debate:
           decision by compromise
```

Figure 12 Ideal Uses of Accounting and Control Systems

In these circumstances decision making should be oriented towards opening up and maintaining channels of communications. Opinions and different perspectives need to be identified and debated in an open and lively fashion. Accounting and control systems can facilitate this by helping managers develop and argue points of view which are conflicting but consistent with the underlying facts, data, and context. In such situations, management accounting and control systems should be used to promote dialogue, to act as a catalyst for debate, and to help participants reach a compromise rather than to provide *the* answer.

They should also be used to bring conflict, power plays, and bargaining over objectives out into the open where discussion and compromise might lead to better decisions. For example, management accounting and control systems can be used to initiate debate during the strategic planning process whereby two quite different but equally feasible strategic alternatives are articulated from the same data base. Dialogues, not answers, are called for.

Similarly, in the situation where objectives are clear but knowledge of how to correctly complete the task is low, management accounting and control systems seldom yield a final optimal answer but they can be of considerable support in figuring out how to get the job done. The need here is for exploration of problems, investigation of the analyzable parts of the work, and the application of judgement and intuition as learning takes place.

In these situations management accounting and control systems can, at best, only suggest a set of feasible solutions, provide data along the way, and help managers assess alternatives thoroughly. Budget variance analysis reports, inquiry systems for probing data bases, computerized models with sensitivity analysis capabilities, and simulations with what-if facilities are examples of systems which help managers learn more about the possible alternatives and their consequences before they call the final judgement. Learning takes precedent over correct answers.

In our fourth situation uncertainty stems from both incomplete knowledge of the task conversion process as well as from unclear objectives or disagreement about which ones are paramount. Accounting and control systems can be used to stimulate and trigger creativity during brainstorming sessions, where any idea, no matter how ridiculous at first glance, is given serious consideration. They can also be used as a supplement to the strategic think-tank sessions where possible critical events are listed and scenarios developed about the consequences of two or more occurring at the same time. It has even been suggested that in extreme situations, semiconfusing accounting systems can be designed deliberately to shake organizations out of rigid behavior patterns in times of changing environmental conditions (Hedberg and Jönsson 1978). While inspiration cannot be guaranteed, it can be given a boost with management accounting and control systems which provide multiple streams of thought to trigger creativity. They can be valuable cohorts for idea generation.

Actual Uses

In each of the four circumstances depicted above, management accounting and control systems can be used to construct the reality of the work situation in the most appropriate way. Yet when these systems are in actual use, they do not always mirror their ideal uses. Instead, they are used in quite different ways, as depicted in Figure 13.

There are no problems when objectives are certain and task knowledge complete. Here management accounting and control systems are used to generate answers. We find, however, that when knowledge of the task conversion process is incomplete, they are still used to provide answers instead of to promote learning. One reason for this may be that accountants are good at providing answers. They are trained and called upon to provide them even when circumstances are in reality fraught with uncertainty. In fact, techniques relying on probability and risk analysis are considered to be rational ways of coping with uncertainty and so work to camouflage the inherently uncertain situation. Either way, the result is that opportunities to stimulate learning and to exploit uncertainty are lost.

Further, we often find that when the degree of certainty regarding objectives is low, accounting and control systems are used as ammunition to win the day. So instead of promoting dialogue, debate, and leading to compromise, they are invoked to support the vested interests and values of specific groups. They emerge from political processes where one

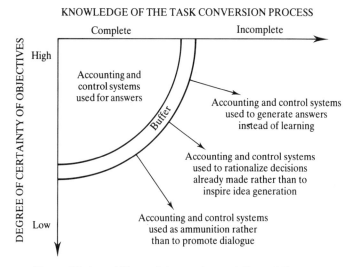

Figure 13 Actual Uses of Accounting and Control Systems

side attempts to prejudice the criterion for selection of a solution and then proceeds to influence the other parties on the basis of that criterion.

Traditional management accounting and control systems undoubtedly are employed as ammunition. Reports containing only financial information are used to reduce multiple objectives to a single financial goal when, in reality, organizational needs also include marketing, engineering, and human relations aspects. By focusing on only one objective, management accounting and control systems can be exploited to further the political ambitions of particular vested-interest groups.[2]

The classic example comes from politics, where various political parties try to influence lawmaking in legislative assemblies. Using the same data base, provided by the Bureau of Statistics, one party develops an information system which supports a reduction in government meddling in the economy while another party develops information which indicates the need for an increase in government planning. Similarly, when accounting systems are used as ammunition, they are dangerous because they override the need for constructive dialogue and compromise.

Finally, when task knowledge is incomplete and objectives are uncertain, instead of the necessary idea machine we often find management accounting and control systems designed to act as rationalization vehicles to justify and legitimize actions already decided upon. Capital budgeting proposals, for example, are put together after the managers have decided that a particular long-term project should be pursued.[3] Task forces, struck to investigate issues which have already been decided upon by top management often use management accounting systems in this way.

Rationalization by mangement accounting systems can have legitimate uses. Sometimes it is necessary to create a rationale to justify decisions to others. Problems arise, however, when they overwhelm the creativity and inspiration necessary for uncovering new and unique decisions. Instead of interactive accounting and control systems, which promote learning and growth, they are used to rationalize and to rubber-stamp the status quo. When objectives and task instrumentality are unclear, accounting and control systems should be used more often than not to facilitate learning; not to rationalize predetermined decisions.

Summary

Under conditions of certainty management accounting and control systems can be relied on to provide answers. But when uncertainty enters the scene they can be put to better use for promoting dialogue, stimulating learning, and generating ideas. But instead we often find them used by managers as ammunition to win the day or to rationalize decisions made unilaterally. These ideas take a radical turn from the traditional perspective. Accounting systems are revealed not as objectively reflecting some preexisting reality, but as constituting organizational reality. They do not merely tell the score after the events, they shape them.

This point is brought home in the well-known story about the two baseball umpires who were asked by a reporter how they saw their job. The first umpire gave the traditional objective realist response that if a pitch was in the strike zone he called it a strike, otherwise he called it a ball. The second umpire, in contrast, replied, "They ain't nothin' till I calls um!" It is the same, more often than not, with accounting. Organizational realities ain't anything until the accounting system and its managers "calls 'um."

SYMBOLIC USE OF ACCOUNTING SYSTEMS

Our next model is based on the observation that organizations and managers gather and process far more information than they can possibly use for decision making (Feldman and March 1981). Recall that the conventional idea depicts information processing as an integral part of the decision-making process. This position assumes that during the decision process managers will secure, analyze, and retrieve information in a timely and intelligent fashion as long as its marginal cost is less than its incremental value. Yet, when they are observed at close hand, managers do not seem to carefully weigh the reliability, precision, and relevance of information. Instead, they collect and process far more information than they could ever reasonably use for decisions. As well, most of the information they collect is totally unconnected to decisions. Organizations, even the best ones, apparently overinvest in a glut of information.[4]

Commonsense Explanations

From the objective realist perspective there are several reasons for this. One explanation is that managers gather a lot of information which turns out not to be suitable for decisions. Another reason suggested is that accounting and information systems are a free good (paid for by accounting, computing, and MIS departments), so managers take almost any formal information they are offered. Yet another commonsense reason is that managers monitor their environment for potential surprises in order to be reassured there are none. Such explanations cling to the version of managers as rational information processors.

Defensive Use

From the subjective interpretive perspective there are quite different reasons for the redundant information processing phenomenon. The first of these is related to the widespread practice in organizations of criticising decisions after the fact. A common version of this is known as Monday morning quarterbacking. On Monday morning, ardent home fans second-guess the decisions made on Sunday by the quarterback in the heat of the battle. Major blunders harmful to the home team are readily apparent. What seems so clear in retrospect was not at all obvious at the time.

Similarly, managers make decisions in the face of uncertainty, knowing only too well that individuals at all levels in the organization will be quick to make criticisms posthoc. The inevitable judgments come in two extremes. One asserts that events occurred which were underestimated, and that the manager should have collected more information. The other concludes that events which did not occur were overestimated; the manager probably overcollected and therefore wasted information processing efforts. The more astute managers recognize that either way criticisms will be forthcoming, but the majority will point to underestimates. The wiser choice is to have more information than is needed. Management accounting and control systems are valuable here as protection systems to head off the inevitable postdecision criticisms. They become the building material for constructing a formidable defense.

Offensive Use

Accounting and control systems are also used in organizations for offensive aims. The dynamics of such use are subtle but elaborate. Managers recognize that information is an important instrument of persuasion, that it can be a source of influence and power. They are, therefore, into a process whereby contending liars, competing in the persuasion game, send and receive mostly unreliable information.

This strategic use of information is a complicated game and is played under conditions of conflicting interests. Wily players do not treat information as if it were neutral and innocent. They must make inferences and discount a great deal of the perverse information they receive, recognizing full well that strategic misrepresentation is a harsh reality of organizational life. Not surprisingly, they counter by circulating strategic information of their own. This subtle game of distorting and misrepresenting information to one's own advantage stimulates an oversupply of strategic information. It is, however, a risky game and managers who are found out are prone to lose their creditability throughout the organization.

Symbolic Use

But perhaps the most pervasive reason managers and organizations collect more information than they can possible process is deeply rooted in a central ideological norm of Western civilization, *the belief in intelligent choice*. Organizations are the central arena for displaying and honouring this paramount social value. Information gathering and the use of accounting data are ritualistic assurance that one does indeed respect this central value. It is a means of reaffirming the value of rational decision making. It is a representation to others of one's competence. And it is a symbol for all to see that one believes in intelligent choice. Managers find value in accounting and control system information that has little or no relation to decision making.

Part of this myth holds that more information leads to better decisions. So, the more information a manager processes, the better he or she must be in the eyes of others. Managers establish their legitimacy by their use of information. The visible and observable aspects of gathering and storing information are not unlike the ritualistic grey suit, white shirt, and dark blue tie of the committed IBM executive of a decade ago. Information is not merely a basis for action, it symbolizes and reaffirms a core value of society. Seeking and collecting information, and this is the striking point, has symbolic value for the manager far beyond its worth in decision making. Managers establish their legitimacy by displaying their use of information. Today, the laptop computer displayed by the busy executive in airport terminals and on commuter trains signals a true believer.

Managers, especially prudent ones, posture accordingly. They diligently ask for and carefully store management accounting and control system reports. They carry them home after work in visibly bulging briefcases. And they assiduously orchestrate decisions to ensure that all believe they take action only on the basis of reasonable and intelligent choice. The command of such information, the access to its sources, and the apparent application of it to decisions all work to enhance the manager's reputation for competency and to inspire the confidence of others. It symbolizes one's broad commitment to reason and to rational decisions. Management accounting and control reports take on value as symbol far beyond their worth as a basis for action.

Summary

It often appears, particularly from the rationalistic, objective realist perspective, that managers gather and process a great deal of redundant information. But from the vantage point of the subjective interpretivist, managers are seen to be using management accounting and control systems to construct a reality that is personally enabling. Sometimes they use them to construct a defensible image of having followed the most competent path by gathering all the information available before making the risky decision. Sometimes they deliberately construct a false reality for other managers and assume that others do likewise. And sometimes they use them ritualistically to create a self-serving image of themselves as believers and active followers of rational choice and decision making. In each case—defensive, offensive, and symbolic—the accounting and control system is used by the manager to subjectively construct meaning.

THE OBJECTIVE ACCOUNTABILITY WAR IN THE WORKPLACE

The final set of ideas in this chapter presents a radical exposé of the *objective accountability* and *bottom-line* mentality so prevalent in today's organizations.[5] The basic premise of these notions is that what gets presented as objective and rational is merely the camouflage over what is really a subjective, invisible war of self-interest that is endemic but inevitable in the workplace. As the developers of these ideas put it, "Each day we march off to an invisible war. We fight battles we don't know we're in, we seldom understand what we are fighting for and worst of all, some of our best friends turn out to be the enemy" (Culbert and McDonough 1980, p. 3).

There are a couple of simple reasons for this state of affairs. First, organizations today feature a complex world of specialization and narrow expertise. Work must be delegated to a variety of people with different skills and interests who can get the job done. So organizations try to specify the type of responsibility and commitment needed in order to strike an agreement with employees who will stand accountable for what has been agreed upon. Then, after the fact, evaluators will be able to objectively appraise whether or not inputs and outputs lived up to expectations.

Second, and just as crucial, most employees have career ambitions. This is particularly the case with managers. They like to have their efforts recognized and their accomplishments rewarded. They like to get ahead and they want to know how they are doing. So organizations need objective accountability and subordinates look for it from superiors. Thus management accounting and control systems loom large; after all, they are thought to contain neutral, objective information regarding plans and outcomes.

So objective accountability becomes a given. But one point is often overlooked, Whose objectivity? As it turned out, getting the product out the door as required (and most people do a pretty good job of this) is necessary but not enough to get a favorable appraisal. Success also depends on getting evaluators to recognize and value one's efforts and accomplishments, particularly to see how they contribute to the organizations overarching mission and purpose.

However, evaluators are also busy getting their efforts recognized and valued. And they have their own cherished ideas about what inputs, outputs, and commitments really count.

Interpretivist Models 157

This orientation, as it is called, usually consists of a self-convenient ideal about global mission, an important accomplishment that matches their own interests and affinities. The evaluator's orientation seldom matches the evaluee's orientation. In consequence, the evaluee's orientation gets discounted and his or her efforts and accomplishments get short-changed by the evaluator's "objective accountability" framework. Evaluators see the world from their own particular orientation, not that of the evaluees.

There are many plausible ways of defining organizational reality, so there are many self-convenient orientations circulating within an organization at any one time. It is no great feat, then, to find an orientation that puts one's own efforts in the best light while devaluing the contributions of others, especially those who are seen as rivals. Thus, even though most people in organizations do a good job most of the time (in fact it is ironic that evaluation systems are not at all necessary to identify those few who perform poorly, everyone already knows) it is easy to trivialize their performance. Evaluators select their own subjective orientation for this task and thus tilt the battlefield heavily in their favor. Evaluation never takes place on neutral terrain.

A specific example may be helpful to highlight these points. In a professional accounting firm most managers have the normal ambition of making partner. The manager who treats client attention and billings (outputs) as the bottom line finds it easy to deprecate the performance of another manager who believes that training juniors (inputs) and service to the accounting profession (impact) are absolute organizational imperatives. Each have a genuine vested interest in establishing their orientations as dominant.

Yet the firm, in all likelihood, is well served by both. There are many plausible organizational realities for a professional accounting firm and this makes it possible for most managers to view performance from some position that most devalues others' contributions while highlighting their own. The astute manager is constantly vigilant for attempts by others to establish grounds of meaning that renders his or her own reality ineffective.

Further, the selection of the particular turf within the evaluation terrain turns out to be pivotal to the outcome of the evaluation. Sometimes inputs are selected by evaluators as the turf for objective accountability. Inputs are the necessary materials, wages, capital, effort, and actions that go into producing outputs. Objective standards are struck for input levels and standard operating procedures defined for effort and action so they can be compared with actual levels and conduct. Evaluators are then quick to point out that input standards were not met, or that essential operations were neglected, or that proper procedures were not followed. Such attacks are mounted regardless of the level of output achieved or the overall impact of effort on global purpose.

This self-convenient logic can be used, for example, by a divisional manager to explain why he fired the marketing manager for neglecting to train salespeople in the proper way to deal with customers, despite the marketing manager having consistently met sales objectives. At the same time, the divisional manager will make much of meeting the standards for inputs and meticulously following all procedures and rules; the manager will deny any blame for unsatisfactory profits (output) and a bad reputation (impact). Similarly, engineers fall back on the standards of inputs rationale when they are on the carpet because the bridge they designed collapsed. Selection of turf is critical to the kind of evaluation given.

When the problems of input and activity accountability are recognized the objective accountability war is shifted to the level of outputs. Outputs are the tangible accomplishments, such as finished product, services provided, and profit and sales volumes, of the effort exerted on inputs. The rationale for output accountability is simple, hold people accountable for achieving agreed output objectives and do not interfere with how they do it; either they make it or a new crew is brought in. This is the essence of management by objectives (MBO). What counts is hard results, the bottom line of performance.

There are, however, major flaws in this way of managing. Agreed profit levels, for example, can be accomplished in the short run by neglecting machinery maintenance programs, reducing institutional advertising, cutting back on customer services, and ignoring the need for training programs for middle managers. When outputs are emphasized, managers can meet them by cutting back on inputs. This is done at the expense of the long-run impact on the health of the organization for future profits and accomplishments. The abuses of bottom-line accountability are legion.[6]

When it becomes difficult to win the day with objective accountability for inputs and outputs, evaluators can shift the evaluation turf to impact. Impact is the effect of effort and output on the higher-order, overarching institutional objectives, such as the long-run health of the organization or its contribution to society and mankind. The television sponsor wants commitment to good taste, and not merely high Nielson ratings. Parents want schools to provide for the social adjustment of their children, not merely high scores on national mathematics or reading contests. High-minded values are invoked as more important than inputs and outputs.

As it turns out, the roughest and toughest objective accountability battles take place over impact. Evaluators try to find out in advance whether or not accountable managers stand for given organizational imperatives. They demand to know whether or not the manager will stand open-endedly ready to do whatever may be called upon as specific situations evolve in order to serve the overarching purpose. These mission standards are treated as absolutes even though in reality they are relatives, as they must be selected on the basis of subjective values and morals which, unlike input and output standards, can never be objective. As with the selection of the turf, objective accountability for impact is a subjective value-laden enterprise.

So the objective accountability game can be played at three levels: inputs, outputs, and impact. Wily evaluators, including rivals, know how to switch from level to level in order to further their own self-interest at the expense of the evaluee. A manager, for example, may be contributing mightily to agreed long-run missions but the evaluator merely points out that input standards, as shown clearly by formal management accounting and control reports, were not met. But if input standards were achieved, the evaluator points out that bottom-line outputs, once again clearly shown by the formal reports, have not been met. When both input and output standards have been achieved, evaluators shift the turf to the dubious commitment by the evaluee to overarching mission.

Thus, it is the ground on which the war of self-interest is fought, and the orientation that best fits the self-interests of specific parties, that decides the winner. The key to winning is to keep the opposition debating within a structure that supports your own position, but be quick to switch the territory when your opponent begins to score well. Formal accounting and control systems provide plenty of ammunition for attacking opponents on either inputs, outputs, or impact.

Most individuals either explicitly recognize or intuitively sense the workaday world is awash with such self-serving definitions of organizational reality and that more often than not, the orientation of significant evaluators is different from their own. In consequence, they tend intuitively to rely on some sort of distinctive survival tactic. Careful research has identified three such ways of coping: framing, fragmenting, and playing it both ways (Culbert and McDonough 1980, 1985).

Framing

Framing is a means of meeting opponents head-on by asserting an orientation that forces others to relate to the entire structure of one's position instead of to the individual parts. It involves subjectively constructing (i.e., framing) a self-serving reality on a foundation that is pretty much unassailable in general but with which others may reasonably differ in terms of its specifics. It means appealing to and aggressively promoting a set of lofty ideals while getting on with business as usual, in terms of specific practical endeavors and outcomes that may or may not be consistent with the overarching frame.

President Bush's handling of the situation in the Middle East just before the Gulf War is an example of a highly successful framing, one that easily swept aside different orientations of various politicians and citizens opposing armed conflict.[7] In getting Congress to vote for a war, the president (accountable to US citizens and their elected politicians) appealed to an interlocking set of lofty goals and ideals. He constructed his frame so as to put at stake many of the cherished imperatives of the modern world, not least of which were the democratic ideals of Western civilization (Saddam was a dictator along the lines of Hitler and with similar ambitions); the humanitarian values of civilized people around the globe (Iraqi soldiers had used chemical weapons in the war with Iran and on its own Kurdish citizens and had inflicted barbarous acts on Kuwaiti civilians, including babies in hospitals); the violation of the international nuclear weapons limitation agreement, (Iraq was developing its own nuclear weapons capability); the economic stability of the entire Western industrialized world (Iraq would get hold of the vast oil reserves in Kuwait and Saudi Arabia and hold the West up to ransom); and the future existence of an Israeli state (Saddam's real plan was to reinstate Israel to the Palestinians as an Arab state). These and other lofty ideals were woven into an almost unassailable frame.

This framing was constructed in the face of and in spite of a host of contradictory specific facts and events (Kaplan 1993 and Friedman 1994). Never mind that Kuwait and Saudi Arabia with their relatively tiny populations were in the hands of enormously wealthy totalitarian family dictatorships at a time when much of the Islamic world existed in a state of poverty and moving towards democratic government; the US itself played a huge hand in supporting and arming Iraq as a counterbalance to Iran; Iraq had never acquiesced to the British carving up of the Gulf region in the 1920s, including separating Kuwait from Iraq when it drew up its arbitrary borders for the Gulf region; the US had the largest stockpile of chemical weapons in the world and had used them in Vietnam; Israel was armed with nuclear bombs and chemical weapons; the continuing settlement of the Palestinian territories was in violation of United Nations declarations; Israel was the biggest threat to any lasting peace in the Middle East; Kuwait had continued to take the lion's share of the oil market with price-cutting tactics in spite of repeated requests from Iraq and other Arab nations to discontinue this practice; US diplomats had indicated to Iraq that it would not intervene if Iraq invaded Kuwait and even seemed to encourage such action; economic sanctions by the US and its allies were beginning to hurt Iraq. These and many other specific facts and events did not count for much against the master and masterful framework constructed by the president.

The strength of Bush's approach did not lie in the facts he presented, many of which later proved to be inaccurate. Instead, he linked his position to the highly principled logic of doing what was needed to rid the US and the United Nations of the biggest threat to their way of life since Hitler and Mussolini. He also made it clear that he called out, not as a single voice, but as the trusted and rightful spokesperson of the president's office and indeed, of the Western free world. His opponents, speaking from only the sundry diffuse facts, didn't have a chance against the highly principled logic of this overarching lofty frame.

Fragmenting

Fragmenting is a survival tactic that becomes necessary when one is on the receiving end, not just of one evaluator's orientation but of several, each of which has been framed with someone else's best interests at heart. It is not uncommon today to be in situations where many people think they have a stake in what we accomplish and how we do it. It is as if we are in a fishbowl being watched and evaluated from a host of subject positions, many of which we are not even aware of until they suddenly appear on the scene to question whether or not our activities and efforts are in aid of their own particular causes. We are constantly in the midst of a host of evaluators and multiple evaluating frames.

In order to cope, it seems necessary and easiest to split the truth into pieces which can be dispensed on self-convenient occasions. This permits us to continue to pursue our own orientation in the face of sundry evaluators, each of whom expect us to be immediately responsive to what they see as absolutes. While each piece (fragment) is more or less accurate, in total they go against the grain of most evaluators' orientations. Unlike framing, where the overall picture works but the details don't, in fragmenting the pieces work but the big overarching story doesn't.

The public accounting profession can be used as an illustration of fragmenting. Shareholders could not fathom how savings and loan companies went bankrupt six months after the accountants had given them a clean audit opinion. Politicians questioned accountants' competence when they had not reported bribes to foreign politicians by US airplane and armament companies. Government agencies and academics wondered how public accountants could be independent and objective in auditing when consulting services to the same companies accounted for nearly half their fees. Financial analysts asked how public accountants could let clients issue favorable, unaudited, quarterly earnings reports which were suddenly wiped out in the annual audited report. And corporate accountants were puzzled by the obvious catch-22 of cooperating fully with the auditors in providing information and explanations which were later used against them in the auditors' reports to the Securities and Exchange Commission (SEC). These are only a few of the concerns by evaluators of public accountants, but they illustrate the point.[8]

The response by the public accounting firms is vintage fragmenting. They simply told different parts of the total story to different evaluators. Regarding the bribes, they invoked the generally accepted accounting principle of materiality as a more important guide for action than ethics. In response to the sudden bankruptcies the answer was that they diligently followed professional standards. As for the conflict of interest regarding consulting, they pointed to their code of ethics and their sacred concern for independence. Regarding the misleading quarterly earnings reports they pointed out there were no generally accepted accounting principles in this regard but they would nevertheless follow SEC guidelines if and when required. And in replying to the corporate accountant conflict, they assured them that they were on their side and only unwillingly used the provided information against them because of pressures from securities commissions and governments. The public accounting profession fragments its total story, invoking professional standards here, ethics there, friendship and collaboration here, and just following government orders there. Each piece makes good sense by itself and is a way of holding off multiple evaluators with different orientations. But the big picture doesn't wash very well, partners on huge salaries who are nonetheless objective, independent servants of the free-enterprise system.

Playing It Both Ways

Playing it both ways, the third tactic for dealing with multiple evaluators, involves telling a different big story to the different evaluators while getting on with the business at hand in the usual manner. It allows the evaluee to say the right things to the right evaluators while keeping the real story about effort and commitment under wraps, thus avoiding embarrassing and unproductive confrontations. In short, it means paying lip service in public to evaluators while going one's own way in private.

University administrators, as a striking example, are highly adept at playing it both ways. The story for legislative funding bodies is that excellence in teaching and turning out graduates who are useful to society is the university's prime mission. The story for the public at large is that the university is a sanctuary for civilization's store of knowledge and wisdom and the wheelhouse for moving society towards a better and more just world. The story for the students who can't find jobs is that the purpose of the university is to develop students' minds so they can enjoy a full and satisfying life. The story for local politicians is that the university pours money into the local economy, provides jobs and service to the community and so should be exempt from property taxes. The story to the support staff is that in spite of professors' high salaries and their occupying all the powerful positions, the university is one big democratic community where all are equal. The story to women's and minority rights groups is that the university is bending over backwards to accommodate their aims and make up for their previous exclusion while at the same time maintaining high academic standards. And the story for the faculty is that the aim of the university is to allow them to get on with their scholarly research and consulting careers with minimal teaching loads, ample support staff, and generous amenities. Each evaluator gets a different big story.

Disorientation and Alignment

While framing, fragmenting, and playing it both ways are common survival tactics, they nearly always lead to *disorientation*. Disorientation occurs when one's activities and efforts become disconnected from one's deep commitments. It happens, for example, when the young trainee accountant starts to acquire the nuts and bolts of computer auditing to please a manager or mentor, instead of following a real interest or aptitude in corporate taxation and tax policy. It also happens when universities move heavily into job-oriented training and privately funded practical research—consulting—when the role of the university is to be critical of the status quo of society and develop knowledge to change it. A fine, almost indiscernible line is crossed and long-term interests and careers get subverted by the structure of evaluators' orientations.

Alignment is a better and more honorable way to engage in the invisible war of workplace self-interest. Alignment involves first being quite clear that self-interest and power games are an inevitable and not necessarily sinister part of the woodwork. It also involves realizing that organizations need many different skills and activities to accomplish organizational goals, so there is a wide variety of valuable orientations. But, most importantly, alignment means being clear about one's own aptitudes, affinities, ambitions, and how they align not only with one's activities, outputs, and commitments but also with the organization's mission and objectives. Finally, it is important to market this alignment to evaluators to see how it plays *before* formal evaluation takes place, not during. Some fine tuning by appreciating the boss's situation may be necessary and valuable, but if the evaluator still doesn't appreciate how one's orientation serves the organization, it is time to look for a new place where it will be not only recognized but also valued.

Summary

These ideas present a radical demystification of the objective accountability and bottom-line discourse so prevalent in the conventional accounting literature and in practice. They reveal how objective accountability is in reality a highly subjective and inevitable invisible war of self-interest. Evaluators subjectively construct self-convenient frameworks of reality and then, treating them as objective frames, use them as the structures to evaluate others. In the process, the evaluee's orientation gets pushed aside. Evaluators also subjectively select the specific turf—inputs, outputs, or impact—for the war and are quick to put the battle on the turf that shows themselves to advantage or which puts opponents in a bad light. They are also quick to shift to a different turf if opponents seem to be getting the upper hand. What is put forth as objective accountability is more realistically subjective framing and maneuvering.

In order to survive when on the receiving end of someone else's orientation and turf, evaluees frequently resort to framing, fragmenting, or playing it both ways. Framing involves constructing a master idealistic framework with an unassailable, highly principled logic while ignoring any specific ongoing activities and outputs which contradict the master frame. Fragmenting goes the opposite way. It involves pointing to different specific activities and outputs for different evaluators while hiding or ignoring the overarching frame. Playing it both ways consists of presenting a different master frame to each important evaluator.

Pursuing such survival tactics, however, frequently leads to disorientation and losing one's way. A better strategy is to make sure one's orientation is consistent with one's personal aptitudes, affinities, and aspirations, and to communicate that orientation to important others before formal evaluation so it is clear how one contributes to the organization's global mission.

AN ILLUSTRATION

We conclude this chapter with a real-life illustration that highlights some of the key concepts outlined above. The situation involves the development and distribution of a cost accounting information system in a private school.[9] The school's eighty faculties taught nearly six hundred tuition-paying students in grades 6 through 12. The school was organized into eight departments each of which offered a variety of courses. The English department, for example, offered 12 courses with projected enrolments ranging from four students in a grade 12 advanced Shakespeare class to 120 students in introductory English literature sessions. As a rule of thumb it was thought that twelve students per class was the appropriate size.

The headmistress, impressed by a demonstration of a sophisticated, computerized cost accounting estimation system developed by an interstate education commission for long-range planning and financial decision making, developed a simplified system for her school. Input data included actual and projected enrolment by department and by course, as well as faculty mix by department and average salary for each pay step. Assuming an average of twelve students per class and forty-five student contacts per day per teacher, the system produced eight exhibits including a schedule of teacher cost by department and per student contact hour, data for current work load per department, and projected staffing requirements. Table 4 includes selected data constructed from these exhibits.

In reviewing the output, the headmistress became alarmed about the apparent large range in costs and loads across departments. Even though she treated the numbers as only preliminary estimates, she distributed the eight exhibits to the members of the board of trustees, department heads, and key employees asking them to comment in writing. To her great surprise, she received a wide range of replies, some giving high praise to the new accounting system and some angrily attacking its accuracy and usefulness. Yet most of the reactions can be explained readily in terms of the frameworks described earlier in this chapter.

Some board members and the business manager saw the system as rendering clear-cut answers, this in spite of the problematic nature of both objectives and instrumentality in the educational process. As one board member put it, "This is the greatest management tool I have seen in this school in ten years. It raises questions that scream to be answered." Another board member stated, "It will help identify areas where excessively large classes are having a detrimental effect on the school's quality of education." And the business manager asserted, "It will serve as a decision-making aid in pinning down areas where our budget is taking a beating because classes are so small." Such responses implicitly assume, wrongheadedly, that knowledge of the teaching task is complete and that objectives of education are clear-cut. Thus, answers are seen as forthcoming when instead the new system should be utilized for dialogue, learning, and debate.

A few replies, however, did suggest that the data produced might be valuable for debate and discussion instead of providing instant answers. The head of mathematics, for example, commented this way:

> Our department's reaction is generally favorable to the ideas that seem to lie behind the model. We think they are potentially useful. But I personally fear that cost/effectiveness analyses are somewhat misleading for measuring the quality of a school and its departments for, ideally, we are not turning out a product for popular demand in quite the same sense as does General Motors. My apprehension is that these figures may eventually be misused at some level. But we are, at least, now getting some concrete data that ought to be taken into account in examining our operations. Also, if the data are used to draw questionable conclusions, we will at least have something definite to argue about.

Comments like this point towards the potential usefulness of the new system for dialogue and learning. Importantly, however, the system depicts mathematics as a relatively low-cost department and so it is not surprising this department head viewed it in a positive light. Further, the well-balanced and carefully reasoned response is suggestive of symbolic posturing at the altar of rationality and perhaps some coyness on the part of the department head, since the model argues for more staff for mathematics.

Other department heads jumped at the chance to use the output as ammunition. The data indicated that social science was understaffed by three teachers and was the lowest-cost

Table 4 Selected Statistics from the Cost Estimation Model*

Department	Faculty		Contact Hours Per Week Per Faculty		Average Salary For Faculty ($)	Cost Per Student Contact Hour		Student Enrolment in Hours Per Week	
Lower school	20	(18.1)	213	(204)	26,550	124.62	(130.14)	4,260	(4,080)
English	14	(11.7)	197	(189)	29,571	150.00	(156.81)	2,760	(2,640)
Mathematics	8	(8.8)	248	(247)	31,500	127.26	(127.26)	1,980	(1,980)
Social science	8	(10.9)	270	(308)	28,125	104.16	(91.49)	2,160	(2,460)
Natural science	8	(8.5)	232	(240)	47,500	161.28	(156.24)	1,870	(1,920)
Fine arts	7	(6.9)	223	(223)	29,142	130.77	(130.77)	1,560	(1,560)
Languages	6	(5.9)	220	(220)	27,481	125.01	(125.01)	1,320	(1,320)
Physical education	9	(9.1)	227	(227)	28,668	126.45	(126.48)	2,040	(2,040)
Total on average†	80	(225)	224	(225)	29,364	130.59	(130.50)	17,940	(18,000)

*Figures in brackets are projections for the next year.
†Figures are weighted averages.
Source: Assembled from data in various Exhibits in the St. Augustine School (A), case in Shank (1981).

department in the school. The head of social science pounced on the data and quickly mounted an impressive offense.

> I knew we were overworked and overloaded, but, up until this point, we haven't had an objective way of demonstrating it. I never would have had the nerve to ask for three additional teachers, but the figures clearly show that that's what we need. I guess enrolment is up because of the new interest in urban affairs and environmental problems, which fall into our domain.

The head of languages, also a low-cost and low-salaried department according to the data, followed suit.

> It will serve as an independent forecast of faculty needs which will provide an objective yardstick for evaluating the hiring needs submitted by department heads. It will bring fairness and efficiency to an area that has previously known the influence of seniority and internal politics.

Thus, as our first model suggests, management accounting systems not infrequently get used as ammunition to support parochial positions, rather than as the raw material for dialogue, learning, and ideas.

The new system also presented an excellent opportunity for some participants to display ritualistic assurance of their belief in rational decision making and intelligent choice. The chairperson of the board of trustees, for example, responded this way:

> I see tremendous managerial applications for the model. It can answer questions I've had on my mind for ages. I don't know how to begin to answer them, but this model provides an analytical framework for doing so.

Even though the data are of little help for any specific decisions facing the school, they provided the opportunity for the top executive to symbolize to the rest of the participants that he was indeed a true believer in intelligent, rational choice.

The situation also vividly illustrates some of the dynamics of the invisible war of self-interest in the workplace. The new management accounting system provides mainly data regarding inputs. This is the turf where departments like social science and languages look best while other units such as English and natural sciences score poorly. Thus, as the framework predicts, these departments were quick to shift the evaluation turf to outputs and impact. The English department head, for example, attempted to divert the struggle to the battleground of impact.

> It is absurd to use a mechanical model to make decisions about a qualitative issue. I am hurt deeply by the prospect of using an adding machine to determine the quality of education that a student at St. Augustine will receive. To suggest that two teachers be dropped from the English Department strikes at the heart of the discipline that made St. Augustine strong, created its reputation, and currently serves as a main source of its pride and dignity. If there are "adjustments" to be made, let us bring the other disciplines up to the standards set long ago by the English Department.

The lofty turf of intangible impacts provides the English department with a much safer position for defending its current staffing levels and from which to attack the more mundane departments such as physical education and social science which look good on input measures.

Similarly, the department head of natural sciences (with the highest cost per student contact hour) attempted to shift evaluation to outputs.

> I suppose the model is correct in pointing out that direct faculty costs per credit hour produced for the Natural Science Department are the highest in the school. The reasons for this are not very subtle or startling: we simply have a greater proportion of senior faculty than other departments. These senior teachers bring more experience and teaching quality to the classroom and therefore command a higher salary. The quality of instruction per credit hour produced in the Natural Sciences Department is higher, and so is the cost. The model yields no decision rules useful for management except that to get more experience and quality in the classroom, the cost is higher. We did not need a model to learn that.

By stressing the quality of the department's output and its more experienced staff, he hoped to shift objective accountability to a more favorable turf.

In summary, as our first framework predicts, instead of using the system to learn why natural sciences, mathematics, and fine arts need smaller class sizes, or to dialogue about the objectives of a private school—whether, for example, to respond to apparently trendy developments like environmental studies or concentrate on the lifelong basic skills like math, science, and fine arts—the system was used as ammunition and to provide answers regarding staffing requirements. The framework also makes it abundantly clear that the system would not automatically give answers but could have been used to advantage to promote constructive dialogue and learning. According to the framework, the headmistress would be wise to keep the data produced by the system to herself and perhaps use it only symbolically at board meetings to demonstrate that she believes in rational and intelligent decision making.

Finally, the new system, as the ideas in the previous chapter suggest, was not well suited for either scorekeeping or control. The appropriate scorekeeping for a private school would be to collect the opinions and beliefs of relevant social groups—parents, alumni, university and college admissions officers—regarding the quality of the school's graduates and its general reputation. As for control, one alternative is to organize a collegium of the department heads and eminent faculty to make key strategic decisions and advise the headmistress on policy and hiring. It also seems likely that building a clan atmosphere for the dedicated faculty would be a good option for control purposes. This might entail regular wine and cheese parties, faculty outings to cultural events, ritualistic dinners for new and retiring faculty, and a well-stocked faculty lounge with coffee, muffins, magazines and newspapers.

Ironically, the headmistress has done the opposite. Her new management accounting system not only precipitated a war of self-interest in the workplace but it also provided the ammunition for the battle. By initiating efficiency scorekeeping she has split the clan into warring factions and will be hard pressed to turn the faculty into a cohesive, proud, dedicated, and loyal social unit.

SUMMARY

This chapter introduced the idea that management accounting and control systems do not simply objectively reflect some preexisting reality. Rather, they are the raw material from which organizational participants subjectively "construct" reality, often a self-serving one. The chapter also showed how such systems can provide the ammunition needed to support a particular parochial position in a debate, how they can be invoked to rationalize decisions already taken, and how they can be used symbolically to advertise to all one's belief in rationality and intelligent decision making. It also demonstrated how management accounting and control systems often provide the ammunition for the objective accountability war during which organizational participants stress input, output, or impact performance, whichever turf favors their own orientation and contribution while degrading those of rivals. Thus the chapter took an important turn away from objective realist thinking to adopt the position of a subjective interpretivist. In the next chapter we look at a more comprehensive framework which incorporates both of these positions.

NOTES

1. This section is based on the work of Thompson and Tuden (1959), Burchell et al. (1980) and Earl and Hopwood (1980).
2. See Dirsmith and Jablonsky (1979), Boland and Pondy (1983, 1986), Covaleski and Dirsmith (1986, 1988), Ansari and Euske (1987) for careful and detailed case histories focusing on the political aspects of management accounting and control systems.
3. See Bower (1970) for a detailed exposition of the politics of the capital budgeting process and Chenhall and Morris (1991) who investigate the interaction of cognitive style and project sponsorship on such decisions.
4. See Schick, Gordon, and Haka (1990) for an excellent if conventional overview and discussion of the information overload phenomenon.
5. This section is based on the innovative and enlightening work of Culbert and McDonough (1980, 1985).
6. See Dearden (1960, 1961) and more recently Merchant (1985, 1990) and Dearden (1987).
7. See Kaplan (1993) and Freidman (1994) for detailed documentation of the Iraq-gate scandal and US involvement in the Middle East.
8. These instances and the following examples of fragmenting are outlined in Culbert and McDonough (1985).
9. The case is described in detail in Shank (1981), pp. 280–295. All quotations are from this case study.

10
A Structuration Framework

In previous chapters we looked at various frameworks concerned with the nature and workings of management accounting and control systems some of which were based on the structural functionalist paradigm and others relied heavily on the subjective interpretivist way of thinking. As a consequence, it may seem that our understanding of the role of these systems has been won at the expense of a certain measure of irreconcilable theoretical differences and an accompanying disagreement about the way to approach this very important aspect of organizations. Moreover, some focused on making the status quo more effective and efficient and others, such as the historical–dialectic model in Chapter 4, were concerned with a radical overhaul of the status quo. In this chapter we present a framework known as structuration theory that incorporates and subsumes both the objective–subjective and the evolutionary–radical change positions.[1]

Another major gain from structuration theory is that issues of power and morality become highly explicit. Management accounting and control systems are seen to do much more than provide information to decision makers for purposes of scorekeeping, attention directing, and problem solving. Many accounting scholars have recently sounded the alarm that to leave relations of domination and legitimacy out of the equation and to view the world only in terms of a collection of individual actors is to miss some very important and essential aspects of accounting systems.[2] In consequence, the degree to which accountants and the systems they design are deeply involved in the social relations of domination and morality is beginning to dawn on most thoughtful academics and practitioners. When this notion really hits home, we will likely see a wholesale revolution in the way we think about management accounting and control systems.

This chapter proposes that the work of Anthony Giddens, the celebrated British social theorist, can be used to develop a comprehensive framework of management accounting and control systems, one that incorporates signification and communication, domination and power, and legitimacy and morality. Thus, the purpose of this chapter is to sketch out a framework based on his theory of social systems and then to use it to explore the way in which it can be valuable for a comprehensive understanding of accounting and control systems in action. The chapter proceeds to outline the main ideas of structuration theory and integrates them into a structuration framework of management accounting and control. The chapter ends with a brief analysis of a case study to illustrate the potential power of this approach.

STRUCTURATION THEORY

Structuration theory distinguishes between structure, agency, and system. Social systems have structures which are the codes for social actions, while agency is the actions of individual members of the system. Agents draw on structures during social action and as they do so they produce and reproduce discernibly similar social practices across space and time. So structuration theory, and this is the point to underline, in contrast with many social theories, includes *both* structures and agency. Thus it subsumes two otherwise fundamentally antagonistic positions: the structuralist position, where social life is determined by impersonal, objective social structures, and the existentialist humanist position, where social life is a product of the individual agent's subjective, existentialist choice-making.

Structures

Structures are the abstract codes or templates which guide our behavior in social settings. They can be thought of as the DNA for agents' social action and interaction. These codes, however, never display the same internal unity as do biological systems, like those of ants and termites. For these and other insects, DNA programs their social roles. Human social systems are not nearly so deterministic. While structures provide for the organizing (structuring) properties of social systems and make for the ordering and binding of similar social practices across space and over time, they can and do change, sometimes gradually and at other times in astoundingly rapid and radical ways.

Structures exist, in a manner of speaking, outside and independently of any particular agent. But they are available to agents as a blueprint for action in specific time–space settings. Another way of putting this is to say that structures exist in virtual time and space and can be instantiated (made an instance of) by agents during social interaction in specific time–space settings. Nevertheless, they are the product of human invention. They are not biologically given by nature, as is the case for ants; rather they are grooved and regrooved by humans in social situations and they are potentially alterable.

An instance may help to clarify the idea. The social codes of the Presbyterian church, for example, direct its followers (agents) to be God-fearing, temperate, law-abiding, hardworking, thrifty, mind-your-own-business Christian folk. Anyone visiting Presbyterian communities in Scotland, California, and South Africa would witness a remarkable likeness in their social behavior, central beliefs, and key values. And, a return visit to one of these churches a year or two later would reveal a great similarity in the pattern of its congregation's social practices. The Presbyterian templates are reproduced through space and across time to institutionalize patterns of behavior. This enables the church as a social system to maintain its unique systemic characteristics.

Agency

Agency is the other major property of a social system. It is defined as the intentional actions of self-conscious individuals as they interact with others in social situations. Unlike our insects, agents can and do make choices in social settings. Agents are not merely social dupes pushed around by structures, rather they are capable of acting existentially. That is to say, they could have acted differently, if they had chosen. The point is that agents take part in social interactions in a way that presents them with the possibility of acting positively in such a manner that social codes are sometimes modified and at other times altered drastically.

For the most part, however, and as a matter of convenience, we respond *reflexively* in social settings, that is to say, without giving them much thought. We simply rely on our implicit store of knowledge about how to behave. In fact, it would be impractical for us to pause, monitor, and reflect about all the choices available. As with the mythical caterpillar who tried to consciously think through the movement of each of its hundreds of legs, we would soon be paralyzed. Nevertheless, agents take part in social interaction purposively and they know a great deal about why they act as they do at what Giddens calls the *practical* and *discursive* levels of consciousness.

At the practical level of consciousness agents monitor their own and others' social behavior by implicitly following the stocks of knowledge about how to act and how to interpret events and the actions of others. For example, individuals chronically apply the laws of grammar and sound production when speaking, writing, or reading even though they could not formulate these laws, let alone keep them in mind at the time. While at the discursive level, agents can give reasons for and rationalize about what they do in social settings. They use their linguistic skills to reflect on their involvement in social interaction. While for the most part individuals respond automatically, they can, if pressed, provide explanations for our own social behavior.

Structuration theory proposes that these two levels are influenced by an agent's primary need for ontological security lodged in the unconscious. Ontological security is grounded in the methods developed during the infant's prelinguistic stage to cope with the anxiety caused by the mother's periodic separation and return. At this stage, the infant has no conception of self and other (i.e., mother). The mentally healthy infant experiences no undue anger or anxiety when the mother is out of sight, develops a general state of trust, and grows confident of the continuity and sameness of outside providers. Conversely, the absence of trust can lead to a lifelong complex of trust anxiety.

Either way, ontological security lodged in the unconscious, is seen as an essential ingredient later in life for action and interaction in social settings. It explains, to a large extent, why agents routinely reproduce social systems, even those they might readily recognize as excessively coercive.[3]

Structuration and the Duality of Structure

Structuration and the duality of structure, are also central concepts in the framework. *Structuration* denotes that structure and agency exist in a recursive relationship. It is the process whereby social systems sometimes function to almost automatically reproduce the status quo, while at other times they undergo radical change. So instead of limiting the analysis to a snapshot, as in many of our previous frameworks, structuration analysis brings the dynamics of history and change onto the scene.

The *duality of structure* idea is closely related to the structuration notion. It denotes that structures are both the medium and the outcome of the agents' conduct which they recursively organize. As actors in social settings we produce (or reproduce) structures, but at the same time we are guided by them. Agency and structure presuppose each other.

Recapitulation

To recapitulate, structuration theory is concerned with the interplay of agents' actions and social structures in the production, reproduction, and regulation of any social order. Structures, existing in virtual time and space, and drawn upon by agents as they act and interact in specific time–space settings are themselves the outcome of those actions and interactions. Agents are not mere social dupes, they are purposive and they know a great deal about why they act in the way they do. They can provide rationales for their actions and interactions. In their reflexive monitoring of action in social settings, agents rely on both their discursive and practical consciousnesses which are motivated by an unconscious need for ontological security. This picture of the agent's psychological makeup and its articulation to structuration, social structure, and agency is shown in Figure 14.

Dimensions of Structuration

Structuration theory also proposes that all social systems have three dimensions: *signification, legitimation,* and *domination*. Signification creates meaning in social interaction, domination produces power, and legitimation provides for the system's morality. These three layers, while separable in the abstract for analytical purposes, are intimately intertwined in reality. Meaning, morality, and power always come along as a bundle. Figure 15 outlines these dimensions within the duality of structure.

Signification

Signification is the abstract cognitive dimension of social life whereby agents communicate with and understand each other. It consists of abstract structures, interpretive schemes, and discursive practices. Signification structures are organized webs of semantic codes. Interpretive schemes are the stocks of knowledge, skills, and rules used by agents to draw on

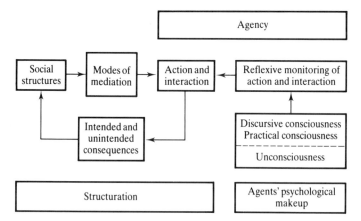

Figure 14 Social Structures, Agency, and Structuration. (1) Social structures are not immediately available to agents but are mediated through modes of mediation drawn upon by agents. (2) Structuration refers to routine reproduction of existing social structures or production of radically different ones during a crisis. (3) Agency refers to actions and interactions of agents in social settings. Source: Macintosh and Scapens.

A Structuration Framework

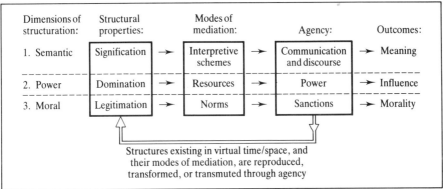

Figure 15 Structures and Agency in Social Settings. Source: Macintosh and Scapens (1991).

the signification structures in order to communicate with each other. They comprise the procedures which mediate between the (virtual) structure and the (situated) interaction. Discursive practices consist of speech, writing, and other forms of discourse used during social interaction. Language, for example, our most important signification structure, consists of the semantic codes (words and vocabulary) which agents draw on using their knowledge of the rules and syntax of language in order to speak, read, and write and thus communicate with other agents.

In day-to-day interactions human agents draw upon interpretive schemes in order to communicate meaning and understanding. Interpretive schemes are the cognitive means by which each actor makes sense of what others say and do. The use of such cognitive schemes, within a framework of mutual knowledge, depends upon and draws from the signification structure, while at the same time, it also reproduces that structure. During an individual speech act, for example, language (the signification structure) is drawn upon through cognitive schemes of syntax and semantics (the interpretive scheme) to create understanding (the communication of meaning). But language itself is the outcome of these individual speech acts.

Management accounting systems can be thought of as a major signification modality. For example, management accounting provides managers with a major means of making sense of the activities of their organization and it allows them to communicate meaningfully about those activities. As such, a management accounting system is an interpretive scheme which mediates between the signification structure and social interaction in the form of communication between managers.

The signification structure in this case comprises the shared rules, concepts, and theories which are drawn upon to make sense of organizational activities. They include the various notions of finance, economics, and management science, as well as central accounting concepts such as income, assets, costs, revenues, and profits. These accounting concepts, moreover, have signification prior to the interpretive scheme. Managers will have shared understandings of their meanings which, although mediated by the management accounting system, are presupposed by that interpretive scheme.

For example, the concept of profit is given specific time–space location through management accounting systems, but it exists outside time and space. It is instantiated only through the use of those systems in practice and it can be changed through such use. Accounting is a major cognitive order shared by executives and managers to make sense of what they and others do. Accounting, as the saying goes, is the language of business.

Legitimation

Legitimation involves the moral constitution of social action. Legitimation structures consist of the normative rules and moral obligations of a social system. They are its collective conscience or moral consensus. They constitute the shared set of values and ideals about what is regarded as virtue, what is to count as important, and what ought to happen in social settings. They also designate what is considered immoral, what is to be trivialized, and what should not happen. Agents draw on these codes during social interaction by means of normative rules of conduct. In doing so, they sanction others (and themselves) in accordance with their compliance or noncompliance with the codes and norms. Compliance, however, does not necessarily mean commitment.

This moral undercarriage inculcates values into the minds of individuals and ensures a fit between the individual and the collectivity. The values and ideals define the mutual rights and obligations expected of agents across a wide range of interactional contexts, including work sites, schools, libraries, discos, sporting events, churches, and shopping malls. The norms specify how to operationalize the values, while sanctions make agents morally accountable for their actions. The legitimation dimension institutionalizes the reciprocal rights and obligations of the members of a social order.

Management accounting and control systems are vitally involved in the moral constitution of managers' actions and interactions.[4] They embody norms of organizational activity and provide the moral underpinnings for the signification structure. In most large complex industrial and commercial enterprises, for example, profit making is the paramount moral ideal.[5] Management accounting systems also legitimate the rights of some participants to hold others accountable in financial terms for their actions. This is widely recognized and accepted as the responsibility center concept.

Accounting systems, seen from the structuration framework perspective, then, are not merely an objective, neutral means for conveying economic meanings to decision makers. Rather they are deeply implicated in the production and reproduction of values and morality. They are an important medium through which the legitimation structure can be drawn upon by managers to produce morally meaningful action for organizational participants.

The structuration framework highlights how these systems can give legitimacy to the actions and interactions of managers throughout an organization. This is because accounting systems set forth values and ideals about what ought to count, what ought to happen, what is deemed fair, and what is thought to be important. They institutionalize the reciprocal obligations and rights of managers throughout the organization. "The giving of 'accounts' of conduct is intimately tied to being accountable for them" (Giddens 1979, p. 85). Accounting and control systems can and do play a critical role in defining the moral constitution of an organization. "Accounting may be seen as a legitimating institution to the extent that it mediates the mapping between action and value...accounting fills this role by structuring relations among actors and acting as the medium through which control is exercised" (Richardson 1987, p. 343).

Domination

Domination concerns a social system's capacity to achieve outcomes, to produce power. Normally power flows smoothly and its far-reaching effects go almost unnoticed in the process of social reproduction. But sometimes its effects are clearly visible and understood, as is the case in warfare or during strikes by unionized employees. While power works to constrain individuals and gain their cooperation, it is also a medium for emancipatory efforts.

Power is generated by drawing on the structures of domination, the blueprints for relations of autonomy and dependency. In feudal times the organization of relations of domination ran in descending order from God, to the king and royal family, to the barons and dukes, to the knights, to the vassals, and finally to the peasants. In a classroom, the power goes from state authority, to principal, to teacher, to student. In contrast to such hierarchical ordering, the domination structure of the US government is one of equal power divided among the legislative, administrative, and judiciary branches. These blueprints are stored in the constitution.

Domination structures are drawn on by means of allocative and authoritative resources. Allocative resources involve the rights of some to hold command over material objects (mines, factories, computers, weapons, farms, etc.) as well as the knowledge of how to operate them (technology, techniques, technical skills etc.). These are the property rights which some individuals hold on these resources. Authoritative resources are different; they comprise the rights of some agents to command others. This is the harnessing of human beings, not physical artifacts. It concerns the rights of some humans to have dominion over other humans. Both power resources provide for the coordination and control of things and people within social systems.

Management accounting and control systems are deeply implicated in relations of domination. They are a vital authoritative resource in the hands of upper management. Responsibility center accounting, a cornerstone of conventional management accounting, involves much more than simply tracing costs, revenues, and capital to the sundry responsibility centers. Management accounting systems also carry the codes regarding who is responsible to whom and for what. They are a major discourse for the domination structure through which some participants are held accountable to others.

Moreover, accounting systems can play a vital role in domination structures. This occurs, for instance, during the ritualistic monthly review of a responsibility center manager's budgetary performances including written explanations of significant variances. It also happens when a superior uses budget performance as an important aspect in appraising the performance of subordinates. It takes place when the responsibility center manager participates in preparing the budget for the center. In each of these instances, the management accounting system is involved in the reproduction of the domination structure, even though for the most part the process goes unnoticed.

In sum, management accounting and control systems are vitally involved in relations of domination and power. Command over them is a key allocative resource used by upper-level executives to hold dominion over the organization's physical and technical assets. The master budget, for example, contains the detailed and all-encompassing blueprint for resource allocation for the entire organization and is a powerful lever in terms of ability to make a difference, to get things done, and to dominate the organization.

The Dialectic of Control

The concept of a *dialectic of control* is also an important building block in structuration theory. It signals that relations of dependence and autonomy are always a two-way affair. While superiors obviously have access to allocative and authoritative resources with which to exercise power over subordinates, the subordinates always have some power resources at their disposal to use over superiors. Superiors are always dependent on subordinates to some extent and the latter are never absolutely autonomous, no matter how much leeway superiors seem to grant them.

The dialectic of control is connected to a social system's primary contradiction, an essential aspect of the constitution of any social system. This is not so much a logical contradiction as a contradiction in the sense that some basic "structural principles operate in terms of one another but yet also contravene each other" (Giddens 1984, p. 193). Each depends on the other while at the same time negating the other. The antinomy between the two basic structural principles is an intrinsic aspect of the social system in that the tension between them is basic to its systemic integration.

A specific illustration may help to bring this point home. Within a private sector organization under the capitalist system of production some members of the social order (the owners) are entitled to the private appropriation of any surplus value created by the system. This is a basic structural principle. At the same time, that surplus is produced by means of socialized (or public) production carried out by the rest of the members. Any capitalist organization, then, is intrinsically contradictory. "The very operation of the capitalist mode of production (private appropriation) *presumes* a structural principle which negates it (socialized production)" (ibid., p. 142). These enterprises only exist as a contradictory form of social entity whereby the two intertwined forces always push against each other.

A striking example is the case of the meganational, conglomerate enterprise.[6] These firms can interject an abstract, impersonal signification structure, the management accounting and control system, into the local environment of a business component halfway around the world. This becomes the means by which executives at headquarters legitimate the appropriation of the profits produced by the social practices of local managers and employees. This paradoxical situation, where some parties produce the profits but others in a far-off land take the lion's share, must in some sense be seen as a double-layered, ambivalent experience at the local level.

The dialectic of control is one secondary contradiction that arises as a consequence of this primary contradiction. Top-level executives, in virtue of the structural principle of private appropriation, have the upper hand over subordinate managers and employees in the sundry business components. While subordinate managers and employees, by virtue of the structural principle of socialized production, have considerable power resources of their own within the dialectic of control.

For example, subordinate managers, such as the divisional general manager in a meganational firm, have authority over the material assets of the business component (an allocative resource) and command over the employees in the component (an authoritative resource). In addition, relative to upper echelons, they have privileged access to a great deal of knowledge and information regarding the component's business affairs, including the possibilities of its production equipment, the technical and administrative skills of its employees, and the strategies and tactics of its competitors. Although they lack authoritative resources with respect to upper executives, upper executives lack allocative resources with respect to component managers. Both parties have access to and can through agency (in the structuration theory sense) draw on their resources of power in the dialectic of control.

One way this is done is through the annual operating budget process. During the budget formulation, subordinate managers participate in the setting of budget targets by submitting their estimates to headquarters for review and approval. Normally, a great deal of negotiation takes place before arriving at a mutually agreeable budget target and component managers are usually able to exercise their power resources to settle for a level they can be pretty sure to achieve.

In fact, evidence is mounting that subordinate managers are able to readily and deliberately build considerable leeway into their budgets. Careful research suggests that in most cases such bargained budgets include 20 to 25 percent slack.[7] Managers accomplish this in a variety of ways: underestimating sales volume and price levels, overestimating expenses, ignoring the effect of known planned improvements, and using discretionary cost estimates in a judicious manner. These actions are taken in virtue of their power resources. Management accounting systems are by no means as one-sided, tightly knit, and inflexible as is commonly portrayed in our conventional textbooks.

Routine and Crisis Situations

Routinization is another fundamental concept of structuration theory. Routine, a basic element of daily social activity, is defined as whatever is done habitually across time–space locations. In *routine situations*, activities undertaken are repeated in like manner day after day. So agents have no need to consciously think or speak about them or to devise and negotiate new social codes every time they meet. Social structures are paramount.

Moreover, in the enactment of social routines, individuals sustain a sense of ontological security. This helps them develop feelings of trust in conducting their daily affairs and staves off what otherwise might be a potentially explosive content of the unconscious. Under routine conditions, social action, including the reproduction of social structures, flows continuously. Routinization is both functional and economical.

In *crisis (or critical) situations*, structuration works differently. Crisis situations occur when the established routines of daily social life are shattered or drastically undermined. Under crisis, agency comes to the fore, often reshaping prevailing social structures. For example, the orderly crowd in a small town general store suddenly turns into a lynch mob, ignores traditional structures, and replaces law and order with vigilante justice. In such a situation members of a crowd can be easily exploited by leaders. Conventional social templates are abandoned, new ones emerge on the spot or repressed ones reemerge in a place of prominence. A striking case in point is the reemergence of sundry Christian religions during the recent breakup of the Soviet Union. In crisis situations, agency often brings different structures, even old ones, into existence. This often involves management accounting and control systems.

This routine–critical demarcation merits further elaboration as it concerns the possibilities for changing social systems. Under routine conditions much, if not most, of reflexive monitoring of action and interaction can be handled at the practical level of consciousness. Here the chances of a significant change in the existing systems through agency tend to be slim. In critical circumstances, by contrast, where there is a radical and unpredictable rupture that affects a large proportion of the individuals in the social system, it seems almost inevitable that the system will change through the actions of individual agents acting at the discursive level of consciousness. If these new social structures continue to be reproduced, a new social order will emerge through the process of structuration.

Change can occur, however, without such extreme critical situations. The intentional action of human agents also taken at the level of discursive consciousness, can bring about gradual social change. Every act which contributes to the reproduction of a structure is also an act of production, a novel enterprise, and as such may initiate change by altering that structure at the same time as it reproduces it. While each of us has the potential to bring about social change, individual acts of themselves will not change social structures until they become institutionalized features of the social system; then new structures will have emerged.[8]

The English language, for example, changes routinely and without much fuss as time passes. The codes (rules) of basketball have changed in an evolutionary, piece-by-piece fashion over the years. Early on a center-jump ball followed each basket, whereas today the defending team simply puts the ball into play from its own end. More recent rule changes include the two-shot free-throw rule, the shot-clock, widening of the key, and the three-point basket. Social structures do change routinely and without radical disruption.

Perestroika (restructuring) in the USSR is a much more striking case in point. In the late 1980s, General Secretary Gorbachev and his supporters faced an economic crisis on the home front and an untenable war in Afghanistan. The aim was threefold. First, to shift the social system of the nation towards more open and democratic processes; second, to reform and revitalize industrial, commercial, and agricultural productivity; and third, to reestablish the credibility and effectiveness of the USSR in international relations and foreign affairs.

Gorbachev adroitly articulated the possibilities for change within the social structures. The substantive changes which have taken place under perestroika are a dramatic example of agency bringing about social change without the emergence of extreme conflicts. This contrasts vividly with the coming to power of the Bolsheviks in 1917 when a revolution proved necessary to secure radical social change.

The Apollo Computers (AC) case, discussed at length in Chapter 5, provides a practical illustration of the role accounting can play in effecting organizational change. The management accounting and control system was a powerful resource in the process of making radical changes in social structures. Prior to the takeover by Dionysus, Apollo featured a centralized chain of command with functional division of duties, and an ethical moral undercarriage emphasizing hi-tech computer and information systems innovations. The new owners effected a radical change in these social structures to transform Apollo into a social system featuring a decentralized chain of command and product group division of duties, along with a value system revering profits, lots of profits.

The new management accounting and strategic planning systems played a crucial role in this transformation. They acted as primary signification systems to show how managers and employees would make sense of organization events and activities. The new accounting discourse reinforced the new domination hierarchy of product center responsibility and accountability. Both control systems served to legitimate profit seeking over scientific creativity and innovation. They played a key role in producing a radically different social order. And once these new structures were institutionalized, the financial controls served to regroove these social templates every time participants drew on the accounting discourse in their daily interactions.

SUMMARY

Structuration theory is concerned with the interplay of structures and agency in the production, reproduction, regulation, and change of social orders. Structure, the codebook for social behavior, exists in virtual time and space, is drawn upon by agents as they act and interact in specific time–space settings, and is itself the outcome of those actions and interactions. This process, known as structuration, denotes the duality of structure.

Agents, however, are not merely social dupes. They are purposive and know a great deal about why they act in the way they do. They can and do provide rationales for their actions and interactions. However, although many of the consequences of agents' behavior are intended and known, other consequences may be both unintended and unknown. In their reflexive monitoring of action in social settings, agents rely on both their discursive and practical consciousness and are motivated by an unconscious need for ontological security.

Structuration takes place along the dimensions of signification, domination, and legitimation. Signification structures involve semantic rules which are drawn on to produce meaning. Domination structures involve resources which are used to produce power. And legitimation structures involve norms and values involved in the production of morality. These three dimensions of social systems are inextricably intertwined and are separate in the abstract for analytical purposes only. In concrete situations of social interactions, agents always draw on these dimensions as an integrated set. In combination, the three structural dimensions influence the social actions and interactions of agents in organizations and institutions. They serve to constrain and coerce agents, but at the same time they function to gain the cooperation necessary to maintain the social order.

Management accounting systems represent modalities of structuration in each of the three dimensions. In the signification dimension, they are the interpretive schemes which managers use to interpret past results, take actions, and make plans. In the domination dimension, they are a facility that management at all levels can use to coordinate and control other participants. In the legitimation dimension, they communicate a set of values and ideals about what is approved and what is disapproved; they justify the rights of some participants to hold others accountable; they legitimate the use of certain rewards and sanctions. Through the modalities described above, management accounting provides for the binding of social interactions in organizations across time and space as well as in some situations for their radical change.

In studying management accounting and control in practice it is always important to recognize the way in which the three dimensions are intertwined. By signifying what counts, management accounting provides a discourse[9] for the domination structure through which some participants are held accountable to others, while at the same time it provides legitimacy for the social processes which are involved. Thus, signification in management accounting terms is implicated in both the legitimation and domination structures, and as such is an important resource in structuring relations of power. However, the structural properties of management accounting are neither wholly explicit, nor unchanging. They can change as they are drawn upon and reproduced in their use by organizational participants.

A SHORT CASE ILLUSTRATION

We close this chapter with an illustration of how the structuration framework might be valuable as a sensitizing device for understanding the full range of management accounting and control systems in action. We will use the Hyatt Hill Health Center (discussed in detail in Chapter 8) for this purpose. Recall that the center had a good financial accounting system as well as a traditional line-item budgetary control system. The new management control system, in contrast, focused on efficient therapeutic medical care (TMC) while ignoring research and preventive, family-centered medical care (PFC). The administrators had begun to use the reports in meetings with the department heads and individual physicians. Reactions to the new system were mixed.

A conventional analysis of the case might suggest that good management control is difficult because physicians see the health and life of patients as their prime responsibility, and concern for costs is a secondary consideration only.[10] It might describe such a professional organization as a hostile environment for management control and then discuss ways of making technical improvements to the system, examine ways of improving the measures currently employed, and suggest additional measures which might be introduced. It would even argue that greater emphasis should be given to efficiency and the profit motive, and that rewards might be based on achievement of budget targets and standards.

Structuration theory, by comparison, provides a means of expanding such an understanding of the role of accounting in this context. The signification structure drawn upon by physicians at the health center is mediated in terms of a medical discourse which emphasizes quality care and preventive medical programs. The new control system employs an accounting discourse, and attempts to integrate it with the medical discourse. However, the accounting discourse has been imposed on the physicians, principally because the administrative director was able to draw upon authoritative resources in the domination structure. The new discourse potentially increases the power of the administrator and his staff, and gives legitimacy to their attempts to control the physicians. Power resources were the means by which the accounting discourse was introduced, but the discourse is itself a source of power to the administrators.

Previously, the participants in the center drew on a set of codes whereby medical language prevailed at the signification dimension, the professional physicians ruled over the administrators, and the effectiveness of PFC and research into the egregious health problems of the slum constituted proper moral behavior. The new control system challenged these codes and precipitated a crisis of domination. Its blueprints for social action and interaction privilege a financial discourse, especially efficiency of medical treatment; a domination structure whereby the administrators have the upper hand over the physicians; and a moral code legitimating efficient TMC, billings for medical services and efficiency of health care delivery.

The physicians reacted to the threat by drawing on resources at their disposal to contest the new discourse. In fact, at their disposal the physicians have authoritative resources—their deep technical knowledge of medicine and their authority as line department heads—and allocative resources—their ability to withdraw their services and leave the center for a more lucrative position at, say, a hospital. This is what the conventional analysis viewed as a hostile environment but which the dialectic of control exposes as a natural form of tension. In the conventional analysis, there seems to be an implied assumption that the accounting discourse is in some way superior to other discourses. This may well be a reasonable assumption for accountants, but it may not appear reasonable to other social actors. This case illustrates conflict in meaning structures, and the way that such conflict can lead to a contest for control that will be worked out within the dialectic of control dynamics.

The structuration framework provides a basis for understanding such contests even though it cannot predict their outcomes. They will depend on the actions of the social actors, in this case, the administrators and physicians, and the distribution of resources among them. Nevertheless, we can expect the administrators and physicians to use the resources at their disposal in attempts to secure control. While the suggestion that rewards should be based on the accounting system points to a resource at the disposal of the administrators, we must not ignore the fact that physicians will also have resources, including withdrawing their services from the center.

Finally, while the conventional analysis can be seen in terms of reinforcing the accounting discourse, the structuration framework provides a much fuller analysis of the social practices involved. Structuration theory teaches us that management accounting and control systems are not only used to communicate financial meaning but also always carry with them the social system's codes for power and morality. Signification cannot be untied from the blueprints for domination and legitimation. In the next chapter we look at the history of two more organizations that illustrate in much more detail the role of management accounting in the process of structuration.

NOTES

1. This chapter is based on articles by Macintosh and Scapens (1990, 1991) who based their research on Giddens (1976, 1979, 1984).
2. See Burchell et al. (1980), Cooper (1980), Tinker (1980), and Tinker, Merino, and Neimark (1982), for pioneer articles making this point. See also Covaleski, Dirsmith, and Jablonsky (1984) and Covaleski and Dirsmith (1988) for the importance of relating accounting to relations of power.
3. See Willmott (1986) for an insightful critique of Giddens' notion of ontological security.
4. See Roberts and Scapens (1985) for an insightful discussion of this point.
5. See Lukka (1990) for an excellent discussion of the socially constructed meaning of profit.
6. See Lowe (1992) for a fascinating, highly readable, and revealing exposé of the power and workings of the meganational conglomerate.
7. See Schiff and Lewin (1968, 1970), Ansari (1976), and Otley (1978).
8. See Giddens (1976), p. 128, for a detailed explanation of this idea.
9. In this sense, a discourse can be said to be the "outcome" of signification, or, in other words, its instantiation in practice.
10. See Anthony and Dearden (1981), pp. 219–221 for a conventional analysis following this line of reasoning.

11
Structuration in Action

The previous chapter presented a detailed account of structuration theory as well as outlining a structuration framework of management accounting and control systems. It proposed that these systems, in addition to providing organizations with a key sense-making apparatus, are also vitally involved in relations of power and morality. The chapter also presented a brief analysis of the new management control system at the Hyatt Hill Health Center. In this chapter, we also put the structuration framework to work in a more extensive way by using it to develop detailed analyses of management accounting and control systems at General Motors and at the US Department of Defense.[1]

In looking at the role of management accounting in production and reproduction of organizational social systems, we will attempt to demonstrate how structuration theory can be used to make sense of the social processes revealed in these companies with a focus on the accounting and control issues and developments. The documented histories of these organizations were not written originally from a structuration framework perspective. They do, however, describe social process that can be understood in terms of structuration theory. The objective is not to provide any ultimate truth about the history of General Motors or the Department of Defense, but instead to illustrate the usefulness of the structuration framework in understanding management accounting and control.

GENERAL MOTORS

The Great Crisis

In 1920, General Motors (GM) was on the verge of bankruptcy and faced a crisis of enormous proportions. William Durant, using the Chevrolet operation as a base, had acquired a huge empire of companies involved in the parts supply, manufacturing, and distribution of automobiles. He had, however, stretched GM's financial resources to breaking point just as a general slump in the economy hit the US automobile industry, reducing sales of new cars from a monthly high of 200 000 in March to about 50 000 in December. GM's shares plummeted on the stock market and Pierre du Pont, representing the majority of shareholders, retired Durant and surveyed the wreckage.

What du Pont and the remaining top executives saw was a crisis of formidable dimensions. First, GM operated under a vague and indecisive product policy. Second, GM's precarious working capital position featured a bloated inventory of raw and semifinished materials, an immense inventory of finished but unsold cars, and a short-term bank borrowing of $83 million. Third, since the production plans of the sundry manufacturing facilities were uncoordinated internally and unconnected externally to sales forecasts, the plants continued to produce parts and vehicles even though dealer lots were spilling over with unsold cars and trucks.

But the most pressing problem seemed to be the fact that headquarters lacked any sort of a systematic management accounting system for control over the far-flung operating units. As Sloan, who was soon to become president, summed up the situation, "The corporation faced simultaneously an economic slump on the outside and a management crisis on the inside" (Sloan 1965, p. 42).

Durant, the consummate entrepreneur, seemed oblivious to the need for a rational alignment of strategy with organization structure and management control systems. Strategic decisions (capital investment, acquisitions, product type and price, etc.) were made unilaterally by Durant, although on occasion he talked with the head of the relevant operating division. Organizationally, there were no clear lines of authority, no effective channels of communication, and virtually no accurate, timely, and systematic information about the performance of the operating divisions.

Further, some divisions were very large and highly integrated while others were small and managed a single function such as spark plugs. The company was little more than a haphazard, ill-coordinated, unsupervised, loosely built federation of operating enterprises. GM was running fast, but it was running blind.[2]

The board of directors appointed Alfred Sloan as the new president. Sloan, already a member of top management, had been president of a roller-bearing company as well as president of United Motors, a GM subsidiary. Sloan formulated GM's first real strategy, realigned the organizational structure, and put in place a sound management accounting and reporting system.

Strategy, Structure, and Controls

The cornerstone of the new strategy was the ingenious realignment of GM's product line. The implicit existing strategy has been to compete head-to-head with Ford in the low-price market segment by making better technological improvements than Ford. Sloan and the other top executives decided instead that GM would aim for quantity production of high-quality cars in all segments of the market, whether or not they were leaders in design within each segment. Sloan banked on attracting customers from below each segment who would willingly pay a little more for the extra quality in the next highest segment. This strategy eventually paid in spades.

GM got another important break. Before being appointed as chief executive officer of GM, Sloan had already worked out in some detail a thorough organizational blueprint to implement these policies. The blueprint, which he labelled "decentralization with coordinated control" involved the realignment of duties and responsibilities by separating the executive from the administrative function. Administrative offices were charged with responsibility for the operations of car divisions including the design, production, and marketing functions. The executive office concerned itself with formulating of long-term corporate strategy, developing critical competitive policies, and measuring and monitoring the efforts of the administrative offices. Key committees, including members from both offices, provided the necessary vertical and horizontal coordination and integration.

The lever to success for this blueprint, however, proved to be GM's new management accounting and control system. It consisted of sound and effective systems of operational controls including well-established techniques for controlling resources and activities such as cash, credit, inventory, production, orders, sales, and shipping. The management control system, built on this solid base, focused on the profit performance of entire responsibility centers, such as the various car, accessories, and parts divisions, each under the jurisdiction of one manager.

Financial Controls and Signification

This new accounting-based signification structure proved pivotal. Financial control was an integral part of Sloan's all-encompassing organizational plan, designed to bring order out of the chaos he saw running rampant throughout the company; a plan incorporating changes which "...though of an emergency nature, coincided with a sweeping reorganization of General Motors, going to the roots of industrial philosophy" (Sloan 1965, p. 15).

In structuration terminology, Sloan's actions reshaped GM's signification structure as well as its domination and legitimation structures into a form that would serve the company for the next half century and provide a model of management for industry at large (Chandler 1962). It was this signification structure which was to provide the dominant meaning system for action and interaction at GM. Upper management at GM, drawing on rights of dominations given to them by du Pont, were able to impose the accounting discourse on the organization in general and on the managers in particular.

The new social order in GM involved a basic realignment of duties and responsibilities. The most important feature was the separation of the executive function from the administrative function. Administrative offices were charged with responsibility for the operations of the car divisions including the design, supply, production, and marketing functions. The executive office, assisted by appropriate staff offices, was to concern itself with formulating long-term corporate strategy, developing critical competitive policies, and measuring and monitoring the efforts of the administrative offices. Key committees, including executive, administrative, and staff officers provided the necessary vertical and horizontal coordination and integration. But the fate of the restructuring of organizational duties and responsibilities, Sloan believed, hinged on the management control issue "... if we had the means to review and judge the effectiveness of operations we could safely leave the prosecution of those operations to the men in charge of them" (Sloan 1965, p. 140).

In structuration theory terms, the actions of Sloan and the executive committee were overturning the prevailing technical engineering signification system and replacing it with an accounting and finance signification. Sloan had appointed Donaldson Brown as vice president in charge of finance. Brown came from the du Pont company where, as we saw in Chapter 6, he had introduced accountants, statisticians, and economists into the finance office and successfully implemented detailed financial controls throughout the organization. At du Pont, financial and economic matters had been the main point of discussion in meetings of the executive committee and of the general managers in charge of the operating field units. Discussions during meetings involved mainly the language of accounting and finance. Executive at all levels used Brown's return-on-investment interpretive scheme to make sense of the operations of the business.

Brown successfully established a financial and accounting discourse at GM as the major means of communication, in place of the prevailing technical engineering discourse. This was no small undertaking. Prior to 1921, the meaning system at GM was that of engineering and technical innovation for the development of better engines, transmissions, suspension, steering, braking, tires, and so on. By 1920, as Sloan observed decades later, "Great as have been the engineering advances since 1920, we have today basically the same kind of machine that was created in the first twenty years of the industry" (ibid., p. 219). The key component of Brown's new signification system was the financial control chart shown in Figure 16 that spelled out the specific relationship of the major elements underlying and articulating return on investment. In terms of structuration theory, Brown's system was instantiated in practice through social action; and in turn it drew on a signification structure of finance, accounting, and control.

In Brown's scheme, standard volume was defined as 80 percent of practical capacity. In order for a division to earn, say, 15 percent per year on operating capital, the standard price for a product had to be set at a level that would yield 15 percent with the plant operating at 80 percent of practical capacity. Standard volume not only guided pricing decisions, but was also used in determining *standard* costs and calculating *standard* burden rates. This concept of standard volume became an essential aspect of the new signification system. For the fixed costs, a burden rate was calculated as the amount needed to operate at practical capacity. For costs which varied directly with volume, a standard variable rate was calculated. Standard burden rates were also developed for cost centers which were partially variable.

Once the degree of variability of all manufacturing expenses was determined, the total expense at the standard volume rate of operation could be established. This made it possible to develop a standard burden rate for allocating fixed, variable, and partially variable costs to products. Any underabsorbed or overabsorbed manufacturing costs were assigned to profit. The scheme, which took into consideration differences in product mix and the quantity and rate of use of machinery and working capital, was used in a very specific way.

> The object was not necessarily the highest attainable *rate of return* on capital, but the highest return consistent with attainable volume, care being exercised to assure profit with each

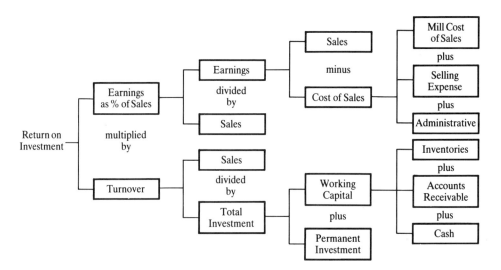

Figure 16 GM Financial Control Chart

increment of volume that will at least equal the economic cost of additional capital required (Sloan, 1963, p. 141).

These new financial controls proved to be the means by which Sloan pulled GM back from the brink of disaster and established control over the far-flung divisions. Differences from standard could be attributed readily to either fluctuations in volume beyond the control of the divisional general manager or to inefficiencies in operations. Financial control arising as it did out of crisis "was not merely desirable, it was a necessity" (Sloan 1965, p. 119). Each segment of the business could now be evaluated in terms of a common yardstick, standard return on investment.

In structuration theory terms, finance and accounting became integral to the signification system for GM's community of managers, with the return-on-investment chart an important element of the interpretive scheme. Managers at all levels in the organization made sense of what each other said and did by drawing on the language of accounting. The importance of financial controls in the meaning structure at GM is evidenced by Sloan.

> Financial method is so refined today that it may seem routine; yet this method—the financial model as some call it—by organizing and presenting the significant facts about what is going on in and around a business, is one of the chief bases for strategic business decisions. At all times, and particularly in times of crisis, or of contraction or expansion from whatever cause, it is of the essence in the running of a business (ibid., p. 118).

The new management control system with its focus on purpose, policies, and operations, enabled Sloan to effectively curtail the freewheeling divisions and to put into effect decentralization with coordinated control. The question which arises, however, is how did the actions of Sloan and Brown successfully introduce an accounting and finance signification structure into GM? Why did managers throughout the organization begin to draw more and more on the new signification structure and less on the previously paramount technical engineering meaning structure? It is precisely for these kinds of questions that structuration ideas can be valuable.

While it cannot produce a definitive answer, structuration analysis sensitizes us to the relevant dimensions of social structure, and particularly to the way that structures of significance are always inextricably intertwined with structures of domination and legitimation. It also alerts us to the way social structures are reproduced, transmitted, or replaced under conditions of crisis and routine. As we will attempt to demonstrate later, Sloan and Brown would probably not have been successful in supplanting the technical engineering discourse without the existing domination and legitimation structures.

But these structures were themselves changed by the new accounting and finance discourse. As structuration theory stresses, the three structures are inextricably intertwined in practice and only separable analytically. The key to understanding the successful supplanting of the old technical- engineering discourse by the new accounting and finance discourse lies in the changes that took place *simultaneously* in the legitimation and domination structures during a crisis of enormous proportions.

Financial Controls and Legitimation

The accounting and finance signification structure at GM was intertwined with the company's legitimation structure. The moral undercarriage for the organization was profit seeking over the longer term. Sloan made the moral assumptions of the business process abundantly clear.

> We presumed the first purpose in making a capital investment is the establishment of a business that will both pay satisfactory dividends and preserve and increase its capital value. The primary object of the corporation, therefore, we declared was to make money, not just to make motor cars (ibid., p. 64).

The new moral code, endorsed by Sloan and the other top executives, was soon adopted as "a soundly conceived theoretical reference to guide us in the practical management of our affairs" (ibid., p. 144). Thus the financial legitimation structure drawn on by the upper echelons of GM was a decisive step for Sloan in putting into effect decentralization with coordination control. As structuration theory states, "The level of normative integration of dominant groups within social systems may be a more important influence upon the overall continuity of those systems than how far the majority have 'internalized' the same value standards" (Giddens 1979, p. 103).

At the time Sloan took over the reins, the prevailing and deeply entrenched competitive strategy called for the development of a revolutionary car with an air-cooled engine to meet Ford head-on in the low-price market segment.[3] Sloan's idea, in contrast, focused on producing a quantity production line of high-quality cars in all segments of the market, whether or not they were leaders in design within their segment, and attracting customers from below each segment who would willingly pay an increment for the extra quality. This product strategy, now recognized as a crucial part of GM's recovery and subsequent domination of the US market, proved decisive. As Sloan put it, "Companies compete in broad policies as well as in specific policies" (Sloan 1965, p. 65).

So profit seeking and growth in capital value came into being as the moral underpinnings for GM personnel throughout the organization. Managers at all levels were morally obliged to strive for, and deliver, a 20 percent after-tax return on their operations. Those who succeeded were rewarded handsomely; while those who failed received well-deserved sanctions. Profit seeking, now formally and informally legitimized, came into being as the moral constituent of GM's collective conscience. The new moral undercarriage was, of course, inseparable from signification and domination.

Financial Controls and Domination

The new financial discourse was also intertwined with the domination structure. The administrative officers in charge of the operating units became responsible for the financial performance of their operations. They were held accountable to the executive committee for making a 20 percent standard volume return on investment. This enabled the executive committee to routinely exercise power over the operating units; power in the narrow sense of providing the medium for the domination of operating managers. It also provided the executive committee with the power to put into effect the new product strategy; power in the broad sense of command over resources to facilitate the transformative capacity of action.

> It was on the financial side that the last necessary key to decentralization with coordinated control was found...if we had the means to review and judge the effectiveness of operations we could safely leave the prosecution of those operations to the men in charge of them (ibid., p. 140).

The financial control system was a critical authoritative and allocative resource in the hands of the top executive team as they put Sloan's organizational model into place and exercised power within the organization.

Recapitulation

Prior to the reorganization, GM's package of social structures featured engineering and entrepreneurial signification codes as the sovereign meaning dimension, reigning over an accounting and finance dimension. While at the domination dimension, the blueprints called for the operating field units to command most of the authoritative and allocative power resources and to operate pretty much independently from the small headquarters office. And at the legitimation level, the morality of high-quality engineering predominated over profit seeking. Sloan, along with Brown and other key individuals, during a period of great crisis and relying on their practical consciousness, were able to replace these structures with a new set of social blueprints, whereby an accounting and finance discourse dominated the engineering and entrepreneurial discourse. The executive offices, now following the dictum of decentralization with coordinated control, ruled over the operating components, and the morality of profit making prevailed over engineering at the legitimation dimension. When the smoke cleared, organizational participants "talked that financial talk," "toed that profit responsibility line," and "walked that money walk." The old social structures were overturned and replaced by dint of reflexive and reflective agency.

At GM, the accounting and finance discourse provided a crucial resource in the hands of central office executives. It allowed them to enact decisions, such as product policy and organizational design, which they favored. It became the key resource through which Sloan and the executive committee routinely exercised power, it regularized relations of autonomy and dependence within the organization, and, through the moral component, it held managers of operating units responsible for their financial performance and accountable to the central office.

The philosophy of profit seeking provided the moral underpinnings of the domination structure embedded in the process of decentralization with coordinated control. Sloan, operating at the level of social practice, drew on the structures of signification, domination, and legitimation to transform GM from a loosely related group of autonomous, freewheeling operating units with a technical engineering orientation into a tightly controlled group of integrated operating units with a financial, profit-seeking orientation.

Financial Controls and Change

GM's history can also be used to illustrate the different dynamics of structuration under situations of critical proportions. It seems clear from both Sloan's and Chandler's accounts that managers at all levels readily accepted and adapted to the new management control systems. At first glance, this might seem to contradict the widespread view that new accounting and control systems tend to meet with stiff resistance. Structuration theory, however, offers a plausible explanation for the swift acceptance of the new controls.

Managers throughout the organization, facing a critical situation—a huge sales slump, a general economic downturn, a cash shortage pushing GM to the brink of bankruptcy, and chaotic organizational arrangements—may have experienced a heightened unconscious need for ontological security and so readily succumbed to the influence of the top executives. As Giddens states, "In such a situation, the members of a crowd are readily exploited by leaders" (Giddens, 1979, p. 125). So the operating managers, responding to their inner feelings, may have followed the mob rather than relying on cultivated reason.

Structuration theory also sensitizes the analysis to the forces which reinforce managers' conscious motivations. At the individual level it seems plausible that the managers, reflexively monitoring the critical economic situation and aware of the changes taking place in the wider social arena, as discussed above, would also have recognized at the discursive level of consciousness the need for radical changes in the accounting and finance systems and the organizational structure. In addition, they would have perceived the power (through du Pont's shareholdings) of the top executives to bring about such changes. These understandings would be legitimated by the economic crisis and the general scientific management movement.

What we see is the working out of the dialectic of control in the context of a great crisis, one that enhanced the already powerful position of top management to effect a radical change in GM's signification structure. This, along with the managers' unconscious motivation for ontological security, reinforced by a working out of the need for change at the level of their discursive consciousness, all favored the rapid acceptance and assimilation of the new system of management control.

Agency and Structure

Finally, the GM case can be used to illustrate duality of structure. GM's new social order was a skilled accomplishment of its upper management, particularly Alfred Sloan, Jr. Facing a crisis of enormous proportions, the top management were able to overturn the existing structures and establish a new social order. According to structuration theory, GM executives were acting as self-conscious agents with discursive and practical consciousness; they put in place a new social order. During crisis agency overcame structure.

The ideas behind Sloan's conception of the new GM did not come out of thin air. His engineer training at MIT, great academic ability, penchant for orderliness, rational Methodist discipline, and experience as president of Hyatt Roller Bearing and United Motors were all drawn upon to conceive of and to put into place the new organizational design. Sloan's own words support this.

> I cannot, of course, say for sure how much of my thought on management came from contacts with my associates. Ideas, I imagine, are seldom, if ever, wholly original, but so far as I am aware, this study came out of my experience in Hyatt, United Motors and General Motors (Sloan 1965, p. 47).

Wider Social Structure

In addition, Sloan drew on structures of the wider social order, which at the time were undergoing marked changes in the US. Sloan's label, decentralization with coordinated control, seems to mirror the metamorphosis taking place in the US in the ideal of individualism. Individualism, one of the cherished and founding beliefs of the American revolution, was imported from France in the nineteenth century. In France, however, individualism had a pejorative connotation. It signified selfishness and blind instinct and denoted persons who had isolated themselves from other individuals and from the collectivity or commonwealth of society. In America its meaning was transformed to denote a condition of social equality. Individualism became an ideal for personal behavior and action. Ralph Waldo Emerson, the great author and philosopher, argued that the enlightened person exists in harmony with nature and has no need for the fetter of state or society.

Individualism, paradoxically, manifested itself in an unpredicted burst of entrepreneurial zeal and led to a world of engines, servomechanisms, levers, and assembly lines in great factories. Harmony with nature as an ideal gave way to the discipline of the machine. The new world of technologies, factories, and machines made organization, the rationalization of activity, specialization, and social interdependence absolutely essential. The ideal of the rugged individual, standing alone, believing no one to be more powerful or spiritually richer than himself, and holding his destiny in his own hands, seemed very much out of place in a social order featuring machines, organization, and interdependence. Public figures, such as the great entrepreneur John D. Rockefeller, began to make public statements to the effect that individualism had disappeared and in all likelihood would never reappear. In consequence, it seemed necessary somehow to recover individualism as an ideal, but change its substance.

The root metaphor for this transformation of individualism was the machine. A machine functions because of its organizing (structuring) principles. The blueprint for the machine articulates the interrelationship of the individual parts. The parts, however, have no meaning or function in their own right. They are subject to the higher order (organization) of the machine. Similarly, the individual, enmeshed in a world of assembly lines, machines, and factories, is subject to the organizing properties of society.

Frederick Taylor, acknowledged as the father of scientific management, saw clearly the nature of the tension between the ideal of the rugged individual and the ideal of the machine.

> Let me say that we are now but on the threshold of the coming era of true cooperation. The time is fast going by for the great personal or individual achievement of any one man standing alone and without the help of those around him. And the time is coming when all the great things will be done by the cooperation of many men in which each man performs that function for which he is best suited, each man preserves his individuality and is supreme in his particular function, and each man at the same time loses none of his originality and proper personal initiative, and yet is controlled by and must work harmoniously with many other men (Taylor 1911, p. 128).

The individual, formerly free from others and the social milieu, is now free to find his niche within a harmonious cooperative network. Taylor had recovered the ideal of individualism by enmeshing it within his metaphor of a cooperation machine.

This new construction of individualism also became embedded in the cooperation machine of organization. Like the parts of a machine, the individuals working in an organization have no meaning of their own apart from the master blueprint that articulates the relationship of each individual and department to the overall collectivity. The ideal organization, like the machine, has no extraneous parts and once set in motion functions as a coherent whole whereby all individuals work toward the greatest possible efficiency. The ideals of the machine became the blueprint for the best possible organization, one where each part became subjugated to the organizing and coordinating principles of the total organization.

Returning to the GM case, it can be said that this new formulation of individualism was an important component of the wider social structure in which GM operated. Durant and his organizing principles seem almost archetypes of the older ideal of the socially unfettered, freewheeling, rugged individual. In contrast, Sloan's organizational blueprint seems to be an exemplar of Taylor's cooperation machine of individualism.

Under Sloan's principle of decentralization with coordinated control, the individual managers of operating divisions were free to run their units as they saw fit with no interference from headquarters, but with the not insubstantial caveat that they mesh their activities with GM's long-run strategic blueprint and policies regarding competitive factors, and that they follow the decisions and policies of the powerful interdivisional coordinating committees.[4] Sloan's organizational blueprint mirrored closely Taylor's ideal of the individual as free within his specialized niche but all the while enmeshed in a cooperative and coordinated social network.

Social Structures Get Cemented In

Once institutionalized, however, GM's structures remained virtually intact for over fifty years. Successive executives routinely reproduced the prevailing social structure through the depression years of the 1930s, the war years of the 1940s, and the postwar expansionary years of the 1950s and the 1960s. Sloan retired as chief executive in 1946 at the age of 71. He remained, however, chairman of the board until 1956, when he became honorary chairman. Even then, he continued to serve as a member of the board of directors and on a few select and powerful committees, including finance and bonus and salary. In 1974 the four top executives of GM had served an average of nearly forty years with the company. So it seems likely that the revolutionary structures of the 1920s were well guarded, preserved, and cemented in.

In fact, they became so entrenched that in the 1970s they resisted the attempts of some executives to change them. As a consequence, GM did not adjust to the very real threats posed by dramatic oil price increases, environmentalists' concerns for clean air, Ralph Nader's exposé of safety needs, and a dramatic increase in sales of smaller, cheaper, fuel-efficient, imported cars. It took almost a decade and a new generation of managers before GM, in the 1980s, slowly began to change the long-standing signification, domination, and legitimation structures.

THE WEAPONS REPAIR ACCOUNTING SYSTEM

Background

Our other empirical study involves the introduction of a system of uniform cost accounting (UCA) for weapons repair by the US Department of Defense (DoD).[5] The UCA was one of a host of new management projects introduced by Secretary of Defense Robert McNamara in the 1960s. Although the UCA concept emerged in 1963, it was not introduced until 1975 and did not become operational until 1979.

In 1985, there were thirty-three major repair depots located in the US and overseas carrying out repair work on the entire spectrum of weapon systems and operated by the four military services: army, navy, air force, and marines. The lines of authority were complicated, but in general ran from the individual repair depots to intermediate command centers, to DoD, to the secretary of defense, and ultimately to Congress.

Motivation for UCA

The prime motivation for the introduction of UCA seemed to be an apparent lack of accountability for funds appropriated to the individual repair depots. Congress demanded some sort of accountability for how maintenance dollars were being spent. The introduction of UCA provided visible external evidence of the DoD's concern for controlling these expenditures: "From the depots' perspective, UCA was a means to demonstrate control over depot operations" (Ansari and Euske 1987, p. 557). Previously, the depots had not been held "accountable" for their spending, they merely spent the monies appropriated to them and requested funding for the coming year on an incremental basis.

Each depot, however, maintained and continued to maintain a management accounting system which was used for budgeting, setting rates for work, and measuring efficiency. Consequently, most of the information gathered for UCA was duplicated in the depots' own accounting systems. Depot personnel viewed UCA, not as a means of controlling their operations, but rather as a way to demonstrate to outsiders that it had control of weapons repair expenditures; whereas DoD personnel viewed it as a way of increasing the visibility of the DoD in controlling depot costs and of enhancing their influence over the depots.

Management Accounting and Signification

The introduction of the new weapons repair management accounting system at DoD parallels the situation at GM, although it was played out over a much longer period of time and under more routine conditions. While at GM the clash of discourse was between the parlance of engineering and the language of accounting, at DoD it was a clash of military and accounting vernaculars. The need for the DoD to demonstrate its control over weapons repair expenditures stemmed directly from the emphasis on economic rationality, quantification, and measurement that McNamara brought to the DoD in the early 1960s.

This emphasis prompted changes in the signification structure that was drawn upon in relations between the DoD and the Congress. The conventional language of accounting—reporting cycles, cost-effectiveness, efficiency, performance standards and capacity utilization—became part of the frame of meaning used to make sense of activities such as weapons repair, which previously had been understood only in military terms. At this level, accounting was seen in purely technical and rational terms.

Management Accounting and Legitimation

The new accounting system, however, played a crucial role at the legitimation level. The prime motivation for the introduction of UCA was an apparent lack of accountability for funds appropriated to the individual repair depots. Congress demanded some sort of accountability for how maintenance dollars were being spent. The introduction of UCA provided visible external evidence of the DoD's concern for controlling these expenditures. From the depots' perspective "... UCA was a means to demonstrate control over depot operations" (Ansari and Euske 1987, p. 557). Previously, the depots had not been held accountable for their spending, they merely spent the monies appropriated to them and requested funding for the coming year on an incremental basis.

Nevertheless, each depot maintained, and continued to maintain, a management accounting system which was used for budgeting, setting rates for work, and measuring efficiency. Consequently, most of the information gathered for UCA was duplicated in the depots' own accounting systems. The depot personnel viewed UCA not as a means of controlling their operations, but rather as a way for the DoD to demonstrate to outsiders that it had control of weapons repair expenditures; whereas DoD personnel viewed it as a way of increasing the visibility of the DoD in controlling depot costs and of enhancing their influence over depots.

The signification structure, as we stressed in Chapter 10, is always implicated in both the structures of domination and legitimation. Signification, in terms of an accounting discourse, was the basis used to legitimate the actions of the DoD in controlling the budgets of its individual activities without the direct involvement of Congress. The introduction of accounting systems, such as UCA for weapons repair, gave legitimacy to claims by the DoD to congress that they were exercising control over the defense budget.

Management Accounting and Domination

The new UCA also played a key role at the domination level. It seems clear that UCA influenced the relations of power between the DoD and Congress. Visible controls, such as UCA, were an important resource in attempts by the DoD to resist direct involvement of Congress in the details of defense budget spending. Signification, in terms of accounting discourse, affected the relations of power between the DoD and Congress.

The effect of UCA on the relations of power between the DoD and the individual depots, however, is not so clear-cut. The personnel at the depots interpreted and continued to interpret, their mission as the provision of facilities for quality repair work. Where this mission conflicted with economic efficiency they would emphasize quality over efficiency. Within the dialectic of control, the depots had considerable resources at their disposal which they could utilize in resisting attempts by the DoD to impose controls through an accounting discourse.

These resources included: the complex command structure that channelled responsibilities through the four military services; the lack of common work among the depots; the absence of clearly defined measurement systems; and the control of local information. These factors gave the depots considerable opportunities for controlling their input to UCA. Furthermore, the DoD had restricted authoritative resources to draw upon, because of the complex command structure. Consequently, UCA was not effective for DoD control over depots, nor was it used by depot personnel to control their expenditures.

UCA, however, proved to have an important role for the depots. Compliance with UCA's information requirements provided a space within which the depots could work without the direct involvement of the DoD. UCA gave a level of visibility to DoD attempts to control depot expenditures without significantly constraining the depots. As such, the new accounting system proved largely symbolic. In contrast with GM, where upper management drew upon the existing structure of domination to impose the accounting discourse on the organization in general and on the managers in particular, the weapons repair system gave individual depots sufficient resources to resist such an imposition, as signalled by the dialectic of control.

Nevertheless, the accounting discourse did impact on depots. Depot personnel became more sensitive to the financial aspects of their activities and they made some technical improvements in their own management accounting practices. Although UCA provided a resource for the DoD to impose controls, because of the resources available to the depots, the introduction of the accounting discourse did not have much influence on the depot personnel.

Summary

In the weapons repair case, signification, in terms of an accounting discourse, proved to be the basis used to legitimate the actions of the DoD in controlling the budgets of its individual activities without the direct involvement of Congress. The introduction of accounting systems, such as UCA, gave legitimacy to claims by the DoD that they were exercising control over the defense budget. These systems also influenced the relations of power between the DoD and Congress. Visible controls, such as UCA, were important resources in attempts by the DoD to resist direct involvement of Congress in the defense budget. Consequently, signification in terms of accounting discourse affected the relations of power between the DoD and Congress.

CONCLUSION

In Chapter 10 we proposed structuration theory as a valuable way to expand the domain of our understanding of management accounting and control systems beyond both a technical efficiency focus as well as many of the ideas in previous chapters to include social, political, and moral phenomena. The analyses in this chapter demonstrate the potential power of the structuration framework to sensitize our understanding of how management accounting and control systems contribute to the *maintenance* of the existing social order, the DoD case, as well as to *changes* in that order, the GM case. At GM, managers relied heavily on the accounting discourse to make sense of the firm's activities. And DoD participants at all levels drew on the accounting discourse to legitimate particular actions. In both organizations, it was an important resource for the exercise of power, both in the broad sense and the narrow sense.

A structuration approach reveals the ways in which accounting is involved in the institutionalization of social relations. It is a focused, integrative, and comprehensive way to analyze case studies of management accounting. Case studies by themselves are not enough to advance one's knowledge of management accounting practice; they need to be informed by theory (Tinker and Neimark 1987; Smith, Whipp, and Willmott 1988). And one such theory is structuration.

NOTES

1. The General Motors analysis relies on Alfred Sloan's (1965) classic book, *My Years With General Motors*, and Chandler's (1962) seminal book, *Strategy and Structure*, on the history of the emergence of the divisionalized structure in US industrial enterprises and Chandler (1977). The Department of Defense analysis is based on Ansari and Euske's (1987) careful, exhaustive, and in-depth study of the new weapons repair management accounting. The analyses are based on Macintosh and Scapens (1991).
2. See Gordon and Miller (1976) for a valuable description of the running blind firm and its accounting and information systems problems.
3. The idea of an air-cooled revolutionary car was not permanently snuffed out. It surfaced forty years later in the form of the Corvair, built by the Chevrolet division. The Corvair, with its air-cooled, rear-mounted engine, went into production in the early 1960s, but proved to be a great embarrassment to GM and the subject of Ralph Nader's famous campaign, "unsafe at any speed."
4. These included the general purchasing committee, the institutional committee, the general technical

committee, the general sales committee, and the operations committee, composed of high-ranking executives from both headquarters and the operating divisions.
5. Ansari and Euske's (1987) study is the source for this case study.

12
Conventional Critiques

It is pretty much taken for granted today that management accounting and control systems contribute mightily to the success of both private and public sector organizations of all sizes. The conventional wisdom tells us they provide three vital functions—scorekeeping, attention directing, and problem solving—and that accountants identify, measure, accumulate, prepare, interpret and communicate information vital to managers in fulfilling an organization's objectives. Management accounting systems are said to play a vital role in planning, organizing, coordinating, controlling, and motivating functions. Many believe these systems are essential and managers rely on them heavily.

Maybe so, yet pieces of evidence rear up to suggest this is not the whole story. In fact, if we pause and take stock, we find an almost embarrassingly large body of research and writing about the darker side of management accounting. And some studies indicate that in certain situations these systems may even be doing more harm than good. Quietly but persistently, careful research keeps uncovering another side to the conventional story. While the criticisms are familiar terrain to academic and practising accountants, for the most part they are put away in a bottom drawer and it is business as usual. Yet, we would be remiss in this book not to discuss them.

To bring some of this dark side into the light, we outline three major critiques that rely on a managerialist philosophy—the *ideology of managerialism*.[1] Before we begin our review, a short detour into the background of the managerialist perspective will help put these critiques into a broader perspective.

MANAGERIALISM

Managerialism can be thought of as a package of ideas, beliefs, and values based on the premise that managers and managerial functions are the essential ingredients of today's organizations. Managerialism came into being in the twentieth century when capitalism's image became tarnished by rapacious robber barons, the worldwide depression of the 1930s, and the effects of large-scale wars between capitalist and bureaucratically controlled nations—the slaughter of millions of human beings and the massive destruction of cities, towns, factories, transportation facilities, and art treasures. While it seemed doubtful that communism was the panacea, capitalism badly needed a facelift.

In consequence, managerialism emerged as an apparently different ideology than the logic of capitalism.[2] It was intended to furnish a more legitimate basis for authority and dominion over society's material production systems, and to validate in the eyes of the general populace the rights of a small group of nonowner executives and managers to hold sway over both private and public sector organizations. As one eminent management scholar put it, "As had occurred when the God of the Western World disappeared 20 centuries earlier, a clergy—management—was created to serve as an intermediary between the workers and 'god' (the owners). The managers knew the will of the owners as the clergy knew that of God, by revelation, and transmitted it to the workers" (Ackoff 1986, p. 17).

The manifesto for these ideas was written in 1938 by Chester Barnard, then President of the New Jersey Bell Telephone Company. In his classic book, *The Functions of the Executive*, Barnard observed that nearly every organization in the history of Western civilization had a short life or eventually failed, and he came to believe that the critical concern must be the organization's continued existence, "Failure to cooperate, failure of cooperation, failure of organization, disorganization, disintegration, destruction of organization—and reorganization—are characteristic facts of human history" (Barnard 1938, p. 5). Survival, he reasoned, must rest with maintaining an equilibrium of forces both inside and outside the organization, forces which constantly threaten to terminate the organization. This is accomplished through control, management, supervision, and administration exercised by "all those who are in a position of control of whatever degree" (ibid., p. 8).

But the real key to organizational survival, Barnard surmised, lay in the willingness of people to put personal effort into cooperative systems along with an intense attachment to the organizational cause. This means "self-abnegation, the surrender of control of personal conduct, the depersonalization of personal action. Its effect is cohesion of effort, a sticking together" (ibid., p. 84). Since willingness and effort is always fluctuating and intermittent, the strategic factor for social integration, he concluded, must be effective executive leadership.

Gaining such commitment and cooperation, however, means walking a tightrope. On one side of the wire, society values highly the ideal of individual freedom with the individual seen as the center of the social universe. On the other side, society firmly believes that subordination and regimentation, under the guidance of the state and society at large, are necessary to achieve social integration. Leadership, the knack of gaining cooperation without stifling individual freedom, becomes the strategic factor. This, Barnard concluded, is *the function of the executive*.

The ideology of managerialism gained full legitimation a decade later when Herbert Simon developed the vocabulary of administration in his seminal book, *Administrative Behavior*.[3] Simon was awarded a Nobel Prize for his highly influential ideas about management and decision making. Most important of all, he declared, managerial decision making is the very heart of organization and administration. He called management a science, a gesture that lent further legitimacy to managerialism. Against this background, we examine three critiques of management accounting and control: goal congruence, relevance lost, and human relations.[4] Table 5 summarizes the main ideas of each one.

THE GOAL CONGRUENCE CRITIQUE

The goal congruence critique, which has been around for some time, is associated with followers of the *management control systems* school. Robert Anthony and John Dearden, professors at the Harvard Business School, and Joel Dean at Columbia, wrote regularly and

Table 5 Conventional Critiques of Management Accounting and Control Systems (MACS)

	Goal Congruence	Relevance Lost	Human Relations
Main concern	The motivational impact of MACS on responsibility center managers	The lack of relevance of MACS to today's manufacturing and organizational environment	The dehumanizing propensity of MACS
Major thesis	MACS have a propensity to motivate managers to take actions that make them look good on the scoreboard but which work against overall organizational purpose	MACS techniques were developed at the turn of this century and are outdated and inappropriate today	MACS tend to be used in a unilateral and coercive manner
Proposed solution	Fine-tune accounting measurements (current cost, annuity depreciation, residual income, marginal cost transfer pricing) and train managers and accountants in ethics	Strategic cost management: widen the scope and focus of MACS; value chain analysis; life cycle costing; and ABC	Train superordinates to use MACS in a way that is more democratic and participatory in order to induce trust, loyalty, and commitment to organizational goals

convincingly during the 1950s and 1960s about the abuses and misuses of these systems (Anthony, Dearden, and Vancil 1965 and Dean 1957). While their work focused almost exclusively on the behavior of top executives and middle managers (in charge of responsibility centers such as divisions, plants, and discretionary cost centers) the problems and issues they identified apply to all levels in the organization. What they discovered is that responsibility center managers almost routinely make some decisions contrary to the overall interests of the organization, but which make themselves look good under the prevailing scorekeeping method.

Such actions, they concluded, violate the goal congruence criterion. Goal congruence holds that a good control system encourages the managers of responsibility centers to take actions that are in their own best interests (in relation to the scorekeeping rules and system) *as well as* in the best interests of the overall organization. That is to say, when the manager puts points on his or her own scoreboard (e.g., an increase in residual income or return on investment) the company also chocks up points on its scorecard (e.g., an increase in earnings per share, cash flows, debt–equity ratio, etc.). In short, goal congruence occurs when the control system induces managers to make decisions and take actions that are good for both parties. But in the case of a conflict, such as when the manager scores but the company scorecard loses points or vice versa, goal incongruence is said to occur. In order to appreciate this line of thinking, it will be helpful to understand the social philosophy upon which these ideas are founded.

Historical Roots

The goal congruence idea can be traced to the thinking that emerged with the advent of the seventeenth century Enlightenment project. According to this new doctrine, the Cartesian philosophy of doubt would erode religious dogma so people's intellectual effort could be used to understand the vision of emerging scientific rationality, instead of trying to fathom the infinite and its wonders and truth. No longer cruel, angry, sublime, and unknowable, God became a benevolent being who wound up and looked after the Newtonian clockwork of the universe. The earth no longer seemed the evil place of medieval times but rather an estate which individuals could and should master for their own betterment. The prospect loomed of a paradise on earth by dint of man's own initiative. God's wonders could be found in the Book of Nature.

Moreover, it seemed to follow that the providential harmony of the natural world also ruled the relationship between man and society. By pursuing their own interests, individuals ensured a harmonious relationship with society. If every individual studied and strove to put personal capital and skills to best advantage, this necessarily led to a preference for what was also most advantageous for society. Previously fettered by church and state, people were now free to improve their lot by dint of their own labor. The invisible hand of the market—as interpreted by prices and choices made in the marketplace—superseded the invisible hand of God—as interpreted by the visible hand of the church and the monarch. Any meddling with this natural order would only upset the harmony of the grander, if invisible, scheme of things. Thus, people pursued the common good under the impression that it was furthering private advantage.

But this thinking had a major hitch. It seemed as if people, instead of being under the guidance of an all-powerful, retributional God, were now under the rule of nature. People had become the slave of the laws of matter and lived only by the law of the jungle. Free will meant the freedom to do whatever people necessarily did and it followed logically that, if people pursued pleasure and derived pleasure from doing wrong, the morality of the system seemed to be undermined. This new world, under the sway of blind determinism, would only lead to survival of the fittest. Something was needed to align private interest with social needs and to recover morality.

A similar problem arose for the economy at large. The economy was seen to be an organism in which the price mechanism guaranteed the optimum allocation of resources. This was fine so long as the act of casting prices adrift on the sea of the market with its invisible hand did not lead to something society at large did not desire, such as war, famine, or plague. While the market mechanism seemed able to accomplish something for all, something that would take an enormous weight of state and bureaucracy to achieve without it, society had always to be prepared to step in if something went wrong. The individual vigorously pursuing self-interest and the invisible mechanisms of the free market had to be kept on a leash.

The way out of this paradox, social philosophers of the time argued, was for the state to somehow align private interest with social customs. The system had to be relaxed enough to allow the government the freedom of action to legislate in such a fashion as to guarantee that individual actions were indeed for the communal good, not for wrongdoing. Public, not private, utility would be the prime source of all moral values. And since laws, now of the legislative variety, would determine social customs, they could be identified with private interests. But when the pursuit of private gain conflicted with the common good, the common good took precedence. The visible hand of the state would reach in to ensure a congruence between self-interest and the overall needs of society.

The divisionalized corporation today faces the very same issues as those confronted by Enlightenment thinkers. During the twentieth century, as companies grew in sales, volume, diversity of products, complexity of technology, and expanded into distant geographic territories, most adopted the divisionalized form of organizational structure. But progress from a one-unit operation to a multi-unit operation meant a significant change in the operating ground rules along with the importation of the classic conflict between individual needs and societal needs. As one eminent accounting scholar noted nearly thirty years ago:

> The reporting of performance of divisions, and the ratings of a division manager and their assistants based on division earnings, created a situation in which a conflict of interest arose. A general manager might make a decision which benefited his division's earnings and his personal compensation but which might not be the best decision from the company viewpoint...other decisions affecting relations between divisions might increase earnings of one division at the expense of another division and perhaps also the company. These questions need investigation and research...[5]

Under divisionalization, the individual business component managers can use reason and rationality to put the resources they command to best advantage. The managers are free to invest and accumulate capital by dint of their own efforts. Any meddling by headquarters executives would only upset the harmony of the scheme. Thus, managers pursue the common good of the overall organization under the impression that they are furthering their own private advantage and that of their particular business component. These are the principles of decentralized management. As with the government and the state, the executives at headquarters need some means of monitoring the various managers to be sure their actions are indeed for the common good of the organization and not for wrongdoing, such as overconsumption of expenses and underexertion of effort.

As it turns out, management accounting systems are exceedingly well suited for this purpose. They are deemed to be neutral and objective. They measure in a quasi-scientific way, resources consumed and wealth produced in a rational way. And as long as the managers performed and were rewarded according to the target levels of the accounting numbers, the manager's private interests are identified with corporate goals. But headquarters must always be prepared to step in if something goes wrong, such as impropriety on the part of any individual manager. Thus, while the managers private interests are deemed to be paramount, when push comes to shove, corporate headquarters have the final say. Goal congruence accounting researchers focus on this basic contradiction.

Managed Profits

A particularly worrisome phenomenon is the way divisional general managers, operating under conditions of decentralized profit responsibility, take deliberate and calculated steps to "manage" their reported profits. Most conspicuously, managers bargain hard at budget formulation time for an easy target level then during the budget year take whatever actions are needed to meet budgeted yardsticks. Further, they can and do manipulate the accounting numbers by means of a variety of cunning tactics. For example, they reduce the investment base of the return on investment (ROI) index by getting rid of temporarily idle production machinery only to buy new the same equipment in a subsequent reporting period. Or they cut back on inventory levels even though the profits on them exceed the corporate average. There are lots of possibilities.

Investigations also revealed that divisional top executives manage reported sales levels. They can, for example, promote early shipment programs whereby goods are pushed out the door at the end of the reporting period before customers really want them. Or they can ship goods by "mistake" to customers that have not yet ordered but are likely to order, record them as sales, and when they are returned early in the new period, treat them as sales returns and discounts. Sometimes they hoard profits when sales for the period are well ahead of budget targets by simply delaying the billing of customers until the subsequent period.

Studies also showed that divisional managers have considerable leeway in meeting budget targets for expenses. One way to accomplish this is by the judicious use of discretionary costs. A manager can simply delay or hurry up spending on accounts such as repairs and maintenance, advertising, or training programs. Another finding is that management accounting and control systems seemed to be a major cause of interdivisional strife and conflict. They add to parochial attitudes whereby divisional managers put the profits and interests of their own component ahead of those of other divisions. This attitude surfaces particularly when managers attempt to negotiate prices for interdivisional transactions. As one prominent academic put it:

> One of the most annoying problems associated with many interdivisional pricing systems is the acrimonious debate and the resulting ill feelings which so frequently accompany price negotiations...Sometimes this animosity between divisions can reach ridiculous extremes. I know of one instance in which the engineers of a buying division were refused admittance to a company plant that was freely accessible to engineers from a competing concern (Dearden 1961, p. 72).

Recent studies along similar lines confirm that such practices are not uncommon today. One researcher reported that managers openly admitted to making decisions that were clearly not in the best interests of their division, such as discouraging employees to be innovative while encouraging them to deliberately manipulate financial performance data (Merchant 1985, 1987, 1990). The managers firmly believed that the company's management accounting and control systems encouraged them to take such actions.

Investigations like this make it abundantly clear, in whatever discomfort, that management accounting and control systems not infrequently induce managers to take actions which they know make them look good on their scoreboard but may not be good for other parts of the company. In short, they "fail to create a community of interests between company and divisional...[they] encourage divisional managers to make decisions that are *inconsistent* with over-all company interests" (Dearden 1961, p. 72), just the opposite of what they should be doing. When pay, promotions, perquisites, and perhaps more importantly, a reputation for one's track record are on the line, it should not be too surprising that ambitious and competent managers engage in such activities. As one expert bluntly put it, "Efforts to maximize any measure of annual performance usually sacrifice future performance."[6] Still, it is no comfort to learn that financial control systems work in this manner.

Remedies

The goal congruence camp is not without remedies for this unpalatable state of affairs. A common suggestion is to introduce better accounting techniques, such as valuing assets at replacement cost; calculating depreciation in a way that records smaller expense amounts early on in the life of the asset and larger amounts later when cash flows are larger; setting transfer prices at marginal cost or market; and using simulated amounts for working capital in the investment base (Anthony, Dearden, and Govindarajan 1992). Another remedy calls for moral guidance, ethical training, and closer auditing (Merchant 1987 and Merchant and Rockness 1994). And one rather ingenious solution suggested desegregating the profit measurement into three parts (Dearden 1987). The investment base, already a sunk cost in any event, would be controlled by means of capital budgeting systems and postcompletion audits (Gordon and Smith 1992), the current operating budget would include only marginal costs and revenues, and a separate budget would be struck for discretionary costs. It remains to be seen whether these remedies will work; to date few comparisons have been reported between companies that adopt remedies and companies that don't.

Summary

Management control systems scholars for some time now have been deeply concerned about the real and potential abuses and misuses of management accounting and control systems. Managers are motivated by these systems to put points on their own scorecard even if the actions they take to do so are detrimental to overall organizational goals. They develop parochial attitudes and manage profits by means of a variety of cunning tactics. Goal incongruence, paradoxically, seems almost to be built into the system. Somehow or other it goes with the territory.

Deep-seated structural forces may work within organizations to render goal incongruent behavior more or less pathological (see later). Nevertheless, it goes against the grain for most of us to know that such things go on. Not many of us would want to go to work each day realizing that if we want to succeed we may have to resort to tactics which we know are devious and distasteful. All of this seems so unnecessary. But as yet, the suggested remedies from the goal congruence camp seem likely only to alleviate without curing.

THE RELEVANCE LOST CRITIQUE

Another group of managerialist accounting scholars have censured management accounting for losing its relevance. The *relevance lost* movement, as it became known, started quietly in 1982 when Robert Kaplan, then dean of the Carnegie Melon Business School, presented an academic paper to small groups of accounting scholars at workshops in several major universities (Kaplan 1983). He wrote persuasively that much of the problem with US manufacturing performance (with respect to Japanese and European competition) could be traced to management accounting techniques and practices which do not match today's manufacturing environment. He concluded, "Traditional cost accounting systems based on an assumption of long production runs of a standard product, with unchanging characteristics and specifications will not be relevant in this new environment" (ibid., p. 688). These criticisms gained momentum in the late 1980s and reached avalanche proportions around the world in the 1990s. As one eminent accounting scholar put it, "Hardly a day goes by without our receiving another paper on the subject, or a solicitation to attend a costly meeting where 'new' and 'better' systems will be discussed" (Ferrara 1990).

Most of the management accounting techniques used today, the argument goes, were invented in the last century and the early part of this one. Manufacturing in those days featured long product life cycles; lengthy production runs of standard products; labor and material intensive factories; and single industry firms. The company that could grind down product costs more than its competitors turned a handsome profit. Henry Ford's Model T was the archetype. At the time, management accounting served the organization's needs well.

But according to the relevance lost thesis, today's manufacturing is totally different. Instead of direct labor intensive, standard products with long life cycles, it features sophisticated, overhead intensive, module-designed, and tailor-made products with short life cycles. Instead of single purpose machinery, there are flexible manufacturing stations and CAD/CAM technologies. Instead of small cadres of technical and clerical employees, manufacturing is planned and controlled by armies of experts in technical specialty departments and administrative offices who manage the actual production operation. And instead of national oligopolistic and tariff-protected competition, the world features relatively unfettered enterprises competing around the globe, and a burgeoning of free trade arrangements. Today's technical and competitive environment would be unrecognizable to the managers of the early 1900s.

This metamorphosis of industry, the relevance lost argument points out, was not followed by a transfiguration of management accounting and control techniques. The old, tried-and-true systems, which focused on reducing and minimizing product costs and emphasized short-run tactical control, have become inappropriate and, more often than not, misleading. Management accountants, including academics, have been asleep for the past thirty years. To alleviate this situation, the proponents of relevance lost offer strategic cost management.

Strategic Cost Management

The strategic cost management idea proposes several ways to restore management accounting to a place of relevance including a radical overhaul of the focus of management accounting and control systems. More specifically, the scope of accounting and control reports would be widened to include information regarding matters such as reductions in product throughput time, rework costs, the number of production interruptions, progress on waste management, levels of work force morale, and indicators of quality in products. And the information base would be expanded from accounting numbers to include measures and statistics of long-run strategic success variables. Such a panoramic focus would direct attention towards strategic considerations, such as customer's changing needs, long-term effectiveness, and quality measures, and away from tactical concerns, such as efficiency and lower product costs.

Strategic cost management also calls for *value-chain analysis.* Value-chain analysis means reconceiving the business as a "linked set of value-creating activities all the way from basic raw material sources for component suppliers through to the end-use product delivered into the final consumers' hands."[7] The aim is to expand the domain of management accounting horizontally to include critical elements external to the company with a particular emphasis on adding value for customers and suppliers. Ironically, the accountant's traditional nemesis, the marketing manager, has long since urged accountants to make this shift.

Strategic cost management also advocates revitalizing management accounting by lengthening the time horizon of control reports by extending cost estimates over the entire life cycle of a product as some Japanese companies are reported to be doing.[8] During the product planning and development stage, these Japanese firms get engineers, purchasing agents, parts suppliers, marketing people, production workers, sweepers, and anyone else involved to estimate the *standard achievable* cost of their part of the new product. Then they establish an *allowable* cost, the difference between the expected selling price and the cost needed to make the desired margin. Next, a *target* cost is struck, usually halfway between the standard *achievable* and the *allowable* cost. These cost estimates are not based on current production methods and equipment but on potentially ideal configurations, taking into consideration learning-curve effects. Finally, during the life of the product, managers relentlessly prod each department to meet the *allowable* cost.

But the strategic cost management initiative that has attracted the most attention is activity-based costing (ABC). ABC advocates argue that many, if not most, of today's large, complex companies rely on traditional absorption cost accounting systems which significantly distort product cost data. This distortion occurs, it is argued, because these systems allocate indirect manufacturing costs using volume-related formulae based on direct labor, direct material, or machine utilization when these resources represent only a small percentage of the total of most products and services. The result, ABC proponents claim, is inaccurate and arbitrary product cost information which, if relied on, leads to poor decisions for vital matters such as product policy and capital investment as well as misguiding management on financial performance and control systems. The ultimate consequence is seen to be a decline in profitability and international competitiveness. The remedy is ABC.[9]

The original ABC method was developed so that system designers could produce a theoretically sound model of the firm's marginal economic cost production function. Initially, the designer identifies and selects cost drivers, activities performed in each factory overhead and service department cost pool and which are assumed to drive the costs of the input resources in each pool. The designer then runs a multiple regression analysis with historical data for each cost pool using the selected activities as independent variables and the total cost of the pool as the dependent variable. The designer must collect data and statistics for enough previous periods (months or years) in order to get a statistically valid regression result. The resulting beta weights for each activity can be treated as the marginal cost per unit of each activity (or cost driver) assuming the factors of production remained constant over this period.[10]

Using these data, the designer derives the total marginal cost of producing each product by combining its activity cost (as per the regression analysis) multiplied by the number of times the activity is performed for the product with the product's direct labor and direct material cost. The result represents the product's marginal economic cost. This amount is multiplied by the normal or average quantity of the product that was produced during the period selected. The designer then combines the results for all products into an equation which represents the firm's theoretical *ex post* marginal economic production function. In the final step, the designer converts this into an *ex ante* model by incorporating identified constraining resources (e.g., production, marketing, and financial) on the *ex post* production function into a mathematical linear programming analysis. The resulting model can be treated theoretically as the firm's *ex ante* marginal economic production function.

It is important to understand that ABC proponents of the theoretical model do not necessarily advocate that firms redesign their existing costing systems along these lines. Rather, they propose that the cost information produced from the theoretical model can and should be used instead of the costs produced by the firm's traditional absorption costing system for important decisions, such as product pricing, production mix, manufacturing equipment investments, and R&D projects.

Even so, the theoretical model is not without its limitations. For instance, designers do not have any rational economic theory for picking the activities to put into the regression analysis and so must rely on intuition and inspection. Thus, the possibility always exists that the set of activities selected does not yield the best (optimal) production function. Further, the model assumes (with no way of knowing for certain) that these activities are being carried out effectively and efficiently. The theoretical model also assumes (implicitly) that the production technology and manufacturing process of the particular firm has remained stable over the time span selected for the observation period. Unless this is true, it cannot be said that the theoretical model reveals the true *ex ante* opportunity cost.[11]

But a more pervasive problem concerns the way that ABC has been popularized in the textbook and popular literature. In contrast to the theoretical model, these versions of ABC merely mimic the theoretical model without any of the latter's elegance and rigorous quantitative analysis. More specifically, the popular version simply calls for the system designer to pick a set of activities (cost drivers) in each factory overhead and service department cost pool, calculate the cost per unit of each activity, and then allocate (assign) these cost pools to one of four cost objectives: units of product, batches of product, production-sustaining resources, and facility-sustaining resources. Finally, the cost objectives are allocated to individual products on the basis of observed cause–effect relationships. Direct labor and material costs are then added to arrive at the final cost figures.

It is important to recognize that the cost data produced by the popular version of ABC are simply the result of a different set of rules for allocating (or assigning) indirect manufacturing costs to products than the rules used for the firm's traditional absorption cost system. This version is, in effect, merely another way to allocate indirect costs to products. It is not a means to calculate the firm's *ex ante* marginal economic cost function. So, the cost data generated are not necessarily more accurate or better than those produced by the firm's traditional absorption costing system. All that can be said is that they are different.

Both the theoretical and the popular ABC model also draw fire from the information economics camp. They point out that ABC proponents do not demonstrate that the new cost data lead to changes in decisions which induce the firm to move from a previously assumed optimal decision to a revised optimal decision. That is to say, they do not assess the value of the new information in terms of the net economic benefits (cost versus value) occuring from implementing the ABC system. In more technical terms, they do not show that the firm has moved away from the original optimal corner solution in the linear (or integer) programming model as a consequence of decisions made on the basis of ABC data. In short, they fail to demonstrate that changes in managers' decisions, if made, result in opportunity lost cost payoffs sufficient to vindicate the added costs of designing and implementing the new system. So they can only hope the new data leads to better managerial decisions. In consequence

> ... researchers who have been wooed over to activity-based costing may be moving too far in the direction of deriving normative implications ... a number of activity-based costing studies

are motivated simply to provide "better" accounting numbers to managers ... better is often assumed to follow automatically from the premise that a selection of additional cost drivers and overhead cost pools will move the researchers close to the "true" cost numbers. This is not the kind of perspective that should guide serious research in this area; it is a perspective one might expect to encounter in a consulting arrangement.[12]

Summary

The relevance lost critique argues that most management accounting and control systems cannot cope with the information demands of the new manufacturing environment, let alone meet the control reporting needs of the robotized, automatic factory of the future. Current systems emphasize short-run efficiencies and cost reduction at the expense of long-run considerations such as quality, morale, innovation, throughput time, and effectiveness. The antidote calls for strategic cost management which includes value-chain analysis, a product life cycle time horizon, and activity-based costing. These remedies, aimed at refocusing controls on long-run, strategic factors and effectiveness, are the means of recovering management accounting's relevance.

Some Concerns With Relevance Lost

The relevance lost movement has stirred up a great amount of excitement around the globe and rejuvenated interest in management accounting issues and problems, largely because it claims traditional management accounting practices may have a lot to do with the declining position of US and European firms in world markets. It is hard to go against its overall aim, to regain a position of relevance for management accounting. Yet, before taking the relevance lost thesis on board holus-bolus, it is important to be aware of some recent concerns.

Advocates of relevance lost might be attributing a great deal more importance to accounting's decision-making role than occurs in practice. Accounting data are part of the decision process, but so are numerous nonquantitative considerations which frequently swamp the numbers. Managers do not simply plug in the cost data and automatically respond to the answer that comes out. Witness the current popularity of strategic considerations in business, with their emphasis on judgment, intuition, and qualitative analysis. And, as indicated in previous chapters, posturing with accounting numbers as well as using them to produce and reproduce power and morality structures are highly relevant aspects of accounting systems which have little or nothing to do with the actual numbers.[13]

Another cautionary note concerns claims made regarding the magnitude of the cost misallocation problem caused by traditional absorption costing systems. While many manufacturing firms around the world are experimenting with ABC, few have yet to embrace it to the extent of tossing out their traditional costing systems. A recent research report, for example, involving eight case studies by leading advocates of ABC revealed that none of these US companies had converted their main cost accounting systems to ABC, although some were using it for selected decisions such as quoting on contracts (Cooper et al. 1992). As the study concludes, "At most of the sites, however, senior operating managers still were contemplating the uses of the newly produced ABC numbers and had not yet acted upon them" (ibid., p. 325). Moreover, some research studies indicate that the product cost information bias in traditional absorption costing is not necessarily significant but rather depends on the firm's production technology heterogeneity, unit input costs, and product mix.[14] While in a revisionist version of the relevance lost initiative, one of its early promulgators has recently expressed serious second thoughts about the efficacy of ABC.

> The belief that activity-based cost management tools will improve business competitiveness, is a dangerous delusion! No accounting information, not even activity-based cost management information, can help companies achieve competitive excellence. Still, that has not kept armies of consultants, software designers, and companies from trying to make gold out of dross. Americans, especially, are big believers in the power of innovative breakthroughs and "quick fixes." The widespread belief that activity-based cost management offers a magic "solution" to American business's flagging competitiveness is no exception (Johnson 1992, pp. 131–132).

Still another study reports that some companies deliberately use less accurate cost systems to their advantage (Merchant and Shields 1993). Some overstate costs to fend off price shaving by marketing personnel. Others deliberately understate costs in order to encourage improvement and innovation in production methods or to stimulate consumption of essential services such as computing and R&D. Still others use very simple and imprecise cost systems to focus employees on critical success variables such as scarce labor or the number of parts in products. Contrary to the ABC thesis, the study concluded, "in many situations the reporting and use of more accurate cost measurements is *not* in an organization's best interests" (ibid., p. 76).

Another worry concerns the proposition that the popular version of ABC produces more accurate product cost data than do traditional absorption costing systems.[15] The reason for this concern is that the textbook version of ABC, as with traditional absorption costing, must come head-to-head with the accountant's Achilles' heel, the arbitrary and theoretically incorrigible nature of any allocation method.[16] Moreover, this version of ABC tends to ignore or gloss over the scholarly debates which took place in past decades regarding the allocation issue and the direct versus absorption costing issue.[17] Thus, in spite of the excitement it has engendered to date, only time will tell if enthusiasm for ABC will wane, as it did for other accounting innovations, such as direct costing, zero-based budgeting, and management by objectives, or whether it will come to replace traditional absorption costing as the firm's main source of costing information.[18]

Furthermore some academics argue that the relevance lost thesis comes over as déjà vu. They point out that eminent scholars, such as economist J. Maurice Clark in the 1920s and accountant Sidney Davidson in the 1960s, advocated different costs for different purposes, taught that allocating costs on the basis of direct labor is not always the preferred method, and argued that businesses need to take a broader, more strategic view than simply managing costs.[19] George Staubus, another highly respected accounting scholar, published his pioneering book, *Activity Costing and Input-Output Accounting*, in the early 1970s.

Other academics are even more sceptical. Some suggest that relevance lost advocates may have set up an imaginary opposition rather than carefully identifying the realities.[20] Still others point out that most of the relevant lost criticisms and remedies have been in the management accounting literature for some time. And some are suspicious that much of the popular relevance lost literature is mainly advertising for consulting services.[21] On the other hand, proponents might come back to say that if these problems and issues have been in the literature for some time, they seem to have been ignored by practitioners and, moreover, that ABC is highly technical and most companies do not have the necessary expertise on hand so they need outside assistance.

Another side of ABC tends to get lost in the flurry of activity over matters of finding new cost drivers and the arguing by scholars (as they should) over matters of theory. This concerns the fact that under an ABC system, service departments, previously treated as a black box to be controlled by means of a discretionary (or managed) cost center budget, now come under the probing, all-seeing accounting gaze which peers into every nook and cranny of the department. All employees are compelled to list the specific activities they perform and to document how each of these contributes (adds value) to specific products (or services) or to product- and facility-sustaining resources. In the process it is possible to do away with redundant and superfluous activities. Since discretionary cost center departments today represent a substantial part of the production component's total costs, the potential for cost cutting is no small matter.

The above concerns notwithstanding, the relevance lost initiative clearly has stirred up and revitalized interest in management accounting practice, research, and theory at a time when a jolt is badly needed. It precipitated a wholesale reexamination of the way manufacturing firms are currently allocating product costs. It gave the technical side of management accounting research a burst of interest. And its aim—to bring management accounting back to a respectable and a relevant position within the accounting academy—is hard to argue against. Assuredly, it is unwise to toss out the relevance lost thesis at this juncture. At base, it has reminded us that management accounting and control is a complicated affair and there is still much to learn.

THE HUMAN RELATIONS CRITIQUE

The third critique within the managerialist perspective looks at management accounting and control systems from the perspective of the human relations school of management. While criticism by the goal congruence and relevance lost camps sees the problems caused mainly by technical flaws in the design of the accounting system, the human relations critique focuses directly on the negative and dysfunctional effects on people working in organizations. First, some background on the human relations school.

The Human Relations Movement

The human relations movement arose as an antidote to the inhumane conditions of US and European factories in the early part of this century. Industrialization had been accompanied, and indeed achieved to a large extent, by the adoption of the principles of scientific management as championed by Fredrick Taylor.[22] Human relations scholars documented the harsh, degrading, and sordid realities of the workplace under the reign of Taylorism.[23] A close look at life in the factory revealed a shocking condition.

Managers controlled the work force by means of harsh autocratic rule dictating to workers the precise way to perform each job and campaigning relentlessly for ever more efficiency. Following the principles of scientific management, most jobs in the factory were deskilled and routinized, thus giving managers control of decisions about how to carry out the work. A worker was no longer an artisan, but more of a wooden automaton. Moreover, working conditions in most factories were appalling; dirt, noise, smoke, little concern for safety, and lack of rest facilities were the order of the day. Unions were ruthlessly put down and, on the least provocation, workers were thrown out the door. Not surprisingly, the workers ganged up to peg production, falsify output statistics, and even deliberately overload machines until they broke.

The human relations (or industrial psychology) school directly confronted and brought to center stage these dehumanizing propensities of rapid industrialization spearheaded by scientific management. The situation was an egregious affront to humanist ideals. Although concerned with fatigue and health problems in the factory, although expressing vigorous concern for humanitarian causes, researchers were clearly proponents of managerialism and on the side of management.

The most famous investigation of this nature is the Hawthorne Study conducted in the late 1920s at the Western Electric plant in Chicago (Roethlisberger and Dickson 1939). When looked at close hand by the researchers, employees in factories were not merely atomistic economic entities with bodies fit for strenuous physical work; they were part and parcel of an informal *social network* with its own values, ideals, norms, leaders, and injunctions. But the big surprise, the one that knocked the researchers back on their heels, was that this social side of the work force seemed to have as much or more to do with productivity than did scientific management. The research discovered, or perhaps more aptly rediscovered, the social person in the worker. The work force had its own social system that existed quite separately from the managerial one.

The human relations movement also championed the idea that the remedy for this dismal state of affairs lay with convincing managers to develop humanitarian administrative practices. This meant invoking a management style featuring democratic instead of autocratic supervision; developing an interest in the employee's welfare over and above a concern for production; and promoting participative decision making instead of bureaucratic rules and unilateral orders. But above all, managers were urged to create a workplace climate of open relationships, trust, and honesty. The result, it was thought, would be beneficial all around— happy employees, greatly improved productivity, and contented managers.

The human relations movement gained momentum in the 1940s and spread throughout the managerial landscape in the 1950s.[24] Corporations hastened to develop a variety of human relations programs for managers and employees. Shopfloor intervention, counselling, supervisory training, innovative pay schemes, participatory problem solving, job enrichment, work redesign, and organizational development initiatives became the order of the day. Managers attended courses which taught the importance of satisfaction, loyalty, trust, openness, self-actualization, and motivation. Prestigious universities, including Cornell, Chicago, and Michigan, opened institutes for researching and promulgating the human relations cause. Consulting firms jumped on the bandwagon as did university-based management schools. By the end of the 1950s, the human relations movement was riding the crest of a wave, sweeping across the nation as well as abroad.

Contradictory Findings

Subsequent research, however, slowly but surely cast a shadow of concern over many of the human relations tenets. The happiest employees, it turned out, were not necessarily the most productive ones. Satisfaction was not always related to higher productivity. In fact, satisfaction sometimes seemed to be related to lower productivity, even though it was negatively correlated with absenteeism and staff turnover. What appeared more likely was that the causal chain went in the other direction; high performance often led to high satisfaction. Moreover, it seemed that high performance could be readily attributed more to good equipment and firm leadership than to human relations factors.

But perhaps the most damaging charge was that, when looked at carefully from close up and with an unbiased eye, sound human relations seemed to be merely another way to extract more productivity out of an already oppressed work force. Human relations practices were a new and cunning form of controlling the work force. In reality, it was business as usual. The Draconian fist of an unsparing managerial class remained under the velvet human relations glove.

Moreover, scholars who looked closely at the human relations prescriptions from the employees' perspective did not see any great benefits for the average worker. Rather, they concluded, human relations practices worked to smooth over and camouflage the fundamental conflict in the workplace between owners and workers. Human relations was simply a clever way to get more effective control over the work force for the benefit of managers and owners.[25] From the late 1960s onward, the human relations movement was still in the running as a dominant theory of management, but it was losing ground.[26] As one eminent organizational scholar summed it up:

> We have tried to show that there is little empirical support for the human relations theory or theories, that extensive efforts to find that support have resulted in increasing limitations and contingencies, and that the grand schemes such as Likert's appear to be methodologically unsound and theoretically biased (Perrow 1972, p. 143).

Nevertheless, the human relations movement did act as an antidote to the dehumanizing tendencies of the scientific management school. Few would disagree that a working environment featuring democratic leadership, widespread involvement in important decisions, and a concern for honesty, openness, and trust has greater value than an autocratic, production-centered system dominated by a bureaucratic hierarchy and impersonal rules. If employees are to be taken advantage of in any event (unless the present social and economic system is overturned) perhaps it is better all around to do it in a way that pays attention to humanistic values. Against this background we now turn to a review of the human relations critique of management accounting and control systems.

Human Relations Studies of Budgeting

The human relations critique got off the ground in the 1930s when a study by the National Industrial Conference Board (USA) indicated that imposed budgets might be causing just as much harm as they were doing good. A survey of managers and supervisors indicated a good deal of dissatisfaction with top-down budgets and recommended that department managers prepare them initially and *then* submit them for editing and revision by executives in the central office. The idea was born that participation in budgeting could remedy some of the dissatisfaction with budgets.

Participation, Group Dynamics, and Management Style

Twenty years later, Chris Argyris, a distinguished human relations scholar of management, conducted a study of budgeting in organizations for the US Controllership Foundation.

Argyris and his research team limited their study to the operating budgets used to evaluate the performance of factory overseers and supervisors in four firms (Argyris 1952). Their findings were highly critical of both the way the financial controls were used and of the attitudes of budget officers and upper managers.

The budget officers saw themselves as the eyes and ears of management. So their job consisted of continuously uncovering errors and weaknesses in operations and reporting them immediately to management. They also believed budgets should include challenging targets to act as a powerful lever for constantly motivating the work force towards greater productivity.

Factory managers, too, believed that budgets had a great impact on the organization. For the most part they invoked them frequently and strongly to maintain their domination over the work force. They used them in an autocratic manner which mirrored their management styles. Behind his back the work force of one factory referred to the top manager as the Whip. Budget officers and factory managers, in their attitudes and use of budgets, displayed very unhuman relations.

Frontline supervisors and overseers, in contrast, hardly ever used budgets. Nor did they mention them to the workers for fear of resentment, hostility, gold bricking, and a consequent reduction in production. They saw a lot of potential problems with budgets: they were geared to results only, with no discussion of the problems and processes used to get the results; they emphasized past instead of future performance; they were based on rigid standards; they were used to apply pressure for ever higher performance; they included unrealistic goals that were almost impossible to meet; and they insulted a worker's integrity rather than offering motivation. Not surprisingly, superiors and overseers thought it best to avoid any reference to financial controls when interacting with the work force.

These observations led to speculations about the underlying behavioral dynamics of budgeting. Workers, it was posited, form cohesive groups to counteract and combat the pressure management exerts through the arbitrary imposition and autocratic use of budgets. This, in consequence, leaves top management in a quandary. When they relax the pressure, the groups do not disintegrate. On the contrary, group cohesion gets stronger and existing negative attitudes are exaggerated so groups continue to do battle with management. Yet, if further pressure is then applied, the result is a head-on do-or-die battle.

As a way out of this quandary, the study recommended that supervisors participate—in a truly *genuine* and not a *quasi* fashion—in making or changing the budget that affects them. Another remedy proposed was to bring all the supervisors together in small face-to-face groups where they would confront each other and their mutual problems; reveal their own feelings, attitudes, and values toward budgets; and then form new ones. Participation and sensitivity training is the standard human relations remedy.

The researchers also concluded that while the pursuit of efficiency is a basic fact of business life, financial controls can easily have just the opposite effect. When used as pressure devices they tend to decrease efficiency in the long run. The report recommended education in human relations for controllers and accountants as well as for accounting students and it alerted accountants to the fact they are dealing not only with individuals but also with groups. It also brought into the open the natural tension involved in their work; success for accountants (uncovering and reporting weaknesses to upper management) means failure for the production supervisors and the work force. Accountants should not expect to be embraced warmly throughout the organization. Respected, yes; but liked, no.

The Argyris study came to be recognized as a seminal work. It stimulated an avalanche of further research into the impact of management accounting systems on the human relations of managers and employees. Most of these studies focused on participation in budgeting, group dynamics, and democratic leadership practices.[27] The results, however, have neither confirmed or proven false that such factors are the keys to successful financial control practices. Nevertheless, some interesting ideas have emerged. We will look briefly at a few of the major ones.[28]

Budget Evaluation Style

One pioneering follow-up study investigated the effect of a manager's *accounting or budget evaluation style* on the subordinate's actions and attitudes. Budget evaluation style refers to the manner in which the manager includes accounting data such as budgetary performance when appraising a subordinate's discharge of duties. The idea is that the way the accounting system is used when appraising performance can have a profound impact on job-related behavior. The study identified three distinct styles: *nonaccounting, budget-constrained,* and *profit-conscious.*[29]

Under the nonaccounting style, superiors are indifferent to budgetary performance information or are unaware of its intended purposes. Accounting information plays a relatively unimportant role in evaluating the performance of their subordinates. Other superiors follow the profit-conscious style treating the budget not as an end in itself but as a means to increase the general effectiveness of operations dictated by long-term goals and programs. Still other superiors rely on the budget-constrained style, the one of most interest to us here.

The budget-constrained style mirrors the way budgets were used by factory executives and budget officers in the Argyris study. Managers taking this approach evaluate subordinates' performance primarily on the basis of whether or not they meet the short-term targets in the accounting reports. They tend to use such information in an unquestioning and overzealous manner. Regardless of the circumstances, they lean on subordinates for unmet budget targets, cost overruns, and sales and profits shortfalls.

From the human relation's perspective, this overemphasis on budget performance to the exclusion of nonfinancial aspects has significant drawbacks. For one thing, the budget-constrained style is a highly task-oriented type of evaluation; there is a deep concern for the task, clear roles and channels of communication, and well-defined patterns of authority. At the same time, it is low on consideration; there is no deep concern for trust, rapport, friendship, open communication, and respect for others. Not surprisingly, budget-constrained subordinates find their superior's evaluations of their performance to be unjust and improper.

Further, subordinates working under the budget-constrained evaluation style respond with a host of negative and dysfunctional activities. They tend to experience widespread job worry and tension. They have poor interpersonal relations with their boss and with colleagues. They avoid innovation. And they adopt short-run expedients at the cost of long-run considerations. In consequence, they dislike financial controls and are not indisposed to manipulate them.[30]

When subordinates participate actively in budgeting, however, the negative effects of the budget-constrained style tend to abate. Participation works to reduce the evaluee's job tension, increase superior–subordinate communication about the problems with management accounting systems, sharpen the evaluee's awareness of the importance of financial performance, and improve the evaluee's interpersonal relationship with peers and superiors. For the nonaccounting style, participation boosts the subordinate's concern for budget attainment and cost control. But more participation has little effect on a profit-conscious style, where budgetary participation is already generally high and accounting information used appropriately in the first place; it is redundant.[31]

A *contagion effect* seems to be involved. This occurs when the accounting evaluation style is transferred from a superior to a subordinate at the next lower level in the hierarchy. In the case of the budget-constrained style, this can be very strong. It is also present for profit-conscious evaluees but they seem capable of resisting it, depending on their own style of using accounting information when evaluating subordinates and on their assessment of the overall needs of the situation. These subordinates are able to deflect the contagion effect, not by reducing the importance attached to meeting the budget, but by increasing the importance attached to other performance criteria. Nonaccounting managers are also successful in diverting it when they think it unsuitable.

Further research has refined these ideas somewhat. Evidence suggests that managers working under a budget-constrained style easily adapt to this narrow use of budget performance information (Otley 1978). They revert to a number of covert practices, including bargaining hard for easy budget targets, manipulating accounting data, and adopting income-smoothing tricks such as early recording of sales, holding over expenses until subsequent periods, and playing with inventory levels. Managers are adept at subverting the budget-constrained style if need be.[32]

Game Spirit of Budgeting

Another interesting idea to emerge from the early human relations studies is the notion of a *game spirit of budgeting*.[33] Managers differ noticeably in the way they engage budgeting systems, the way they play the budgeting game. Some ignore the budget, some become overly concerned with it and carefully weigh every move in terms of its effect on meeting standards, and some treat it in a positive but not pathological manner. Upper management can establish this spirit to create an atmosphere where the budget process is seen as a game, not an end for its own sake.

The logic is that people play games for the game itself. Players become highly involved and enjoy the challenge of competition. Play also involves certain rules which the players accept and self-consciously follow. It is this attitude that has the potential for creating team spirit. Similarly, budgetary planning and control, essential factors for any organization, can be seen by managers as a game. The trick, then, is to get managers to approach the budget as if it were a sporting game, in a positive and high-spirited way. A well-played budget game means involvement, cooperation, team spirit, excitement, and a positive contribution.

One key to promoting a game spirit, the study concluded, is budget participation. Participation by itself, however, is not enough to get managers to live with budgets and be motivated by them. Sufficient communication, correct target levels, judicious performance appraisal, and appropriate superordinate behavior are also necessary. But even then, only with a positive game spirit will healthy budget motivation emerge. Consequently, accountants and top management must work hard to instill a game spirit into the budgeting process in their organizations. After all, they too are part of the budgeting game.

A Path–Goal Theory of Financial Controls

Another important idea that emerged from the human relations perspective is a path–goal theory of financial controls.[34] Studies in leadership behavior had found that initiating structure (clear-cut roles, formal communication channels, and detailed job instructions) as well as consideration efforts (trust, respect, democratic leadership, and good rapport) have an important relationship to job satisfaction. This relationship, however, varies with one's level in the organizational hierarchy. At high levels, initiating structure is positively associated with satisfaction while consideration efforts are not. At low levels, the pattern is reversed.

Financial controls are a vital part of most organization's initiating structures. They can, however, be put into practice with consideration efforts. It follows that at upper levels, where role ambiguity runs high, financial controls will be welcomed, since upper management jobs are rife with uncertainty, ambiguity, and often conflict. Here, financial controls, such as a budget, serve to delineate the budgeted manager's relationship with the superior, provide an important channel of communication with subordinates, and help define patterns of authority and responsibility. Financial controls help managers at this level identify their goals as well as the proper paths to reach them.

For most managers, the job is inherently ambiguous. But by referring to the budget, they can assess decisions in terms of their impact on planned versus actual financial outcomes. Budgets give structure to the job and serve to reduce sharply role ambiguity. They also provide motivation and satisfaction if the budget target is met. Not surprisingly, then, careful surveys usually report that managers hold very positive attitudes towards financial controls. Consideration efforts, such as participation in budgeting and a profit-conscious budget evaluation style, however, are not related to job satisfaction. They are not necessary, since the job is inherently challenging and satisfying.

At lower levels in the organization, the path–goal theory suggests that financial controls work the other way around. Here, financial controls, such as budgets and cost standards, may tend to be perceived as redundant parts of the unnecessary and overloaded initiating structure. The well-defined path, often void of intrinsic satisfaction, is at least familiar; and if upper management view the financial controls as important to operational control, they will be thought of negatively by the lower-level employees. So financial controls will not be as welcome at lower levels and may even correlate negatively with employee satisfaction. Under these conditions, consideration efforts, such as active employee participation in financial plans and standards, may mitigate against these negative attitudes.

Summary

The human relations perspective of management accounting and control systems began with a deep concern about the negative effects of budgetary pressure and a promise that budgetary participation would soften, if not remove, the rough edges of these systems. Many insights emerged, particularly a growing understanding of the social dynamics of budgeting, the way different styles of using accounting information by superiors affects subordinates, and the need for a game spirit to make budgeting systems fly.

A large number of studies followed these pioneering efforts. Many investigated the interaction of performance with human relations variables, such as participation and leadership style. Others looked more closely at the vexing problem that while participation seems to be required, it can also lead to budgetary slack. And some began to look at the effect of different environmental and technological factors on the human relations aspects of financial controls. But these studies did not produce any grand human relations theories of management accounting and control. Instead they seemed merely to confirm the contradictory findings which appeared earlier in the organizational behavioral literature. While undoubtedly research along these lines will continue, the odds are against any huge discoveries appearing.

CONCLUSION

Investigations with managerialist underpinnings have provided rich critiques of management accounting and control systems. Goal congruence research brought to the surface the fact that these systems frequently induce responsibility center managers to take actions which are not in the best interests of the overall organization. Relevance lost advocates precipitated a deep concern for the tendency of management accounting systems to focus on efficiency, cost cutting, and short-run performance at the expense of long-run strategic considerations. They also brought the cost allocation issue, the accountant's Achilles' heel,[35] on to the front burner once again and motivated many organizations to take a hard look at the possibilities of adopting activity-based costing to revamp their accounting systems. And the criticisms from the human relations perspective have given us deep insights into the social dynamics of management accounting systems. Bringing this *dark* side of management accounting into the light seems all to the good.

These critiques are rooted in managerialist premises. The goal incongruence critique, for instance, gives center stage to managers and management accountants. It sees the problems as a consequence of the interaction of responsibility center managers with financial control systems whereby the misdeeds of managers are motivated by improper messages and misdirected motivations inscribed in the controls. The proposed remedy, to adjust the controls so that they give proper messages and inducements as well as to train managers in ethical behavior may, however, merely relieve some of the symptoms without changing any of the deep-rooted, structural causes (Macintosh 1994).

The relevance lost critique also puts managers and management accounting systems into the forefront. The problems are seen to arise since management accounting systems focus on the wrong variables, have a short-run myopic view, and feature cost allocation systems which are outdated and inaccurate. The recommended antidote is to expand the domain and scope of management accounting to include factors such as quality, morale, a longer-time span, just in time (JIT), and to shift to activity-based costing. More and better management accounting, it is believed, will lead to wiser management decisions.

Similarly, the human relations critique puts managers and management accounting systems at center stage, although frontline supervisors and employees are not left out of the picture. The dissatisfaction is that accounting-based controls provoke managers and management accountants to engage in poor human relations. The remedy calls for all parties to follow sound human relations, such as participative budgeting and the profit-conscious evaluation style, when delivering and using financial controls. In short, management accounting systems can be the ship that delivers the human relations cargo to the organization.

These critiques are founded on managerialism; they ascribe a great deal of importance to managers, the functions of managers, and to management accounting systems even though in most of today's organizations managers represent only a tiny fraction of the total number of employees. At General Electric, for example, only a dozen or so managers enjoy top executive privileges including lucrative bonuses and stock options; while another thousand or so managers at headquarters, group, divisional, strategic business unit, and product department levels (out of the total of about 300 000 employees) spread around the globe, also enjoy privileged status including a chance to join the inner circle of top executives. Within the scope of the three critiques, the vast majority of people working in organizations and the work they perform are downplayed while the work of the upper echelons and other managers in the hierarchy is decreed to be paramount.

Managerialist critiques are descriptive and prescriptive so they do not add much to our theoretical store of knowledge about management accounting and control systems (Smith, Whipp, and Willmott 1988). Furthermore, they do not bring into consideration the broader social, economic, and political setting in which management controls operate; they take the status quo of organizations and society for granted.[36] Nevertheless, we do not wish to downgrade managers, management accounting, and managerialism. They are extremely important regardless of whether the social order is underpinned by capitalism, socialism, or a mixed economy. Now that the problems caused by financial control systems are out in the open after a period of neglect, maybe management accounting scholars and practitioners can get busy mending them and setting standards of excellence worthy of professionals.

NOTES

1. See Miller and O'Leary (1987, 1990) for an insightful critique of managerialism and the way accounting constructs the individual as practical and rational.
2. See Anthony (1977), Burrell and Morgan (1979), and Clegg and Dunkerley (1980) for discussions of the ideology of managerialism.
3. Simon (1960). See Kilduff (1993) for a critical reading of March and Simon's (1958) classic book on organizations.
4. See Birnberg (1992) for a review of these three strands of management accounting which he labels, respectively, managerial, economic, and behavioral.
5. Solomons (1965) p. v. This classic book remains a masterpiece in terms of succinctly pinpointing the measurement and control problems in the widely diversified divisionalized organization and the solutions available at the time.
6. Ackoff (1986), p. 35. See Macintosh (1994) for a review of the profit manipulation phenomenon and the problems of the ethical approach to it.
7. Shank (1989), p. 50. See also Shank and Govindarajan (1992) for a detailed field study of the use of value chain analysis in a paper manufacturing company.
8. See Hiromoto (1988) for a detailed account.
9. See Kaplan and Akinson (1989) for a concise and clear summary of this argument.
10. See Kaplan and Atkinson (1989) especially Chapters 3 and 4.
11. I am indebted to Dan Thornton for explaining the theoretical model to me and for pointing out its limitations.
12. Dopuch (1993), pp. 618–619. Goldratt (1992), p. 46, echoes this concern: "The advocates of ABC, when confronted with a request to show exactly how all these data, once collected, will be used to judge decisions, are basically avoiding the question."
13. See Macintosh (1985) for a description of the way managers use accounting information and reports strategically as a symbol and sign of their belief in and ability at rational decision making and see Macintosh and Scapens (1990, 1991) for an analysis of the power and morality dimensions of accounting systems. Also, see Feldman and March (1981) for the way managers use information systems strategically and symbolically to signal their belief in rationality and reason. See also Preston, Cooper, and Coombs (1993) and Covaleski, Dirsmith, and Michelman (1993) for the way "new" management accounting systems have been used in power and domination struggles by

administrators in health care organizations as well as Hirst and Baxter (1993) for a case study of the way managers in an Australian firm used accounting systems strategically and symbolically to obtain funds for capital expenditures.
14. See Kaplan (1993) for a review of some of the research testing the assumptions underlying ABC systems. See also Gupta (1993) who argues that aggregating costs may not always be bad and that the costs of disaggregated ABC systems must be weighed off against simplicity, parsimony, uniformity, compliance with GAAP and strategic considerations inherent in aggregated costing systems.
15. See Noreen (1991) who demonstrated mathematically that the conditions under which ABC provides more accurate and relevant data than traditional testing are "quite stringent and include, among other things, that all costs must be strictly proportional to their 'cost drivers' " (p. 159). See also Noreen (1987) for a trenchant critique of basic ABC presuppositions regarding fixed and variable costs.
16. See Thomas (1969, 1974, 1975a, 1975b, 1978) as well as Eckel (1976) for discussion of the arbitrary and theoretically incorrigible nature of accounting allocation in general.
17. See Ferrara (1960, 1961, 1963), Fess (1963), Horngren and Sorter (1961), Fess and Ferrara, (1961), Fremgen (1962), and Weinwurm (1961) for detailed discussions of the cost allocation problem.
18. This is a testable proposition, as Robert Kaplan kindly pointed out to the author in private correspondence.
19. Frank (1990) provides convincing evidence that most of the issues raised by relevance lost proponents are not new at all and were indeed addressed by Clark (1923) nearly sixty years ago.
20. See Reider and Saunders (1988) who comment, ". . .management accounting education does not share a large portion of blame for the problems being faced by management decision makers. . .it does not teach the exclusive use of a short-term management focus or the making of decisions using allocated costs. To the contrary, experience indicates that management accounting education teaches just the opposite" (p. 59).
21. As Noreen (1987, p. 116) in a commentary on the relevance lost thesis wrote, the recommendations "read more like an advertisement for consulting services. And, like any advertisement, I suggest these latter chapters be read with a healthy dose of scepticism."
22. For a detailed exposition of scientific management as well as an insightful discussion of control in the workplace from a labor process perspective see Braverman (1974) Chapter 4, "Scientific Management."
23. See particularly Mayo (1933, 1945).
24. See Barley and Kunda (1992) for a concise and insightful review of the human relations movement in the US.
25. See Perrow (1972), Braverman (1974), Burrell and Morgan (1979), and Clegg and Dunkerley (1980) for detailed expositions of this view and reviews of the research supporting it.
26. See Barley and Kunda (1992) for documentation of a swing away from the human relations approach at the time.
27. Not all the studies, however, supported the human relations imperative. For example, Cherrington and Cherrington (1973) found that managers, operating under conditions where budget targets were either imposed or revised upward immediately after they had set them, outperformed those whose budgets were accepted on submittal. Becker and Green (1962) pointed out that participation can work either for you or against you depending on the attitudes and cohesiveness (or lack thereof) of the work group. Stedry's (1960) study showed that in some situations participation in budgeting was less beneficial in terms of productivity than when upper management unilaterally set budget targets. DeCoster and Fertakis (1968) unexpectedly found that budget pressure seemed to induce, not autocratic leadership, but rather ideal managerial behavior including an increase in considerate and humane leadership practices. Milani (1975) reported that while budgetary participation may lead to more positive job attitudes it was not associated with performance. And Kenis (1979) reported a positive relationship of budget participation and goal clarity on budgetary performance as well as on job and budget attitudes.
28. See Burgstahler and Sundem (1989) and Birnberg and Shields (1989) for useful overviews of behavioral accounting studies and methodologies over the past couple of decades.
29. See Hopwood (1973) whose seminal study produced these ideas; surprisingly, nearly half of the

departments included in his study reported that their superiors followed the nonaccounting style. See Macintosh (1985), particularly Chapter 2, for an overview of the accounting evaluation style research.

30. A good deal of research followed looking into leadership and budgeting. For example, Ansari's (1976) data showed leadership style and information structure (standards) interacted to affect performance and satisfaction. Collins (1978) reported that an authoritarian versus a flexible style moderated budgetary response attitudes. Collins, Munter, and Finn (1987) found that leadership style (positive versus punitive) influenced subordinates' devious budget behavior. Other studies have looked at the effect of sundry personality variables on budgetary behavior; see Dermer (1973), Driver and Mock (1975), San Miguel (1976), McGhee, Shields, and Birnberg (1978), Birnberg, Shields, and McGhee (1980), Brownell (1981), Otley and Dias (1982), Gul (1984), Chenhall (1986), Chenhall and Morris (1991). Other researchers found relationships between participation, job difficulty, attitudes, and performance, see particularly Mia (1988, 1989).

31. Several researchers followed these leads. For example, Brownell (1983), found that high budgetary emphasis in conjunction with high budgetary participation enhanced managerial performance. Brownell and Hirst (1986) reported that while budgetary evaluation style interacting with budgetary participation affected job tension it did not have an effect one way or the other on performance. In contrast, Dunk (1989) found that budgetary emphasis by supervisors interacted with budgetary participation to influence performance (e.g., high participation and high emphasis led to lower performance). See also Dunk (1993).

32. Many subsequent studies investigated the way managers utilize budget participation to negotiate slack budgets, especially when budget performance is perceived to influence their reward system. See Lowe and Shaw (1968), Schiff and Lewin (1968, 1970), Onsi (1973), Collins (1978), Christensen (1982), Baiman and Evans (1983), Penno (1984), Antle and Eppen (1985), Brownell and McInnes (1986), Lukka (1988), Chow, Cooper, and Waller (1988), Waller (1988), and Dunk (1989). The results of these studies generally support the idea that participative budgeting, reward systems, and slack budgets are associated.

33. This insightful idea originated with Hofstede (1967) as a result of his exhaustive study of human relations (and other) aspects of budgeting in six different companies in the Netherlands.

34. Macintosh (1985) and also see Ronen and Livingston (1975).

35. See Eckel (1976) and Thomas (1969, 1974, 1978) for concise and articulate discussions of the arbitrary and theoretically incorrigible nature of any method of cost allocation. Also, see Zimmerman (1979) who argues that while any allocation is essentially arbitrary, cost allocation might be useful for control and as a proxy for long-run (unobservable) marginal costs.

36. See Burchell et al. (1980), Hopper and Armstrong (1991), Armstrong (1987, 1991) and Macintosh and Scapens (1990, 1991) for arguments to expand the scope of management accounting research to include power and morality dimensions.

13
A Radical Critique

This chapter presents a more radical critique of management accounting and control systems.[1] It is based on work by Michel Foucault,[2] the renowned French social historian whose research and methods have been the source of inspiration for scholars around the globe in a wide variety of disciplines, including accounting.[3] Foucault's[4] major effort involved documenting the rise of a general carceral-like disciplinary drive circa 1700, a drive which became widespread throughout Western civilization in the next two centuries. The result is that today "we are much less Greeks than we believe. We are neither in the amphitheatre, nor on stage, but in the panoptic machine. Is it surprising then that prisons resemble factories, schools, barracks, hospitals, which all resemble prisons?" (pp. 217–218).

As this disciplinary drive gained momentum, various human sciences, such as psychiatry, psychology, criminology, pedagogy, and sociology, came into existence. Each of these developed its own discursive formation and practices. A discursive formation (or mini-regime of truth) refers to a cluster of concepts, assertions, and propositions about the human individual which makes the individual knowable and correctable according to its particular package of knowledge. Each of these regimes of truth also has its own unique set of discursive practices consisting of techniques and apparatuses for applying its knowledge. Discursive formations appear to be scientific and so are accepted as legitimate by society. So this knowledge regarding the individual is power; power imposed on the individual.

Freudian psychology represents a striking illustration of a discursive formation. Highly regarded a few decades ago, but now pretty much in retreat, Freudian psychoanalytic theory casts the human individual as a three-tiered psyche—id, ego, and superego—in which the three parts struggle over control of the individual's personality. The strong ego of a normally healthy psyche acts as the controller of both the id's inclination to blindly follow its instinctual, natural, libidinal urges and the superego's demands to conform to the moral dictates of society. In contrast, an unhealthy psyche stems from a distorted working through of the Oedipus complex, whereby the infant envies the father's sexuality but fears castration. This psychoanalytical knowledge is dispensed in discursive practices, such as university lecture theatres, textbooks, and popular magazines, as well as during lengthy confessional therapy sessions where Freudian psychiatrists aim is to normalize clients.

The discipline of management accounting and control systems can also be regarded as a discursive formation. It represents a peculiar way of knowing the manager and it has its unique set of discursive practices. As a discursive formation, it knows the manager as the person with the authority to make decisions, with command over subordinates and resources within a designated responsibility center, and with responsibility for its financial performance. The manager's function is to achieve the financial targets incorporated in the controls. Discursive practices include, for example, the procedures for preparing, submitting, revising, and approving budget plans; meetings to discuss results; variance reports; and incentive schemes linked to achievement of financial targets. This accounting regime of truth is legitimated by virtue of its seemingly scientific status as neutral, objective, and quantitative as it is depicted in accounting textbook and journals. The manager is knowable in accounting and financial terms and is produced as a docile and obedient automaton who generates financial returns for the organization.

The management accounting discursive formation is put into practice in accordance with the general principles of discipline and control described in the next section. These principles are illustrated by way of the large body of literature available regarding Harold Geneen's reign as chief executive officer at the International Telegraph and Telephone Company during the 1960s and 1970s. The chapter concludes by speculating about the possibilities for resistance to accounting regimes of truth.

GENERAL PRINCIPLES

The general principles of discipline and control which follow were extracted and systematized from Foucault's celebrated book, *Discipline & Punish: The Birth of the Prison*, in which he detailed the emergence from the classical era of an all-encompassing disciplinary drive that became ubiquitous during the modern epoch. Foucault identified three general principles underlying the way the disciplinary society functions: the principle of enclosure, the principle of the efficient body, and the principle of the disciplined mind. These principles achieve the maximum effect within the architectural arrangements of the panopticon.

The Enclosure Principle: Disciplining Space

General Enclosure

Discipline proceeds initially by the careful distribution of individuals over general purpose, self-contained places of confinement. These include monasteries, poorhouses, prisons, schools, universities, factories, hospitals, office buildings, military bases, asylums, and so on. Once accumulated in the general enclosure, the monk, the pauper, the criminal, the pupil, the scholar, the worker, the sick, the clerk, the soldier, or the mentally deranged can be controlled in a sheltered, monotonous disciplinary state.

Partitioning

General enclosure by itself, however, is not sufficient to achieve disciplinary spaces. It is also necessary to partition the enclosure into smaller, self-contained locations in which it becomes possible to know, master, and make useful each and every individual. This cellular principle can be traced back to the monastery of the classical era where each monk had his own cell. Partitioning also makes it possible to effect the rule of functional sites, whereby each location is defined in terms of the specific, regular, useful function to be performed therein. In a factory, for example, each workstation is assigned a particular task. A university library contains various rooms, each with a particular function: reserve reading, periodicals, archives, and study cubicles.

Useful Spaces

Partitions are also arranged horizontally and vertically. In the first instance, each partition is serialized within the general enclosure in a perfectly legible fashion so that its usefulness is identified in relation to all the other partitions in the functional chain. In a factory, work moves from one workstation to another in serial fashion. Similarly, in a university, students are shunted from one building to another and from one class to another. Each partition is also defined in terms of the rank it occupies in the hierarchy and by the space that separates it from the partitions immediately above and below it. The result is the formation of a relatively permanent grid of functional, useful, serialized, and ranked spaces. The crucial consequence is that each individual becomes defined by the physical space he or she occupies.

When put into practice, the cellular principle produces a living picture, *a tableau vivant*, of useful spaces which is both *real* and *ideal*. It is real in the sense that it governs the disposition of material objects (machinery, inventory, furniture, cabinets, etc.) and ideal in that it defines the function, serial relationship, and rank of each space. With each space having a fixed identity, individuals can be distributed in and circulated around a network of real and ideal locations. The spaces retain their identity while the individual is identified by the space he or she occupies. This art of distribution with its cellular power transforms a mass of individuals, otherwise confused, useless, or dangerous, into an orderly assembly of live subjects arranged in a purposeful useful grid in which the individual is dominated by the partition he or she occupies.

The Efficient Body Principle: Disciplining Time

The enclosure principle disciplines space and paves the way for the efficient body principle, which disciplines the individual's time within any specific partition. The efficient body principle works according to three practices: the timetable, which programs the individual; the maneuver, which defines the precise timing of body movements, and dressage which produces automatic responses to signals.

The Timetable

The timetable has a long heritage as a disciplinary practice. Religious orders such as the monasteries of the Middle Ages employed it to great advantage to establish a meticulous timetabling of the monk's daily life. Vespers, lessons, bible study, chores, and contemplation time were ritualistically scheduled to ensure regular cycles of sanctimonious activity. The timetable later came into use in schools, hospitals, poorhouses, prisons, and workshops where it established rhythms of actions, regulated cycles of repetition, and effected a clockwork repetition of useful activities. According to Foucault, since time was now measured and paid for in the factory, it "must also be a time without impurities or defects! a time of good quality, throughout which the body is constantly applied to its exercise" (p. 151). The individual was enmeshed in a constraining chain of detailed minute actions.

The rules of one factory, for example, required all personnel to start the day by washing their hands, thanking God for their work, and making the sign of the cross. Such pious exercises gave legitimacy to the timetable. Sanctions were invoked for being fifteen minutes late for work, for talking or joking with coworkers, or for leaving one's workstation. Even during meal breaks, no conversation was permitted that might distract workers from their duties. Every attempt was made to "...assure the quality of time used: constant supervision, the pressures of supervisor, the elimination of anything that might disturb or distract: it is a question of constituting a totally useful time" (p. 150). Time penetrated the worker's body rendering it docile, obedient, and efficient.

The Maneuver

The maneuver, which also became widespread in early modernity, emerged as a technique to intensify time even further by articulating the body with the work object. The maneuver links the individual's body and the pen, rifle, wagon, machine, or tool into a man–object–machine by specifying the precise way to perform the task as well as the exact timing required to complete the job. In schools, for example, the correct technique for handwriting was spelled out in detail. The position of the feet and arms, the movement of the hand, fingers, eyes, elbow, and even the chin were rigorously prescribed. Each movement was assigned a direction, a range, and a duration within a prescribed sequence until handwriting resembled microgymnastics. The result is a systematic and meticulous meshing of the body with the object.

Along similar lines, the French army prescribed the precise body movements for various maneuvers such as marching and shooting. In the sixteenth century, the orders simply called for the troops to march in file, raising their feet in unison to the rhythm of the drum. In contrast, one hundred years later, the regulations detailed four different sorts of marching steps: short, ordinary, double, and oblique. The oblique step had to be eighteen inches, measured from one heel to the other, and had to take slightly longer than one second. The instructions for firing fusils contained precise details for various stages of preparing, aiming, and shooting. The user-friendly instructions and software of today's ubiquitous personal computer, although more subtle, can be seen as a modern day maneuver, tying the individual to the computer keyboard and screen.

Dressage

Dressage also emerged as a highly effective mechanism for disciplining time. While dressage today refers to the habitual training of an animal such as a horse in obedience and comportment, it was applied to human beings in schools, the military, and the factory at the beginning of the modern era. In the case of the disciplined soldier, a hand gesture or a whistle from the officer called for immediate and blind obedience to carry out a specific action. The least delay was considered a crime. And in the school, at a particular signal from the teacher (ringing a bell, a clap of the hand, a nod, or a glance) the student became instantly attentive to its implicit but unambiguous command—recite the catechism, open the workbook, put down the pen, and so on. Today the factory whistle and the coach's signal give similar messages to workers and baseball players. The signal elicits a reflexive response from the disciplined body.

Dressage is not applied as a celebrated triumphant power; it works in a quiet modest way, exercised gingerly so as not to weigh too heavily on the individual. It places the body in a world of signals, each with a moral imperative and each requiring instantaneous response. It also works on groups of individuals, linking them together to multiply their usefulness, as with a team of horses or a pack of sled dogs.

The intensive and exhaustive use of time by means of timetables, maneuvers, and dressage received the warm endorsement of society. In premodern times, the principle of nonidleness prevailed and to waste time was deemed economic dishonesty in the eyes of one's fellows and a mortal sin; God counted time and man paid for it. So the secularization of these practices in modernity came easily. The efficient body principle became an important strand in society's disciplinary fabric.

Disciplining Minds: The Correct Comportment Principle

While the enclosure principle organized and controlled space, and the efficient body principle organized time, the disciplined mind principle emerged to hold sway over the mind. It works according to three mechanisms: hierarchical surveillance, normalizing sanctions, and the examination.

Hierarchical Surveillance

Hierarchical surveillance refers to a disciplinary practice which spread across the social landscape in the late eighteenth and early nineteenth centuries. Previously, architecture was concerned with designing buildings which permitted their occupants to look *out* on a spectacular view, or to be seen for their splendor, or to protect the inhabitants from attacks. The new configuration, in contrast, called for a geometry of enclosure that permitted the establishment of a constant gaze which looked *in* on each of the inhabitants. Hospitals, schools, prisons, poorhouses, and factories adopted similar architecture to become human observatories.

The geometry of the military camp served as an ideal. An artificial, short-lived city populated by armed men, the military camp required intense disciplinary power. The officers' tents looked onto the main gate and the arms depot; the captains' tents faced rows of tents of their own companies; and the noncommissioned officers' tents fronted the lanes of the soldier's tents. This layout produced a relay of observation and information with no missing links. It brought into play an all-encompassing network of hierarchical surveillance whose constant gaze disciplined as it watched.

The emerging factories also called for intensive discipline and control. Previously, the master in the workshop worked alongside the apprentices and helpers keeping a close eye on them. And government inspectors checked occasionally to see that laws and regulations were followed. But, as the size of the work force, the complexity of the work flow, and the sophistication of machinery grew significantly, a different kind of surveillance was necessary. This gave rise to a new cadre of supervision composed of overseers, supervisors, and clerks. While they also kept a close watch over inventories, machines, tools, and quality of output, their major function was to provide an intense and constant surveillance on each worker's skill, zeal, promptness, and comportment on the job.

Thus, a new regime of surveillance came into existence alongside the physical system of production. Although running parallel to the actual work flow of machines, inventories, locations, and workers, it remained separate from the latter. This cadre of watchers, who looked on from raised stands over the heads of the workers, treated the employees with severity and contempt (the origin of the overhead account). Hierarchical surveillance worked as an uninterrupted, anonymous, automatic, and indiscreet disciplinary gaze which played out over the entire organization.

Normalizing Sanctions

In order to make their presence felt more deeply, incumbents in the supervisory hierarchy relied on systems of normalizing sanctions. Normalizing sanctions refer to the practice of meting out various rewards and punishments in accordance with a set of arbitrary rules (like the confessional in the Catholic Church). These rules and sanctions are created without reference to any philosophical ideals; they constitute, in effect, a private system of justice which operates outside of the state's legal judicial system. Together, hierarchical surveillance and normalizing sanctions functioned as a miniature penal mechanism.

The practice of normalizing sanctions spread throughout society in the eighteenth and nineteenth centuries. Schools, military establishments, prisons, hospitals, factories, state bureaucracies, and so on, all developed their private systems of justice. Regulations covered timeliness (lateness, absences, task interruptions), attentiveness (negligence, laziness, lack of zeal), comportment (disobedience, impudence, rudeness), speaking (idle chitchat, insolence, rudeness), appearance (dress, cleanliness, gestures, posture), and sexuality (lewdness, impurity, indecency). Even the smallest departure from correct behavior, or not measuring up to a certain required level, became the basis for a subtle range of penalties, including petty humiliations (standing in the corner or wearing a dunce cap), minor deprivations (no recess), or light corporal punishments (a tweak of the nose). The individual was caught in a network of disciplining sanctions.

Sanctions were not meted out so much for expatiation or repentance as to induce the individual to conform with established norms of correct behavior. They were designed to motivate the individual to strive for a higher level of competence, as in the case of punishment in the form of repetitive, intensified learning exercises. They functioned to reduce the gap between actual behavior and the norms. Offenses became opportunities for correcting and training. The individual became constantly available for normalization.

The use of penalties was to be avoided whenever possible. The teacher, master, jailer, overseer, or reformer tried to dish out positive recompenses much more frequently than painful ones. Favoring rewards over penalties followed the principle that a deviant individual could be stirred up more by the desire to get rewarded in the same way as were the diligent ones, than they could by the fear of punishment. The positive use of sanctions was much preferred. Many institutions devised intricate regimes based on this principle.

In the Christian Schools during the eighteenth century, for example, the Brothers gave out merit points to students for correct catechism answers. These points could be inventoried and used later to gain exemptions for penances imposed in the future. And military schools in France introduced sanctioning systems of color-coded epaulettes. A silver epaulette signalled very good, a red silk and silver one good, a red wool one mediocre, a brown wool one bad, and no epaulette but a sackcloth jacket identified shameful cadets. Moreover, the silver epaulette holder could make military arrests and mete out punishments to other cadets, including solitary confinement in dark cells. These systems of quiet sanctioning induced each student or cadet to constantly strive for a better ranking until reaching the top.

Along similar lines, in the early 1800s Robert Owen, the famous utopian social reformer, replaced whippings and strappings by overseers in his experimental Scottish cotton mill with a silent system of ranking the daily conduct of each worker (Walsh and Stewart 1993). He hung a four-colored small block of wood (called the telegraph or silent monitor) over the head of each worker whereby black denoted bad, blue indifferent, yellow good, and white excellent. Overseers turned the block each morning according to their judgment of the worker's conduct the previous day. By simply walking through the factory, the telegraph informed Owen of each worker's general diligence.

Such judicious mixtures of gratifying and negative sanctions became widespread as an integral part of repeating cycles of knowing and correcting the individual. A carefully managed mixture of sanctions worked automatically, silently, and without physical force to induce individuals to want to correct themselves and conform to standards of comportment and improvement. Quiet sanctioning produced self-normalization.

The Examination

So that sanctioning appeared objective, masters, overseers, sergeants, wardens, and so on, came to rely on systems of examination. Previously, the scholar or apprentice worked alongside the master and after a long period of tutoring, presented a masterwork for examination. If the master deemed it worthy, the novice became a fully fledged member of the academic community or guild. This intimate ritual changed when the examination emerged as a constant and pervasive procedure in schools, workshops, prisons, military establishments, hospitals, asylums, and professions. The examination combined with normalizing sanctions as an integral part of the principle of correct comportment.

The new examination process featured a system for testing, interrogating, and inspecting each and every individual. The Brothers in the Christian Schools, for example, examined their pupils every day of the school week on a particular subject and twice on Wednesdays. They also administered a monthly examination to select students deserving to be examined by the government inspector. Examination results guided and legitimated the sanctioning process. The student had become an object offered up for compulsory visibility and for perpetual examination and normalizing sanctions.

An important part of the examinatory process was the writing and numerical grading of each individual. The keepers (teachers, sergeants, physicians, psychiatrists, wardens, supervisors, etc.) administered the examination, assigned a numerical grade, and made out a written report, thus accounting for the individual. Moreover, each individual became written in a permanent archive of records, results, files, report cards, and so on. Collectively, this array of documentation formed a grand register and a total field of comparison, thus making it possible to calculate averages, create categories, designate classifications, and to establish norms and stages of development. Anyone and everyone could be defined in terms of normalcy and located at a particular stage of development. Compulsory objectification and perpetual examination became a natural part of the social fabric. Embedded in a cumulative system of writing and grading, the individual had no place to hide.

National Data Bases and Case Studies

These records and archives also made it possible for the various human sciences to accumulate large and ever growing fields of knowledge, each of which made the human being knowable in terms of the special attributes, characteristics and capabilities measured by the examinations. Moreover, any individual could be treated by the expert professional of a particular discipline as a case study, an object to be measured, described, compared, and judged according to the norms and averages of the general population. The individual existed as a thing to be corrected, normalized, and treated in accordance with the discursive practices of that particular discipline. Knowledge about the individual became power, power over the very individuals from whom it was extracted. Hierarchical surveillance, normalizing sanctions, and the examination were most effective when applied within the panopticon.

Panopticism

Panopticism refers to the unique architecture of Sir Samuel Bentham's (1757–1831) famous panoptic prison. The geometry of the prison called for a central tower in the middle of a peripheral ringed building which was divided into solitary cells each one facing the tower. Every cell had two large windows, one at the rear to light up the cell from the outside, and one in the front, facing the tower. Thus, the prisoner stood out against the backlighting of the peripheral ring while the side walls prevented any visual contact with fellow prisoners. The cells acted as tiny theaters, putting each inmate on the stage, alone and individualized, but constantly visible from the central tower. Unlike the dungeon, which hid prisoners in a dark hole, the panopticon brought them out into the light.

Panopticism also called for a constantly visible but unverifiable gaze. The central tower, always clearly in sight from each cell, was designed so the occupant could never tell whether or not someone in the tower was gazing in. This was accomplished by installing venetian blinds on the windows of the tower and by arranging its interior into intersecting partitions and zigzag openings so that any small noise, movement, or ray of light in the tower seemed to indicate the watcher's presence. A petty clerk, a janitor, an inspector, a visitor, or even a tourist moving in the tower was enough to instill on the occupant's mind the feeling of being constantly watched. Under the power of an all-knowing, all-seeing gaze, the prisoner's anxiety rose, making him or her amenable to any normalizing prescriptions.[5]

The panopticon design had further advantages. Various treatments of correction could be administered to the prisoner and their effects readily observed. Criminologists could discreetly experiment with different punishments, work regimens, and drugs. Moreover, wardens and administrators could easily monitor the guards and watch correction workers with an eye to devising more efficient methods. The panopticon, in its ideal form, was a highly efficient and effective laboratory of power. It could be put into place in schools and military encampments and, importantly, in factories and business enterprises.

DISCIPLINE AND CONTROL AT ITT: THE GENEEN ERA[6]

The relevance of these disciplinary principles for management accounting and control systems can be illustrated by reference to International Telephone and Telegraph Company (ITT) during Harold Geneen's tenure as chief executive officer in the 1960s and 1970s. By applying management accounting principles, Geneen constituted ITT's managers as amenable, docile, obedient bodies and enabled him to establish ITT as a vast international conglomerate. The principles of discipline and control, it is important to realize, do not fully capture all the important factors germane to this period in ITT's history and its aftermath. Rather, they complement other perspectives, many of which have been covered in previous chapters.

Geneen: A Disciplined Disciplinarian

Geneen's childhood started him on the road to becoming an archetype of Foucault's disciplined subject. At five years of age he was put in a convent boarding school

> ...where the discipline was strict and the nuns loving. I can still picture in my mind Sister Joseph, who more than once whacked my outstretched hand in punishment, not for any infraction of discipline, but because I had misspelled a word. It instilled in me, I think, a serious appreciation of my responsibilities as a student; certainly I learned to do my homework (p. 56).

Each summer Geneen went to a boy's camp where he encountered the discipline of the camp counsellors. And at school he was content during holidays such as Thanksgiving and Christmas to sit by himself reading books. Although no more than a small child, Geneen was well on his way to becoming a model of the disciplined–disciplinary individual.

During his adolescence his parents enrolled him at Suffield Academy, a military-style college preparatory school based on the West Point model. The academy had a balanced mixture of boarding and day students as well as a good proportion of scholarship students who set high standards for the others. During his stay there, Geneen waited on tables and worked in a local bakery. Although the general atmosphere was democratic and egalitarian, "...within that academic freedom there was a discipline and a value system and a reverence for life that I absorbed, which would stand me in good stead all the days of my life" (p. 57).

Geneen continued in summer camps until the age of 15 when he took a job as an errand boy for a lithograph company on New York's west side. His duties entailed carrying heavy packages all over the city, often travelling by subway and frequently working until 9:00 at night:

> The sharpest memory of that summer was a crisis and furore over a lost piece of copy that I had delivered to the printing room. A major client was furious. I was called into the manager's office to explain what I had done with the papers involved. I explained and he agreed that I had left the documents in the proper place. He thanked me and sent me back to work. For many years, that man's polite and fair treatment of me remained in my mind as a lesson (p. 57).

Discipline, Geneen learned, gets recognized and rewarded. When he graduated from Suffield he got yet another lesson in discipline. His diploma was unsigned. However, the headmaster assured him that the moment his outstanding boarding and tuition bills were paid the diploma would happily be completed.

Instead of university, Geneen had to go to work. He took a job as a floor page at the New York Stock Exchange, but also attended evening classes, where he studied accounting. Six years later, he witnessed the great crash of October 1929 and watched in wonder as great fortunes disappeared in a few hours. But he was more impressed by the ways fortunes were being made by cunning short-sellers. After the crash, things took a turn for the worse. His salary was cut. He lost his $200 of life savings in a bank failure. And he had to live on day-old bread and taffy. A series of jobs followed: a book salesman, newspaper advertisement salesman, bookkeeper for a small investment firm, and paraprofessional for a large public accounting firm where

> My job was temporary and seasonal and I was promptly assigned to help in the audit of Floyd Odlum's Atlas Corporation of Journal Square in Jersey City, which had become notorious and highly profitable in buying up bankrupt and near-defunct mutual funds and companies. Atlas

> Corporation owned a grab bag of securities representing hotels, barge lines, frozen foods, utilities, as well as a wide variety of other assets. The job was conspicuous, most of all, for its overtime. We worked on that audit until ten o'clock at night five and sometimes six days a week, and in the final three weeks until three or four o'clock in the morning. Considering that I had to take the tube back to New York, sleep at home, and return to work at 9:00 $_{a.m.}$, I think I slept about two hours a night those last three weeks. When the audit certificate was signed, I requested an audience with Mr. Lenhart. "What's on your mind?" he asked. "Well, I haven't had but nine hours sleep all week and I can't help worrying about this job," I said, "I just want to know whether I am permanently on the staff or not, so I can go home and get a good night's sleep." I got the job and a good night's sleep (p. 64).

As before, discipline saw him through and got its just rewards.

Geneen revelled in the work of a public accountant. He got to see inside many companies. He liked his colleagues and coworkers. But best of all, he liked the training, "Public accounting taught me analytical approaches to business problems, objective reasoning, and the highest order of discipline in making factual presentations" (p. 65). He stayed six years and successfully took the certified public accountant (CPA) examinations. Yet professional life, as with stockbroking, somehow seemed empty. He missed the active side of being on the firing line of commercial and industrial life.

As a consequence, he took a series of jobs in industry, serving with American Can, Bell and Howell, Jones and Laughlin Steel, and Raytheon Electronics. While these were mostly in finance and controller positions, he was also involved in getting new plants and ventures up and running, and, in some cases, divesting existing ones. After being bypassed at Raytheon in favor of a relative of the owners, Geneen became chief executive officer (CEO) at ITT with the avowed aim of transforming it into a disciplinary organization. He did this, as we shall see, by following almost to the letter, the principles of discipline and control articulated above.

The Principle of Enclosure at ITT

The principle of enclosure and its counterpart, the principle of responsibility center accounting, is illustrated vividly in Geneen's story. Once installed as CEO, Geneen moved quickly to reorganize the company. He replaced the old functional–geographic structure with one featuring decentralized profit centers. Managers became fully responsible and accountable for financial performance. By 1977, ITT's line operations consisted of nearly 400 000 employees enclosed in 250 profit centers. Following the principle of enclosure, each space had its own manager; and each had his or her own space.

Having neatly partitioned the company into profit centers, Geneen made each responsibility center analyzable through what he called "the discipline of the numbers." For most people, he postulated, numbers are much more easy to read than words. They are unambiguous symbols which measure the tasks and operations of the organization and, most importantly, they inform upper management about what is going on:

> The difference between well-managed companies and not-so-well managed companies is the degree of attention they pay to numbers, the temperature chart of their business. How often are the numbers reported up the chain of command? How accurate are those numbers? How much variation is tolerated between budget forecasts and actual results? How deep does management dig for its answers? (Geneen 1984a, p. 80).

The financial control system provided Geneen with continuous, functional surveillance of each enclosed responsibility center. For Geneen, this was absolutely essential if ITT was to become a disciplined and productive company. Previously, ITT had been disciplined by technical, engineering, and electronics discursive practices. The new accounting regime of truth replaced it. It seems clear that Geneen, deeply influenced by his own educational experiences, his knowledge of control systems at General Motors, his attendance at courses at Harvard Business School, and his own accounting background, relied heavily on the principle of enclosure inherent in the management accounting axiom that organizations should be divided up into responsibility centers headed up by an accountable manager.

The Efficient Body Principle at ITT

Geneen's financial control system at ITT bears more than a little correspondence with the principles of disciplining time. In terms of timetabling, each profit center manager and staff divisional head submitted their annual budget and business plans in February for review and revision at both the local level and at headquarters. Then, in November and December, Geneen and other key headquarters officials met face-to-face with each manager and his or her own staff to discuss, review and settle on the plans and budgets. The finalized budget, now carved in stone, became the benchmark for performance in the ensuing year.

> ...each division manager and his own management staff had negotiated an agreement with headquarters on his budget and business plan for the following year. He had made a firm commitment to ITT. His subordinates down the line had made their commitments to him for the integral parts of his budget. He would hold them to their word as we would hold him to his commitment—or know the reason why (p. 92).

Geneen also required each manager to sketch out two-, three- and five-year profit plans of the profit center as well as anticipated capital expenditures. He did not, however, put a great deal of stock in long-range, qualitative strategic plans but instead focused on the current year: "The budgets and business plans for all our divisions, bound in loose-leaf books, occupied more than thirty feet of bookcase shelves. But those books were the bible we lived by" (p. 93).

The cornerstone of the financial control systems, however, was the monthly operating report. Each profit center manager submitted to headquarters, by the fifth working day of each month, reports which included pertinent and detailed information on sales, earnings, inventory, receivables, employment, marketing, competition, and R&D along with any current or anticipated problems. The managers also reported on the current economic and political situation in their territory. Divisional comptrollers also made a monthly financial report to the headquarters comptroller. Moreover, all headquarters staff division heads (engineering, accounting, marketing, R&D, etc.) sent Geneen a monthly report about the situation in their specialized area, as did the product line managers. Geneen, and his headquarters staff, personally scrutinized each and every report. He summed up his surveillance network this way, "Information flows up the chain and orders flow down. Everyone knows his or her own place and responsibilities in the hierarchy. Logic and order are supposed to reign supreme" (p. 85).

The profit center philosophy also served to train managers to act like individual entrepreneurs. Geneen selected each manager carefully to ensure that only those persons who fitted his predetermined mold got the job. He did not want geniuses who could not communicate with ordinary, hardworking people. Nor did he want people who got by on their good looks, smooth talk, or family connections. Instead, he looked for people who shared his enthusiasm for hard work. Intelligence, knowledgeability, and experience were necessary but not sufficient characteristics. Each manager also had to display an enthusiasm for laboring. Geneen's normalizing mold was clear for all to fit into.

More specifically, the information in the financial controls became the basis for the dressage-like training of ITT's managers. The on-site, monthly meeting with 150 European general managers and forty headquarters staff managers quickly became Geneen's training grounds.

> Soon after I came to ITT I saw the advantages of meeting face-to-face with our European directors, rather than trying to solve problems over the transatlantic telephone or telex systems. The look on a man's face, his tone of voice, his body language made a difference in the decisions I was making. We started out in Europe in small, smoke-filled hotel rooms, but as the company expanded and built its own European headquarters in Brussels, the monthly General Managers Meeting usually consisted of 120 to 150 managing directors. Every month I flew to Europe with about forty headquarters staff, and we sat down together and went over the monthly operating reports. The pertinent figures from the comptrollers' reports and the managing directors' reports were flashed on giant project screens in three corners of the room. Everyone on the headquarters staff had read every monthly report to be reviewed. We were informed. In going through the two large brown leather-bound loose-leaf books of reports, I made it a practice to jot down my queries in red ink, and turn down the corner of the page to mark any item I wished to query at the meeting.
>
> We sat around a large U-shaped table, covered in green felt, facing one another, and I asked questions based upon the notes I had made on their monthly operating reports. Why were the sales down? Was he sure of the reasons? Had he checked it out? What was he doing about it? What did he expect in the month or two ahead? Did he need help? How did he plan to meet or outdistance the competition? (pp. 96–97).

Geneen came armed with a series of "red-ink" queries. He exhorted the others to do likewise.

> Not only I but anyone else at the meeting could say anything, question anything, suggest anything that was pertinent. Each man had a microphone in front of him. With the figures on the screen, we could all see how each profit centre measured up to its budget commitments, its last year's performance and whatever, in sales, earnings, receivables, inventory, etc. (p. 96).

The financial control system and the mandatory monthly examinatory meetings provided the means for training ITT managers in the correct moves. The signals from Geneen and the financial control system automatically triggered the required proper behavior. ITT managers performed in unison as highly efficient bodies.

Sampson[7] captures the dressage image in describing ITT's takeover of Avis Rent-a-Car. Many people at Avis complained that they now spent more time dealing with ITT than they did renting cars. The new president (who replaced Townsend, the consummate artistic, antibureaucratic manager) "had the task of fitting the company into the hard shafts of the ITT harness" (p. 79). The metaphor of shafts and harnesses captures well the spirit of dressage, a key ingredient in Geneen's control system.

The Principle of Disciplined Minds at ITT

Geneen's account of his financial control system at ITT also typifies many aspects of the principle of disciplined minds. One of his first, but perhaps most critical, moves in reorganizing ITT was to restructure the comptroller's organization to establish his own accounting hierarchic surveillance. Under the old system, field comptrollers reported directly (solid line) to the field managers and only indirectly (dotted line) to headquarters. Geneen changed this to a solid-line reporting to headquarters *and* a solid (but weaker) line to the field general managers. Initially, this move met with stiff resistance. Line managers, fearful that their controllers would turn into home office spies, submitted formal protests. Yet as Geneen explains

> They wanted complete control over their domains and the absolute loyalty of their financial men. But I wanted an independent check on their activities by comptrollers who would be personally responsible for the figures they submitted to headquarters. It is all too easy to fudge or cover up the facts with numbers as well as with words. The temptation is always there. Even without conscious lying, different men honestly interpret events and situations differently. Company or division managers can exaggerate anticipated sales or underestimate costs or whatever and the men under them will go along because their jobs depend on it. I wanted the comptrollers to feel free of that pressure and be able to give the home office their honest opinion. If the division manager and his comptroller could not agree, we would settle it at a higher level after a full and open hearing (p. 90).

Geneen completed his system of hierarchical surveillance by setting up a cadre of technical staff and product managers in his office headquarters "panoptic tower." Technical personnel, experienced and proficient in all aspects of ITT's activities (such as telecommunications, electronics, consumer goods, engineering, accounting, marketing and personnel), were organized into specialized headquarters staff offices. These managers also reviewed and analyzed the monthly reports after which they were free to go to an ITT location without an invitation to investigate anything within their area of expertise. On the site, they asked questions, got answers, and reported their findings back to Geneen. Before reporting to Geneen, however, the staff had to tell the local manager involved, as well as their own boss, exactly what they were doing and what findings they came up with. This way the manager had a chance to correct (normalize) the situation. Geneen describes it as follows:

> These staff men out of headquarters cut through the structured rigidity of the formal organization, monitoring each of the subsidiaries. The accounting staff man monitored the profits, the engineering staff monitored the engineering department, and so on with marketing personnel, legal, etc. The staff people worked very closely with the men out in the field, and they made their reports and recommendations, and they were held equally responsible for whatever went well or poorly in the unit they were monitoring (p. 88–89).

This part of the surveillance hierarchy played out its normalizing, disciplinary gaze over the surface, lines, and fibers of the entire line organization.

Any manager or employee who had a suggestion for an improvement anywhere in any aspect of the company's operations, regardless of geographic location, was encouraged to write a signed report, have his or her superior initial it, and send it directly to Geneen. If Geneen had a question he talked directly to the person who wrote the report, "I really wanted to know what was going on in the company. I thought it was essential" (p. 88). He also appointed a dozen or so senior, product-line staff managers who

> ...roamed over complete product lines, representing competition. He was monitoring the competitive ability of an ITT subsidiary in the marketplace. The product line manager was paid to look cold turkey at an ITT company and its competition and raise questions as he saw fit ... They had in effect a licence to speculate on what could be done differently and better (pp. 89–90).

Product line managers did not, however, have any authority to give orders to line managers. Instead, they had to sell them their ideas. But they had to report any disagreements directly to Geneen who "quickly settled them."

Geneen held strong opinions about organizational hierarchies. Although he believed in the conventional, textbook version of organization—pyramid, layers of supervision, a labor force at the bottom supporting the pyramid, and a regular chain of command—he was convinced that such a structure could impede information flows. As a consequence, he demanded that all important information came directly to him and his staff. ITT utilized accounting as its disciplinary gaze; but it was a highly personalized one.

Geneen also initiated a system of ranking for each and every responsibility center in ITT. A typical example is the rating system he set up for the field comptrollership units.[8] In the late 1970s, ITT employed nearly 23 000 people in the comptrollership activities, including 325 corporate headquarters staff. Each field comptroller was examined and rated by an effectiveness score based on thirty identified areas of comptrollership including, for example, intercompany accounting, budgets, cost accounting, capital expenditures, payables, debt management and foreign exchange, the comptroller's monthly operating and financial review, and the comptroller's interface with both the unit general manager and the director of financial controls.

A questionnaire was developed for each comptrollership area with some thirty to sixty yes/no questions, depending on the area. Each unit controller answered nearly sixteen hundred self-evaluation questions in completing the examination while the divisional financial controller (DFC) answered over 150 questions. The self-evaluation items, based on a 0–5 point scale, were weighted 1 and the divisional controller's items carried a weight of 5. Thus, the final score was weighted about 70 percent on self-evaluation and 30 percent on DFC's ratings. The two ratings then were combined to get an overall score where a perfect evaluation equalled 100. The scores for each of the 250 field units became the basis for a color-coded ratings timetable (blue, green, yellow, and red) where blue (the highest rating) indicated satisfactory performance and another rating in two years, while red (the lowest rating) meant unacceptable performance and another rating in three months.

These ratings were displayed on a massive color-coded comptrollership grid. The grid listed each of the 250 comptroller field units on the vertical axis and each of the thirty areas of comptrollership on the horizontal one. As a reslt, Geneen and other top-level executives could see at a single glance how well any particular field controller was performing as well as get the picture for any specific function. Newly acquired units and units featuring a high situation complexity (unfavorable business environment, inadequate staff, degree of multiple operations, troublesome governments, or tricky foreign exchange transactions) frequently received poor ratings. Here, the measure of the unit comptroller's effectiveness was the time he or she took to remedy the situation.

So the comptroller's exact actions were detailed in minute fashion. Like the color-coded epaulettes and sackcloth uniforms for the "shameful" French cadets, and Robert Owen's four-colored block hanging over each worker, the comptrollership grid provided an exhaustive, automatic examining, ranking, and sanctioning of the comptrollers. The result was obedient, disciplined, and willing comptrollers.

Such disciplinary practices prevailed throughout the company. Each of the line and staff managers received a similar dose of surveillance, discipline, and sanctions. Every operating and staff maneuver included in their monthly report a brief description of any significant problems they were facing, a clear statement of the action recommended, the reasoning and numbers used to analyze the problem, and a brief opinion statement regarding the resolution of the problem. These problems remained "red-flagged" until they were solved. They also became part of the examination at the monthly face-to-face meetings.

> If the man knew (of a problem) and was reluctant to put the facts in the open, my questions would force him to admit what he was trying to hide. If the man did not know or understand his own lines, which was often true because he had not written them, then my questions, doubly embarrassing, would force him to do his homework (p. 100).

The stigma of the red flag induced managers to get up to par.

While Geneen's disciplinary grid provided for hierarchical surveillance, the numbers were the most critical part of it. In fact, "the numbers make you free" became his famous credo. Managers were required in formulating their budgets to put down on the paper the "...whole gamut of costs of the product, supplies, production, labor, plants, marketing, sales, distribution—and also anticipated income from sales based upon market share, back orders, and what have you" (p. 191). These numbers, however, were not merely to be pulled out of the air. Nor were they to be based on hopes and whims. Rather, they were to be carefully gathered by the line managers and based on the best possible available figures and facts.

As the budget year unfolded, a similar set of numbers flowed into headquarters each month, or weekly in the case of red-flagged units. Geneen scrutinized every piece of information searching for anything that might be off plan. He believed fervently that unshakeable facts, along with hardheaded, hard-hitting, cross-examinations, were essential in order to instill the requisite degree of discipline into the organization:

> It is discipline that is built into the credo management they must manage. Part of that discipline is recognizing that the first answer you receive is not necessarily the best one. That is why I put so much emphasis upon probing for unshakable facts (p. 123).

Monthly reports and the face-to-face meetings provided for regular and compulsory examinations and ratings. Geneen acknowledges the importance of examinations in his education upon his systems of management.

> Upon reflection, I can see that it probably goes back to my days as a student at Suffield Academy. I worked conscientiously because I liked to get good marks and it bothered me if I got bad ones...So at ITT I instinctively sought to instal something of the same in the company (p. 134).

Geneen preferred to examine the managers of the operating units himself. The notorious monthly meeting in Brussels of 150 of ITT's top executives served as Geneen's examination.

> The invited 150 officers hear introductory remarks by Geneen, an operations report from Dunleavy, and reports on such matters as inventory levels and receivables. Then the action begins. The heads of the bigger companies and the line group vice-presidents responsible for the others track the performance of their operations against their budgets. Anybody attending can ask questions and make suggestions.
>
> Some former employees complain that the big meetings reek of Kafkaesque courts, of volleys of verbal invective fired at underachievers. "Many of us have frankly left the organization for having been put upon publicly," says a former European unit manager. Geneen, by contrast, views the meetings as open, business-like forums at which participants try to help one another.[9]

This examinatory practice, featuring an alphanumeric inquisitional process of reading, examining and rewriting each manager was seen by the managers not so much as a help session, but a hell session.

Geneen also believed ardently in the exhaustive use of time and the perpetual struggle for improvement. He tells of numberless meetings which started early in the morning and continued until 10:00 P.M. or midnight, or even in some cases, all night. He writes extensively about his own working habits when he constantly travelled back and forth to Europe, took home huge cases of office work each weekend, and normally worked twelve- to sixteen-hour days for seven days a week. He did so, he recalls, not simply to get his own work done competently, but more importantly to establish standards of hard work and commitment for the entire organization, to set the norm.

> I worked as long and hard as any man at ITT and they knew it...I did set an example, an honest example, which travelled down the ranks of management and, to an extent, established a standard of performance for the whole company...If I could do it, so could the next man (p. 136).

Along with gruelling standards for working hours, Geneen constantly harangued everyone for more productivity. He set what he thought were challenging and competitive goals of steady, stable increases in earnings of 10 to 15 percent each year. Any chance he got he talked about growth. His expectations were for everyone to stretch beyond the ordinary. He never let up and he did not hesitate to reward those who followed his dictums.

> I wanted to create that kind of invigorating, challenging, creative atmosphere at ITT. I wanted to get the people there to reach for goals that they might think were beyond them. I wanted them to accomplish more than they thought possible. And I wanted them to do it not only for the company and their careers but also for the fun of it. I wanted them to enjoy the process of tackling a difficult piece of business, solving it, and going onto bigger, better, and tougher challenges. I wanted them to do this not for self-aggrandizement, but as part of a greater team effort, in which each player realized his own contribution to the team, knew that he was needed and appreciated, and took pride and self-satisfaction from playing a winning game (p. 135).

Geneen's ultimate aim was self-normalization for every manager.

Geneen also believed that raising standards beyond what most managers thought was possible was one of his major personal contributions to ITT. He was convinced that the levels of achievement he insisted upon penetrated the entire company, "We stretched and stretched, we reached and reached, we managed, and we achieved our goals. And we felt good about it" (p. 129). Standards for long working hours and constantly increasing goals put into effect a system that exemplified the principle of the exhaustive use of time.

Geneen was also a master of dressage. Each manager had to at least posture as a sophisticated user of accounting information, and more importantly, a believer in managing by the numbers. They were trained to meet face-to-face, to look into each other's eyes, to listen carefully to the tone of other's voices, and to pay attention to their body language. Telephone or telex would not do. You had to see the other person's reactions.

Geneen's management accounting and control system also mirrored the ideals of the panopticon. From the central headquarters office, the accounting system cast its constant normalizing gaze into every responsibility center throughout the organization. At a glance, it could monitor any part of ITT. It effected a continual flow of both formal and informal information into Geneen's office. Individual managers, however, never knew at any particular moment whether or not Geneen was gazing directly at them through the window of the numbers, or if not Geneen, then some other member of the anonymous headquarters staff. Within this accounting and control panopticon, the line organization anxiously conformed to the prescribed normalization of the numbers.

But Geneen's position, "the numbers will make you free," is the antithesis of Foucault's. For Foucault, numbers would be a critical part of the prison which incarcerates managers in their responsibility centers. "For the disciplined, as for the believer, no detail is unimportant, but not so much for the meaning it conceals within it as for the hold it provides for the power that wishes to seize it (p. 140).

Reflections

ITT's control systems under Geneen's reign provides a striking illustration of how management accounting practices can resemble many of the principles of surveillance, discipline, and punishment. This case study reveals how accounting controls, reinforced by direct personalized celebrations of accountability, are akin to a panoptic gaze. Their effects upon managers were marked and, in some instances, disturbing. They produced a feeling of perpetual observation not dissimilar to the victim of a panopticon. In Sampson (1974) an ITT manager recounts, "You'd realize that being an ITT manager is like living in a room with closed circuit television and a bug up your ass" (p. 126).

There is little doubt that Geneen's methods intensified managerial work. While the financial rewards for executives who complied could be high, they carried heavy personal costs, as was reflected in high rates of managerial turnover (Menzies 1980) as well as fear by those who remained.

> For a newly joined manager... the ordeal can be terrifying: there are stories of one man fainting as he walked in, and another rushing out to get blind drunk for two days. For the hardened ITT man it is not more than a routine test of sangfroid; "You have to be prepared," said one of them, "to have your balls screwed off in public and then joke afterwards as if nothing had happened (p. 96).

Geneen advanced and refined the principles of disciplinary power through the aggregative abstractions of management accounting controls. Supplemented by interrogation, control could be extended over time and space, and, most importantly, it came to be self-policed by creating within the minds of managers a conception of what being an effective manager entailed.

The ITT case indicates how the images and metaphors of Foucault are analogous to controls in a large business organization. Moreover, ITT and Geneen became revered for awhile as representing a model of best practice. Geneen succeeded in creating a public image as a great manager. When he left Raytheon in 1959 its market value decreased $18 million overnight. ITT was dubbed Geneen University, due to its export of managers to lead many large American corporations. By the mid-1970s Geneen was perceived by many as one of the great managers of his time, the Michelangelo of Management. Geneen achieved this in no small way by applying accounting principles learned in the classroom to the governance of corporate conglomerates.

The ITT history, however, goes further than just providing an illustration of the way management accounting and control systems reflect the principles, discipline, and control.

It also reveals how Geneen's achievements were of much greater significance, not only for ITT, but for industrial development and capitalism at large.

The 1960s were the years of the conglomerate. Large, single-industry, nationally based firms realized that fast growth was impeded by government anticombines actions and the slow and steady growth in their generic industry. Early movers, particularly those with high price-to-earnings ratios, could build spectacular growth records in profits and earnings per share by acquiring low price earnings companies in exchange for cash and shares of the acquiring company. It is not surprising, then, given Geneen's accounting wisdom and skill, his experience with sorting out the Atlas corporation, his background in getting acquired companies up and running and divesting segments with unlikely prospects, and his abilities to arrange financing, that ITT was one of the most active conglomeratizers during this decade, not only in the US, but internationally.

The important upshot of these events, and this is perhaps Geneen's most remarkable but unnoticed achievement, is that he invented (or at least stumbled upon) and put into place a new social institution, the widely diversified global meganational firm. As one ITT watcher astutely observed, "For in swallowing a succession of companies, and imposing his ingenious system of control, he helped to create a new kind of industrial animal, a conglomerate that was also multinational" (Sampson, p. 73). Geneen demonstrated that management accounting systems could discipline and control giant corporations of immense dimension and power.

While today these controls are enhanced by advances in electronics and computers, and are more elaborate, elegant, and subtle, the basics were assembled by the Geneen Machine. A mere two decades later this new form of corporate organization has resulted in a "brave new world of corporate might" in which many of these corporations are larger and more influential than most of the countries in which they operate.[10] They all but rule the world.

Yet it would be wrong to conclude that Geneen's accounting regime of truth produced only docile, obedient managers. According to Sampson, many rebel managers "kicked against the traces of discipline" and there were many instances of outright "revolt against the tyranny of facts" (p. 128). As one successful executive replied when asked why he left ITT, "I decided to join the human race" (p. 129). And European managers opposed Geneen's disciplinary practices by using his devotion to timetables to delay discussion of key topics and embarrassing results until shortly before his return flights. Undoubtedly, many managers resorted to covert practices such as manipulating spending, adopting profit-smoothing techniques, building slack into budgets, and resorting to creative bookkeeping techniques. While clearly a tyranny of the numbers existed, there was also considerable resistance.

Other managers resisted more overtly. For example, Robert Townsend, president of Avis Rent-a-Car before and after ITT took it over, publicly flouted Geneen and his numbers machine. Townsend described in his best-selling book how his success was due to business methods the exact opposite of Geneen's. Reported in Sampson, Townsend calls Geneen's control system a joke that "consumes ten pounds of energy to produce each ounce of misunderstanding...they are just playing a numbers game and couldn't care less if they make zombies out of people" (p. 78). Other Avis managers complained that fitting into the "hard shafts of the ITT harness" meant spending more time dealing with headquarters than renting cars.

RETHINKING MANAGEMENT ACCOUNTING

In many respects the current management accounting textbooks also contain overtones of disciplinary surveillance, punishment, and normalization. While they may be idealized caricatures of organizational practices, as case study research has increasingly revealed, they are not without empirical foundation and they help shape reality, as the ITT case confirms. Management accounting is not a carbon copy or an exact correspondence in form and conduct, but it does reflect some of the central ideas of Foucault's principles of surveillance and discipline. It may be valuable to rethink management accounting and control in Foucault's terms.

In the first instance, the principle of enclosure is very much evident in the management accounting concept of responsibility center accounting. The organization, the general enclosure, is partitioned into responsibility centers (divisions, strategic business units, investment centers, profit centers, responsibility centers, discretionary cost centers, engineered cost centers, departments, and workstations) each of which is defined in terms of the useful function to be performed therein and in terms of its serial relationship with the other enclosures. The manager and employees in each center are defined by the place they occupy.

Thus, the cellular nature of the organization is captured and reinforced by management accounting reports and practices which idealize each partition in terms of its use, serial relationships, and hierarchical ranking. As one popular textbook states, "A management control system is a *total system*. The plans developed encompass the whole organization, and one important aspect is that plans for each part of the organization must be so coordinated with one another that the various parts are in balance" (Anthony, Dearden, and Govindarajan 1992, p. 12).

The principle of the efficient body and the disciplining of time are also manifested in management accounting and control systems principles and practices. Industrial engineers use the techniques of management science (stopwatches and analysis of body movements) to determine the normal time and precise bodily movements required to perform jobs in factories and offices. Accountants combine this information with data collected on labor time cards and tickets and payroll records to develop standard cost systems. Scientific management and standard costing establish a rhythm and regularity to the work, blending the worker's body movements with the machine, tool, or filing system into a "man-machine." These regimes of truth produce docile and obedient production workers and clerks. The strict timetabling, systematic procedures, and precise rhythm of the management control process bears more than a little resemblance to *dressage*.

> In the management control process, decisions are made according to procedures and timetables that are repeated year after year. The first step in sequence is formulating a strategic plan or program; next, one year of this program (the "budget year") is translated into a budget; next, operations take place, usually more or less guided by this budget and by prescribed policies and procedures; and finally, actual results are compared with this budget and evaluated, and corrective action is taken if necessary. These steps are labelled programming, budget preparation, execution, and evaluation (ibid., p. 12).

Most large corporations also follow precisely timetabled reporting procedures. It is not unusual in these organizations that on the same day of the year, the sundry managers spread around the globe enact the budget formulation ritual for budget approvals, submission of results, and sending in reports regarding variances and outlooks for the rest of the year.

And the practice of rolling budgets, whereby each month "continuous budgets compel managers to think specifically about the forthcoming 12 months" (Horngren et al. 1993, p. 245). A silent signal from the calendar triggers prompt obedience from managers around the entire organization to perform the budgetary dressage.

Management accounting and control practices also mirror in the abstract the principle of correct comportment. When managers present their budget plans, upper-level executives grill them closely about the assumptions, trends, and calculations in the budget targets. Managers also undergo a rigorous examination when presenting their strategic plans. And they are examined again by upper executive and financial officers when they report their monthly results. These examinatory practices work to normalize managers according to the mold of hardheaded, results-oriented, entrepreneurial executives who produce financial results for the organization.

Management control systems also function as an important apparatus for upper-level executives to measure, describe, judge, compare and rank each responsibility center manager. They are an important apparatus for diagnosing managers and classifying them as "normal" or in need of closer, more detailed examination (as in the case of Geneen's examinatory control system). Furthermore, each manager leaves behind a perpetual accounting track record and archive which accounts for his or her working life. Managers are constantly examined, sanctioned, and normalized by means of management accounting and control systems.

Finally, management accounting and control systems provide for a panoptic, all-seeing, all-knowing, constant gaze over each and every manager. According to one widely used textbook

> ... the management control system pervades an organization and consists of four parts ... an observation device that detects, observes and measures ... an assessing device that evaluates the performance of an activity or organization, usually relative to some standard or expectation ... a behavior modification device for altering or changing performance ... [and] a means of transmitting information within and among the three parts (Anthony, Dearden, and Bedford 1989, p. 7).

Management control systems, as with the panopticon, put into play surveillance, evaluation, normalization, and constant communication throughout the entire organization. "The accounting eye is indeed a significant and omnipresent one" (Hopwood 1987, p. 223).

GOAL INCONGRUENCE AS RESISTANCE

The foregoing picture of the carceral nature of management accounting and control systems assuredly is not a cause for celebration. There seems to be little to approve of let alone cheer about. Yet, paradoxically, this disconcerting but compelling reversal of the conventional, scorekeeping, attention-directing, and problem-solving orientation opens up some space for a fresh look at the goal incongruence phenomenon, which has troubled management accounting practitioners and academics for decades, from a postmodernistic point of view.[11]

A postmodernistic perspective, in the first instance, sees an organization's sociocultural landscape as an open field on which it is possible for managers and employees to act and comport themselves. However, sundry discursive formulations, including the management accounting formulation, crisscross and penetrate this field in an attempt to structure and organize it. It is a matter of the actions of some individuals (the experts of the various discursive formations) acting upon the actions of the other individuals (who may well be experts of other formations).

It is important to recognize that this depiction differs radically from the conventional picture. The conventional picture relies on a biological metaphor and portrays the organization as an organism whose various parts work together so the entity can adapt to its constantly changing environment in order to survive, grow, and prosper.[12] In contrast, a postmodernistic portrayal relies on the metaphor of the Roman coliseum and depicts the organization as a combative amphitheatre in which various competing power–knowledge regimes come at managers and employees alike from all sides. No one regime ever achieves a final victory over the other. So the manager's field of action is perpetually agonistic.

If we pause for thought, this should ring true. The individual manager is literally bombarded with mini power–knowledge regimes of truth regarding "normal" actions and comportment. Planning departments construct the manager as a general at war, cunningly deploying resources to defeat competitors. Financial offices construct the manager as an economic forecaster of long range balance sheets and profit statements. Policy and standard operating procedure manuals construct the manager as an obedient rule-following automaton. Personnel staffs construct the manager as a democratic, employee-centered leader who delegates decisions to subordinates. MBO trainers construct the manager as someone who sets clear goals and gets results, almost regardless of the means. MIS departments construct the manager as one piece of a rational expert decision system. And today, the manager is constantly watched, examined, and corrected according to the discursive formation of the politically correct police. There seems to be nothing individual left for the individual manager to manage.

A postmodernistic view, however, does not stop there. It contends that wherever such power–knowledge regimes operate, we will also find resistance. Power and resistance are inseparable. Moreover, power is not something some parties or institutions hold an inventory to be used at their discretion. The very basic and immutable condition of power is its perpetual relationship of provocation and the struggle for freedom it incites in those it attempts to control. Nor is it a matter of any abatement of power on one side as a consequence of either consent or violence on the other side. Power does not exist where insubordination of some kind is absent. Wherever power is exercised there will always be resistance (Foucault 1980, p. 95).

Two characteristics of resistance from a postmodernistic perspective stand out. First, resistance occurs locally. Individuals in a community struggle against local exercises of power such as a school busing regulation, a plant closedown, a reduction of television time in a prison, or a proposed landfill site in the neighborhood. They do not aim their opposition at some central power base such as the area school board, the multinational headquarters in New York, the government department of correction services, or the city hall. They take a stand at the local community hall rather than marching to the nation's capital. Resistance, except in situations of great crisis, occurs in the immediate vicinity of the exercise of power.

The second major aspect of resistance is that it is transversal. Transversal refers to the idea that the same kind of struggles take place across institutions and organizations. For instance, we find students protesting against tuition increases at universities around the world. Similarly, environmental groups rally in dissent at local industrial pollution sites in all parts of the world. And, similar confrontations by women, gays, and minority groups occur across organizations and nations. The nature and concerns of such struggles are similar wherever they break out.

Foucault's own words on the resistance–power relationship are worth quoting at length.

> Where there is power, there is resistance, and yet, or rather consequently, this resistance is never in a position of exteriority in relation to power. Should it be said that one is always "inside" power, there is no "escaping" it ... the strictly relational character of power relationships ... depends on a multiplicity of points of resistance: these play the role of adversary, target, support, or handle in power relations. These points of resistance are present everywhere in the power network. Hence there is no single locus of great Refusal, no soul of revolt, source of all rebellions, or pure law of the revolutionary. Instead there is a plurality of resistances, each of them a special case: resistances that are possible, necessary, improbable; others that are spontaneous, savage, solitary, concerted, rampant, or violent; still others are quick to compromise, interested, or sacrificial; by definition, they can only exist in the strategic field of power relations. ... Are there no great radical ruptures, massive binary divisions, then? Occasionally, yes. But more often one is dealing with mobile and transitory points of resistance, producing cleavages in a society that shift about, fracturing unities and effecting regroupings, furrowing across individuals themselves, cutting them up and remolding them, marking off irreducible regions in them, in their bodies and minds. Just as the network of power relations ends by forming a dense web that passes through apparatuses and institutions, without being exactly localized in them, so too the swarm of points of resistance traverses social stratifications and individual unities (ibid, pp. 95–96).

These ideas gesture towards an alternative conception of the goal incongruence phenomenon. A postmodernistic perspective would contend that managers' goal incongruent behavior is a form of resistance to the power–knowledge effects of management accounting and control systems. Moreover, wherever these systems are used there will always be resistance. Building slack into budgets, bargaining for readily achievable targets, maneuvering accounting records, manipulating reported profits, and even accepting financial controls in a docile manner are ways of resisting these systems. And, the greater the force of power–knowledge exercised, as is the case for the budget-constrained accounting evaluation style, the more will managers be incited to resist. Yet, and this is the point to underscore, goal incongruent behavior is not necessarily a sinister part of the management accounting woodwork; it just goes with the territory.

In fact, there is a lot of evidence around to support this conception. For example, one extensive and pioneering study of budgetary slack in large, multidivisional firms concluded that goal incongruent behavior is endemic in these organizations, "... management can and does create slack to achieve attainable budgets and to secure resources ... This behavior seems universal among managers; it occurs in profitable and unprofitable companies, whether stable or growing." The study also found that top management were aware of such actions and, at least implicitly, condoned it. Subsequent research over the past three decades has consistently supported this picture.[13]

The ITT case also bears witness to this fact. According to Sampson (1974), while clearly a tyranny of the numbers existed, there was also considerable resistance by rebels. Such pockets of resistance indicate that opposition to the accounting disciplinary drive is not only possible but also real. For others that stayed, however, the price in human terms was high: the sense of strain is very noticeable in the Park Avenue skyscraper, not only the tension of worry and work but no satisfaction of personal achievement. It shows itself in heavy drinking, broken families, and a dazed, bleary look that seems characteristic of ITT staff (p. 127).

This postmodernistic depiction of managers' goal incongruent actions differs radically from the conventional textbook picture. It suggests that a basic and immutable condition of exercising discipline and control by means of management accounting and control is the struggle it incites in the managers it attempts to control. Moreover, managers will resist locally rather than confronting head office. And the kind of goal incongruent actions taken will be similar across organizations around the globe. So, regardless of any amount of fine tuning of the accounting measurements and regardless of how many doses of ethics training managers receive, goal incongruent behavior on the part of managers will continue to be part of the woodwork. We need not see it as sinister, rather as the inevitable and necessary way for managers to cope.

NOTES

1. This chapter is based largely on Hopper and Macintosh (1993).
2. Foucault's obituary in *Le Monde* (June 1984) described him as the most important intellectual achievement of the twentieth century.
3. See Loft (1986), Hoskin and Macve (1986, 1988), Hopwood (1987), and Miller and O'Leary (1987) for seminal accounting studies, drawing on Foucault.
4. Unless otherwise stated Foucault references are to Foucault (1979).
5. Foucault did not believe that such practices worked to rehabilitate the hardened criminal, only that the panopticon designers believed it would.
6. Unless otherwise stated Geneen references are to Geneen (1984b).
7. Unless otherwise stated Sampson references are to Sampson (1974).
8. See Alleman (1985) for details of the controllership grid.
9. *Business Week*, November 3, 1973.
10. See Lowe's (1992) expose of the world's largest meganationals.
11. This section is informed by Foucault's afterword in Dreyfus and Rabinow (1983) and Smart (1985), particularly "The Question of Resistance," pp. 132–136.
12. See Daft (1992) for an outline of the organic perspective.
13. Schiff and Lewin (1968), p. 62. Since Schiff and Lewin (1968, 1970) published their study, their findings have been consistently replicated by other researchers. See Onsi (1973), Otley (1978), Merchant (1985), Merchant and Manzoni (1989), and Dunk (1990, 1993).

14
Employee Accounting Systems

This penultimate chapter addresses a recent proposal to empower employees by means of accounting systems. It is not surprising that this idea has found its way into the accounting domain since empowerment is the most recent management initiative aimed at getting more effort, commitment, involvement, and output from employees throughout the organization.[1] A chief proponent of employee empowerment through accounting is Thomas Johnson who, in his important book, *Relevance Regained: From Top Down Control to Bottom-Up Empowerment,* parts company with relevance lost proponents such as those who believe the answer lies in more accurate cost allocation and a more strategic perspective for management accounting systems. Against this, Johnson argues that the solution must come from a real human relations initiative whereby lower-level employees are given detailed accounting information which they will use, in combination with their deep understanding of production processes and customer concerns, to revitalize their organization.

Remote Control Management

Johnson argues persuasively that in the past couple of decades top executives in large, diversified US firms have adopted management by the numbers as their main means of controlling managers and employees throughout the organization. He points out that a new breed of accounting, finance and law-oriented top executives have appeared in the past couple of decades who run their companies by "remote-control management." Johnson describes the problem this way:

> Attention of management at all levels focused on costs, profit, ROI, leverage, and other financial results. Not shop floor organization and satisfaction of customer wants. Managers began to manipulate processes to achieve accounting results, instead of monitoring well-run processes by occasionally checking accounting results. Process management—intuitively understood by most managers before the 1950s—was quickly replaced by "managing by the numbers". Worse, top managers increasingly identified and defined "process" in terms of whatever it took to achieve desired accounting results. And that view of things was telegraphed throughout every level of almost every business in the nation (Johnson 1992, p. 26).

These executives are not at all knowledgeable in either the production processes or the marketing techniques required to meet customers wants. Even more disturbing, they are out of touch with the values and needs of lower-level employees who actually produce and market the products.

Remote control management, Johnson observes, impairs performance rather than enhancing it. With its prime focus on return investment yardsticks it, "misleads managers into chasing false imperatives...always sustain output at a level to cover all costs, and always persuade customers to buy output at prices high enough to earn the market's required rate of return" (ibid., p. 10). Remote control by the numbers merely encourages lower-level managers "to *manipulate* (or tamper with) processes in order to achieve accounting cost and revenue targets dictated by 'top down' command and control information" (ibid., p. 10). Such actions are the opposite of the responsiveness and flexibility needed in today's global markets. It is, for Johnson, like putting the cart before the horse.

An Earnest Proposal

Two major initiatives are proposed to remedy this egregious situation. First, the key performance goals of today's organizations must be overturned. Jobs, a stable society, and profits must replace the market price of stock, dividends, and steady growth of return to shareholders. Second, lower-level employees must be empowered by means of accounting information, "Empowerment management means simply giving people 'bottom-up' problem-solving information and asking them to continuously improve the output of processes" (ibid., p. 10). This means overturning the idea that only managers know how to use information to improve products and processes and replacing it with the idea that "*everyone* in a company understands how to translate information into competitive actions. To wait for instructions from above is inimical to flexibility" (ibid., p. 11).

Empowerment will ensue, the argument goes, by giving employees the ownership of accounting information. This will enable them to respond faster to customers, to increase process flexibility, to reduce production lead time, and to improve morale. But most importantly of all, it will let employees develop a new commitment to regaining global excellence for US firms. Empowerment can be the dominant initiative to overturn remote-control management by the numbers and thus regain relevance for managerial accounting. The irony of this proposal should not be overlooked. Recall that the labor process perspective reminded us that during early capitalism, employees were dispossessed of the financial knowledge of the production process. Yet, today, experts are calling for measures to give it back to them.

Nevertheless, this earnest proposal has a great deal of appeal today. It comes in the wake of Clinton's sweep of the 1992 US presidential election, Ross Perot's urgings for citizen empowerment of the political processes and institutions, the steady loss in the 1990s of manufacturing jobs in the US, Canada, and Europe to developing nations, and the public's growing concern about the industrial world's degradation of the natural environment. The proposal that employees' jobs and social stability should rank ahead of stock market winnings for shareholders plays well into the hands of resurgent populist politics.

Pseudo Versus Real Empowerment

Yet to put into effect *real* empowerment, as opposed to only *pseudo* empowerment, means putting employees on an equal power footing with managers and shareholders. This would more than likely require not only the reversal of organization goals that Johnson argues for, but also some kind of overturning of the current power structures of today's organizations. This can only be accomplished, however, by privileging employees and establishing a social order in which employees are legally on at least a level playing field with managers and shareholders.[2] Until this issue is resolved, Johnson's proposal would likely be perceived by employees as only *pseudo* empowerment.

As Chris Argyris warned half a century ago in the case of introducing real budgetary participation into the firm, pseudoparticipation is

> ... participation which looks like, but is not, real participation. True participation means that the people can be observed to be spontaneous and free in their discussion. Participation, in the real sense of the word, also involves a group decision which leads the group to accept or reject something new. Of course, organizations need to have their supervisors accept the new goals, not reject them, but if the supervisors do not really accept the new changes, but say they do, then trouble will exist. Such an acceptance is characterized by the necessity of the person who induced the change to be always on the "look out", to request signatures of the "acceptors" so that they cannot deny they "accepted", should they later complain, and to constantly apply pressure (through informal talks, meetings, and "educational discussions of accounting") upon the "acceptors". . . In other words, if top management executives are going to use participation, then it should be used in the real sense of the word. Any dilution of the real stuff "will taste funny" and people will not like it (Argyris 1952, p. 28).

Similarly, if top executives are going to truly empower employees, it must really be "new and better wine." Otherwise it will taste bitter and employees will not swallow it.

Empowerment, when looked at closely, however, seems similar to other nostrums in a long list of similar remedies that have come and gone over the past sixty years (e.g., participative management, democratic leadership, theory Y, sensitivity training, Likert's System 4, organizational development, team building, collaborative management, human asset accounting, cultures of excellence, and, more recently, quality control circles, just in time, and total quality management.[3] Each of these initiatives, as with empowerment, holds that the prevailing mode of bureaucratic, depersonalized, and mechanistic administrative practices (including power relations of fear and coercion) must be replaced with organizational arrangements featuring "a humanistic existential orientation and a new concept of power, based on collaboration and reason" (Bennis 1966, p. 188). Yet the employee empowerment initiative somehow tastes a little bit like a wolf in sheep's clothing, the manager pretending to be just another employee.

The *pseudo* nature of managerialist-based empowerment schemes can be illustrated by unpacking a recent article that makes a case for federalist empowerment management (Handy 1992). The article argues that today's massive multinational firms must abandon (restructure) *centralist* management structures for *federalist* governance. Under federalism, top executives give power away even though, curiously, the article also argues, following Chester Barnard, that power actually "lies at the lowest point in organizations and it cannot be taken away."[4] However, the center, the argument goes, retains the ultimate power even though it should be kept small, since it can now rely on the possibilities of today's awesome information technology to keep track of the operating units.

This curious contradiction is repeated later in the example of British Petroleum. The company "went Federalist" in 1990 when it devolved authority and responsibility to its separate businesses but had to decide which powers the center would retain. "The center came up with a list of 22 'reserve powers' but, after discussions with the separate businesses, these were pruned down to the 10 most essential to the future direction of the company. In a federal system, the center governs only with the consent of the governed" (Handy 1992, p. 64). At British Petroleum, however, the center, clearly retained powers that were *not* given to them by the governed. So where is the federalism whereby the people (the employees) elect the rulers?

So all of this seems like merely *pseudo* federalism. The operating units will still be accountable to central executives who will judge them "against their type two accountability." Did they seize every opportunity? Did they make all the possible improvements? Further, the center will intervene if an operating unit makes a decision that "would substantially damage the organization" (ibid., p. 65). Moreover:

> Autonomy means managing empty spaces. Subsidiarity and signatures both imply a lot of individual discretion. Yet unbounded discretion can be frightening for the individual and dangerous for the organization. Groups and individuals therefore live within two concentric circles of responsibility. The inner circle contains everything they have to do or fail—their baseline. The larger circle marks the limits of their authority, where their writ ends. In between is their area of discretion, the space in which they have both the freedom and the responsibility to initiate action. This space exists for them to fill (ibid., p. 70).

In spite of this accountability to the center, the real possibility of central intervention, and the limiting of the space of lower level's freedom, Handy curiously concludes that federalist governance is ultimately democratic.

The argument that corporations should be like democratic federalist nations and govern themselves in the same manner is readily demystified. A federalist nation, such as the US, is founded on the fundamental notion that the president and the legislature are democratically elected (one person, one vote) and that all citizens are free, equal and entitled to the basic rights inscribed in the Declaration of Independence and the US Constitution. Abraham Lincoln captured the fundamental premise of a federalist state of affairs in his famous and authoritative Gettysburg Address that has come to be admired by peoples around the globe.

> Fourscore and seven years ago our fathers brought forth on this continent, a new nation, conceived in Liberty, and dedicated to the proposition that all men are created equal. Now we are engaged in a great civil war, testing whether that nation or any nation so conceived and so dedicated can long endure. We are met on a great battlefield of that war. We have come to dedicate a portion of that field, as a final resting place for those who here gave their lives that that nation might live. It is altogether fitting and proper that we should do this. But, in a larger sense, we cannot dedicate—we cannot consecrate—we cannot hallow—this ground. The brave men, living and dead, who struggled here, have consecrated it far above our poor power to add or detract. The world will little note nor long remember what we say here, but it can never forget what they did here. It is for us, the living, rather to be dedicated here to the unfinished work which they who fought here have thus far so nobly advanced. It is rather for us to be here dedicated to the great task remaining before us—that from these honored dead we take increased devotion to that cause for which they gave the last full measure of devotion; that we here highly resolve that these dead shall not have died in vain; that this nation, under God, shall have a new birth of freedom; and that government of the people, by the people, for the people, shall not perish from the earth (Lincoln's Address at Gettysburg, November 19, 1863).

The parallel seems clear. Until corporations are truly democratic in Lincoln's terms and permit the employees (who actually do the work, produce the goods, and make the corporation possible) to freely elect their managers and top executives (just as citizens in democratic federalist nations truly elect the president and legislature), initiatives such as empowerment and federalist governance will remain as *pseudo* empowerment and *mock* federalism. Federalist governance is merely managerialism in new robes. Attempts to empower employees by means of accounting and information systems would seem to be equally ill fated. Employees will quickly recognize the old fist of managerialism under the new velvet glove of empowerment by accounting information.

Furthermore, a recent historical study of the managerialist ideology puts these empowerment themes into a more sobering and broader historical perspective (Barley and Kunda 1992). The analysis indicates that managerial discourse in the US since the 1870s has been elaborated in long waves that have alternated between a systems–rationalism control rhetoric (a managerialist-based concern for imposing scientific discipline on both the physical production equipment and the work force) and a normative human relations control rhetoric (a high concern for the employees' well being and welfare). The study provides convincing evidence that currently the US is in a normative human relations wave in which the ideology of mergers and acquisitions

> ...may actually have fuelled interest in the rhetoric of corporate culture. It is certainly the case that even as they implemented their hyperrational vision of industry, proponents of this point of view also sought justification and legitimation in the rhetoric of values, motivation, and morality characteristic of the currently surging normative ideology (ibid., p. 395).

The study concludes that the current emphasis on normative control will be "followed by a resurgence of rationalism...in conjunction with a long-term expansion in the economy and the rise of a new paradigm of automation" (ibid., p. 395). From this historical perspective, both the empowerment and federalist proposals come across as merely part of yet another wave of pseudo concern for employees, one that will be followed in turn by a new wave of managerialist systems rationalism.

Such critiques of empowerment and federalism suggest that while Johnson's well-intended and admirable initiative goes a long way towards democratizing management accounting and control systems by putting employees' concerns, jobs, and a stable society above the stock market's demands for steady growth in reported profits, it does not address in a satisfactory way how this radical reversal of the organizational status quo might come about. This chapter concludes by considering three possible choices for initiating such an eventuality.

SOME CHOICES

There lies before us a vital choice, one in which accountants and accounting academics alike might choose to be actively involved. The choice is not *whether* to change the hierarchical social order of today's organizations, but rather *how* to change it in a way that is truly democratic and which will result in real employee empowerment. In making this choice we can look for guidance from three strands of critical social theory: the historical dialectic materialism of Friedrich Engels and Karl Marx, the radical humanism of Jurgen Habermas of the Frankfurt School of Social Philosophy, and the postmodernistic analysis of discourse and discursive practices of Michel Foucault. These approaches are representative of three different possibilities for changing power arrangements in organizations and society at large.

Modernistic Approaches

In 1845 Engels wrote his classic book, *The Conditions of the Working Class in England*. Sent from Germany at the age of 24 by his father to open a branch of the family textile machinery business, young Engels documented the horrifying picture of the plight of the employee in early capitalism in England. As discussed in Chapter 2, overcrowding, abject poverty, child labor, sexual exploitation, dirt, and drunkenness were the order of the day for the working class.[5] Engels' aim was to warn the German citizenry that unless they took steps to prevent it, the same conditions would come to prevail in Germany and that a workers' revolution would ensue in which

> ....the proletarians, driven to despair, will seize the torch which Stephens has preached to them; the vengeance of the people will break out with a wrath of which the rage of 1793 gives no idea. The war of the poor against the rich will be the bloodiest ever waged (Engels 1987, p. 298).

Engles hoped, however, that "there will be enough intelligent comprehension of the social question among the proletariat to enable the communistic party, with the help of events, to check the brutal element of the revolution and prevent a 'Ninth Thermidor'."[6] Either way, the eventual outcome for Engels was a classless society in a communist state.

In contrast, for Habermas the solution lies not in a bloody revolution but in the development by society of an enlightened critical theory with which to confront the cult of material progress that fuels the social order of late capitalism.[7] Habermas maintains that we can reach a consensus regarding the validity claims of capitalist ideology by means of "ideal speech situations" in which discussion would ensue under conditions of debate conducted free from domination of any kind. The result, he argues, would be a widespread consensus regarding how to shift the ideological steering mechanisms of society from the cult of material progress to a true moralization of motives for action both materially and spiritually.

> ....social roles are based on the intersubjective recognition of normal expectations for the possibilities of sanction situationally available to a role occupant...this change (from the animal status system based on the occupant's capacity to threaten) means a *moralization of motives* for action (Habermas 1979, p. 136).

If we would only pause, reflect, and debate on the human condition, society can evolve, Habermas insists, into a social order free from distorted communication, exploitation, and hegemonic domination by the tainted ideologies of the elite.

It seems unlikely, however, that either of these two alternatives have much of a chance today. The possibility seems remote that the proletariat will get organized, undertake a revolution, and overturn capitalism to form a classless society where property is owned by all. It also appears unlikely that the hopes of the radical humanists will be realized for the development of a widespread normative, idealistic shift in social consciousness that might replace late capitalism's cult of material progress. Moreover, critical theory of this kind has been denegrated for its primacy of subjectivity, its excessive idealism, its exclusion of the historicity and economic materiality of social life, and its excessively philosophical (metaphysical) tenor.[8] In consequence, we focus here on the third choice.

Discourse Analysis: A Postmodernistic Approach

Postmodernism, simply put, is the apprehension that late capitalism has coalesced into a new epoch that has superseded modernity just as the latter emerged to replace premodernity.[9] Modernity, spawned by the eighteenth century Enlightenment social philosophical movement, challenged and overturned the old emphasis on handed down wisdom and traditional modes of thought based on religion and mysticism.[10] The central and at the time flagrantly radical idea underpinning the Enlightenment project was that humankind could progressively build a paradise on earth for all human beings by relying on an intellectual landscape founded on science, rationality, and individual freedom for all. Similarly, postmodern social philosophers maintain, modernity has been displaced by a new social, cultural, and intellectual terrain.

Proponents of postmodernism read modernity, including the ideas of Engels, Marx, and Habermas, as a failed attempt to complete the Enlightenment project. Instead of a paradise on earth, the twentieth century has witnessed the two world wars among industrialized nations, which resulted in the death of millions on millions of civilians and soldiers (the latter mainly civilians in uniforms); the massive destruction of most great cities of Europe and Japan; the egregious degradation of the natural environment; the abject poverty of three-quarters of the world's population, especially those in developing countries; the development of weaponry capable of annihilating the entire world many times over: and the appearance of the deadly virus, HIV. Instead of a paradise, the vast majority experience life as a heinous perdition; while for most of the lucky minority, life consists of mindless, television-driven mass consumption of material and immaterial products in the global mall.[11]

Moreover, postmodernists call into question the grand theories of modernity, such as those of Smith, Marx, Weber, Freud, and Habermas, whose totalizing accounts are seen as bankrupt attempts to erect some final, permanent, and untouchable vision of the ontological nature of the human condition or a final version of the social order.[12] The argument is that such metanarratives must claim to be built on some kind of unassailable foundation, first principle, or unimpeachable ground. From a postmodernist perspective, they are merely neat, tidy, ossified stories which suggest a coherence that is either unreachably utopian or absent from the world they claim to describe (Sarup 1993).

Postmodernists also object to the notion of any ultimate sign or final true word-concept. Such signs (or transcendental signifieds as they are technically called) work as an anchor to give meaning to all other signs in the metanarrative. So they act as an unquestionable source of meaning to which all other signifiers in the system can point.[13] Against this, postmodernists point out that for such a signified to found the whole system of thought, it must somehow not be a part of the signification system for which it acts as the unshakable foundation. In other words, in order to act as the linchpin or fulcrum for the whole thought system, the sign around which all others revolve, it cannot be entangled in that system.[14] It must somehow be outside of itself. Such a position, postmodernists argue, is an impossibility.

So metanarratives and transcendental signs, postmodernists contend, can only be elevated to such a place of privilege by social ideologies. Yet, these in turn are constructed by humans (who use the signification system) and who construct them from some ideological power position. So metanarratives and transcendental signs are not, and cannot be, outside of the meaning system. Thus, they are merely the human longing for the presence of some ultimate reference, center, or authorizing voice. For the postmodernist, the era where such axiomatic systems, totalizing metanarratives, and transcendental signs ruled has melted away.

Rather, postmodernists see the life world and its institutions as being penetrated by a host of discourses and discursive procedures. Discourses are particular ways of writing, speaking, and thinking; while discursive practices are the special means of communicating particular discourses. Importantly, a primary condition of discourse is dialogue, which means that discourse is always social and so is always inextricably intertwined with power and domination (Giddens 1976, 1979, 1984). In consequence, it is vital when analyzing discourse to identify the speaker's (writer's, thinker's, artist's, accountant's, etc.) position including class, occupation, social role, political stance, and so on. Discourse is always and everywhere political.

Thus, it is crucial to understand how any discourse takes its stand in relation to its related but opposing discourses (Macdonell 1986). Any particular discourse can only appear to be a unified, coherent, and solid piece of meaning because of its difference from what it excludes or marginalizes through its strategy of prioritization and hierarchization of what are really only opposites.[15] To put it slightly differently, a discourse only achieves its unique identity (its coherent meaning) by what it excludes or relegates to a lower position, what was necessary for it to come about in the first place. Snow White, for example, obtains her identify as beautiful, young, pure, innocent, and good only by virtue of her alterity (or self same other), the Wicked Witch who is ugly, old, tainted, contaminated, and evil. But the Wicked Witch was once young, pure, and innocent like Show White who, in turn, will one day be old, contaminated, and jaundiced. Discourses do not mirror the world as it is; rather they reflect the wishes, ideals, attachments, and power positions of their promulgators.

Thus, any discourse, and this is the point to underscore, can only achieve a central, coherent meaning by dint of awarding center stage to certain objects, ideas, and ideologies while simultaneously marginalizing or ignoring other objects, ideas, and ideologies. Michel Foucault's own words on this important point are worth quoting.

> In earlier studies, we were able to isolate a distinctive level of investigation among all those approaches which permit the analysis of systems of thought: the analysis of discursive practices. This context discloses a systematic organization that cannot be reduced to the demands of logic or linguistics. Discursive practices are characterized by the delimitation of a field of objects, the definition of a legitimate perspective for the agent of knowledge, and the fixing of norms for the elaboration of concepts and theories. Thus, each discursive practice implies a play of prescriptions that designate its exclusions and choices.
>
> These principles of exclusion and choice, whose presence is manifold, whose effectiveness is embodied in practices, and whose transformations are relatively autonomous, are not based on an agent of knowledge (historical or transcendental) who successively invents them or places them on an original footing; rather, they designate a will to knowledge that is anonymous, polymorphous, susceptible to regular transformations, and determined by the play of identifiable dependencies (Foucault 1979, p. 199).

Postmodernists believe discourses and their discursive practices can be confronted and challenged. Just as nature does not grow on trees, discourses are not natural, neutral, and untouchable; they are human creations and so are contestable terrains.

Poststructuralists believe discourses can be undermined from within by following a deconstructivist strategy of reading any text. They can be dedoxified by showing how any text has a political side that inherently organizes the discursive turf in favor of some special interest group, as is the case for the managerial accounting discourse which tilts the playing field towards upper executives and managers and away from employees. Postmodernists aim to reveal how discourses and discursive practices are put together (constructed or packaged) to prioritize and privilege certain players and meanings at the expense of others. The important point is that discursive formations with their discourses and discursive practices are socially constructed and thus available for deconstructing (unpackaging), dedoxifying, and debunking.

Management Accounting as Discourse

Management accounting discourse can be seen to be closely aligned with that of the financial accounting discourse of upper executives. It relies on the rhetoric of market capitalism to construct the goal of the corporation as that of maximizing the long-term wealth of stockholders by means of steady growth in earnings per share.[16] Yet, from a postmodernistic perspective, today's market capitalism is a far cry from the real capitalism of the last century. It is merely a simulacrum of capitalism, capitalism in its pure form, the *idea* of capitalism but not its *substance* (Baudrillard 1988). The capitalist who owns the means of production, manages the enterprise, and hires workers as wage labor commodities has disappeared from our major corporations. Instead, today's organizational world is caught up in the web of the ideas of capitalism but without the spider, the capitalist.

More to the point, most of us work in organizations in which the discourse of capitalism (markets, competition, finance, stock owners, workers, capital markets, etc.) abounds. Within these organizations we are workers—accountants, professors, information technology experts, machine tenders, managers, truck drivers, secretaries, etc.—yet at the same time we are all minicapitalists in that a large part of our wealth is tied up in pension funds managed by investment dealers who buy and sell stocks and bonds on our behalf. We are at once worker and capitalist; the owner capitalist has all but disappeared. The *image* of capitalism remains but not the *substance*.

Within the accounting world, even though lip service is paid to a heterogeneous set of stakeholders among the users of financial accounting information, the shareholders' aims are usually given first priority. Thus, the discourse and discursive practices of financial accounting marginalize the aims of the other stakeholders (Cyert and Ijiri 1974). As two accounting scholars aptly put it:

> ...few would dispute that the profit-maximizing objective of the firm is merely a shorthand way of stating the objectives of the *shareholders* of the firm under the assumption of homogeneous shareholder preferences; it does not represent the specific objectives of the managers, employees, creditors or of any other parties inside or outside the firm (Dopuch and Sunden 1980, p. 14).

It seems clear this privileging of shareholders over other stakeholders is merely a "cover story" to hide the reality that maximizing shareholders' wealth is a narrow, parochial, exploitive, and self-serving discourse of an elite group. Upper executives normally hold substantial stock options, to say nothing of huge salaries and handsome bonus plans based on reported profits, sufficient to make themselves very wealthy if they deliver the necessary steady growth in earnings per share.[17]

The 1992 compensation scheme for top executives at Chrysler Corporation provides a striking case in point. Lee Iacocca, chairman and CEO, took home a staggering fortune of $20.3 million, not counting his pension benefits of $700 000 per year and $500 000 per year for serving as a consultant. Stock options and a bonus based on Chrysler's reported net earnings constituted most of this total. New Chairman Robert Eaton received about $4.7 million in compensation for his first ten months, about 60 percent of which came as stock options with the rest in cash equivalent income. President Robert Lutz received $2.7 million in compensation nearly $900 000 of it in stock options. Chief Financial Officer Jerome York received $1.5 million, $462 500 of it in stock options while Chief Administrative Officer Thomas Denomme received $1.4 million, $473 000 of it in stock options.[18] With such huge sums at stake, there can be little doubt that the "maximize the shareholders' wealth" discourse plays nicely into the hands of upper executives in today's large corporations. Shareholders' wealth is a euphemism for top executives' wealth.[19]

The management accounting discourse also relies on the rhetoric of market capitalism. It constructs the role of managers as scorekeeping, attention direction, problem solving, and decision making in the interests of organizational efficiency and goal congruence in order to maximize shareholders' wealth. This discourse is also narrow, self-serving, parochial, and exploitive since these managers have reasonable chances to rise to the level of the upper executives and, in the meanwhile, have interesting and remuneratively rewarding work. The reality is that for the most part, the manager's job is one of surveillance, watching, controlling, and evaluating the performance of the rest of the employees in the corporation by a variety of means, not least of which are master budgets, detailed operational control systems, standard costing reports, and product costing systems. In short, the manager's job is to see that the employees deliver the required profits.

GESTURING TOWARDS REAL EMPLOYEE ACCOUNTING SYSTEMS

While the accounting discourses of upper executives and managers circulate as good currency throughout organizations and society, no similar accounting-based discourse exists for the rest of the employees. In consequence, the latter are without a strong voice within the very organizations in which they design, produce, market, and keep track of the products and services which produce the profits. So it could be argued that Western civilization has reached a curious landmark whereby the vast majority of its citizens have a say in electing politicians and presidents but are disenfranchised in the workplace. Without some equivalent of the upper executive and management accounting discourse, it seems that the employees' aims and needs are likely to continue to play second fiddle to those of the upper executives and managers.

Perhaps the time has arrived when academic management accountants devote some of their research efforts towards developing real employee accounting systems instead of merely reinventing management accounting systems using different cost allocation algorithms or training management accountants in ethics. When factories and marketing divisions are closed down in the "interests of profits" legitimated by financial and management accounting systems, unless employees have some sort of accounting discourse of their own, it is likely they will remain disenfranchised at work, if they still have a job.

Yet, assuredly it would be a challenge and great fun to be involved in designing employee accounting systems. They might report, for example, on the number of hours worked, the wages and salaries earned by employees, the value added by their work, the number of new jobs created, as well as the progress made by the corporation in pollution controls and environmental improvements and the results of affirmative action programs. And who knows, *real* employee empowerment in combination with employee accounting systems and discursive practices could well lead to greater earnings-per-share growth than will continuing with the current state of affairs.

Undoubtedly, many will argue that we are not ready yet for an experiment of this size and shape. After all, we have only really tried capitalism, in one form or another, for about 150 years out of a total of thirty or so thousands of years in the recorded history of humankind. Yet it seems to me that such an experiment is almost inevitable. The people that do the work, produce the results, and devote their emotional and spiritual lives to organizations should take their fair share of the rewards, rather than only a miniscule minority getting the lion's share. With time, and some luck, things could well turn in this direction.

NOTES

1. The Concise Oxford Dictionary (1982) defines *empower* as to authorize, licence (person to do); give power to, make able, (person to do).
2. This proposal may not be as far-fetched as it seems. In the 1970s in Sweden, Bill Starbuck and Bo Hedberg (in an unpublished working paper) designed a new steel plant including a social system in which the employees democratically elected the managers and the prime ministers every four years. The plant, however, was never completed as the bottom dropped out of the world steel market just as construction was to begin, causing the government to abandon the project. This idea is not entirely new. Jönsson and Grönlund (1988) present a detailed case study of a Swedish subcontractor organization where accounting information was aimed at enabling employees on the shopfloor to learn how to handle new production processes and to enhance team building, thus going a long way towards unofficial but real employee empowerment.
3. See Kanter (1992) for a review and extension of the managerialist empowerment concept, as well as Handy (1992) for a neoconservative managerialist version of employee empowerment and corporate federalism.
4. Handy (1992), p. 61. See also Barnard (1938) and Chapter 12 on managerialism.
5. Conditions which are mirrored today in most developing nations.
6. Engels (1987), p. 292. The thermidor metaphor refers to the French Revolution of 1789.
7. See Laughlin (1987), Arrington and Puxty (1991), and Broadbent, Laughlin, and Read (1991) for excellent papers using some of Habermas' key constructs to inform accounting practices. See also Macintosh (1990).
8. See Guess (1981) and Bottomore (1984) for excellent critiques of this line of social theory.
9. Postmodernity is by no means a single, coherent concept. It includes descriptors by various postmodern scholars including "the information society" (Lyotard 1984); the "disciplined–disciplinary" world (Foucault 1979); the "cultural logic of late capitalism" (Jameson 1984); and the "consumer society" (Baudrillard 1988). For excellent and highly readable books about the territory of postmodernism and poststructuralism see: Hebdige (1988), Harvey (1990), Norris (1992), Sarup (1993), Poster (1990), Hutcheon (1989), Macdonell (1988), Rose (1991), Wakefield (1990), and Weedon (1987).
10. For example, Adam Smith in his momentus 1776 treatise, *An Enquiry into the Nature of the Wealth of Nations*, attacked the dominant economic theory of the day, promoted especially by the French physiocrats, which held that land was the only source of wealth. Smith demonstrated, to the contrary, how the rational division and use of man's labor could be the key ingredient for a nation's wealth to experience exponential growth.
11. See Herman and Chomsky (1988) who indicate how the mass media today frames and sanitizes these matters on behalf of an underlying small elite percentage of the population.
12. Well-known metanarratives of modernity include Adam Smith's (1723–1790) vision of a world progressively unfolding according to the laws of the market; Marx's (1818–1883) historical materialist formulations of the inexorable unfolding of capitalism into a classless egalitarian

society without private ownership of the means of producing society's economic wealth; Weber's (1964–1920) image of society dominated ever more by the general rationalizing process of bureaucratic state oligarchies. Jurgen Habermas' conception of a reflective critical theory capable of guiding humankind towards enlightenment as to its true interests and towards emancipation from unsuspected forms of social and self coercion; and, in the accounting world, agency theory's construction of the organizational world as a web of atomistic principal (owner) to agent (employee) contractual relationships whereby all agents pursue self-interest with guile. For the postmodernist, these ideas are just as unbelievable as the metanarratives of premodernity such as the Koran, the Talmud, the New Testament, and, more recently, the Book of Mormon.

13. Examples of transcendental signifieds are: Truth, God, Essence, the Invisible Hand of the Free Market (Smith); the Iron Cage of Rational Bureaucracy (Weber); the Dialectic Contradiction of Historical Materialism (Marx); and the Unconscious Oedipal Libidinal Drive (Freud).

14. As Ryan (1983), pp. 16–17 explains, "To use Godel's terminology, the system is necessarily 'undecidable', because it generates elements that can be proved both to belong to the system and not belong to it at the same time. The axiomatic system is necessarily incomplete."

15. Ryan (1982, p. 9) explains how standard practice in constructing meaning such as in a discourse is "...to understand the world using binary oppositions, one of which is assumed to be prior and superior to the other... The second term in each case is inevitably made out to be external, derivative, and accidental in relation to the first, which is either an ideal limit or the central term... The reason why this is so, according to Derrida, is that the second term in each case usually connotes something that endangers the values the first term assures, values that connote presence, proximity, ownership, property, identity, truth conceived as conscious mastery, living experience, and a plenitude of meaning. The second terms usually suggests the breakup of all of these reassuring and empowering values, such terms as difference, absence, alteration, history, repetition, substitution, undecidability, and so on." See also Derrida (1976, 1978). See Arrington and Francis (1989) for a deconstructive reading of agency theory in accounting.

16. Bedford (1974) and Beaver and Demski (1974). See also Thornton (1988) for the rhetoric of accounting as metaphor.

17. Research by Posner and Schmidt (1986) supports this view. Their survey indicated that US executives rank themselves first (after customers) in terms of stakeholders. They also ranked shareholders eleventh out of twelve stakeholders. This finding "casts doubt on the assumption that corporations operate solely, or even primarily, for the benefit of their shareholders" (p. 215).

18. Source: The Associated Press, March 26, 1993.

19. Chrysler executives also put forth a public notice proposal (required by a recent rule of the Securities and Exchange Commission) for approval at the annual meeting that sought to allow directors to continue awarding up to 8 percent of Chrysler's annual earnings as incentive bonuses to management. The proposal also sought to exclude from the calculation of annual earnings (for purposes of calculating these bonuses), the effects of new accounting rules, which require booking expense allowances for the value of stock options given to officers and employees as well as postretirement health benefits. Professor Emeritus Graef Crystal at Berkeley, an outspoken critic of excessive compensation, described the proposal as "cherry-picking the income statement. What's next? You want to take out depreciation or the cost of buying steel? Where does it end?" Source: The Associated Press, March 26, 1993.

15
Conclusions

This book has presented a collection of what I believe are valuable ways to think about the crucial nontechnical aspects of management accounting and control systems, particularly as they influence the organization's sociocultural landscape. This selection brings out the redeeming features of these systems, but not at the expense of turning a blind eye to important critiques of their darker side. Either way, such systems are essential to the functioning of nearly all organizations of any size and shape. So it seems almost mandatory that professional management accountants, students, and academics get beyond a narrow, technicist perspective. The major aim of this book is to help these people develop a deeper understanding of the social side of these systems than is customary.

The frameworks discussed come in five different basic brands: structural functionalist, radical structuralist, interpretivist, radical humanist, and poststructuralist. At the outset we looked at three competing universal frameworks: agency theory, nerve center, and labor process. While each had something important to say, each also seemed to have a blind side in that it left out important elements included in the other two, or it seemed excessively biased towards either owners, managers, or workers. Moreover, they did not provide much leverage on the way their universal elements might vary systematically, depending on the peculiar circumstances facing the organization. In contrast, the structural functionalist models which followed focused on how environment, strategy, technology, and uncertainty lead to patterned variations in universalistic elements of management accounting and control systems. The central idea was to achieve an appropriate match of circumstance and system characteristics.

Next, we introduced some frameworks which indicated that while management accounting-based controls work well when task knowledge is well understood and the goal is unambiguous, they lose their robustness when instrumentality is only poorly understood and goals are ambiguous. In the latter circumstances, results tend to be unmeasurable and unpredictable; so organizations look to other types of scorekeeping and control to supplement, complement, and even usurp accounting-based controls. These frameworks also served to expand the intellectual domain of control beyond that of accounting systems.

The interpretivist frameworks looked at another side of management accounting systems. They showed, in a systematic way, how managers do not simply react to the systems but not infrequently engage them in a proactive way. Sometimes they use them as ammunition to convince others of a particular course of action, they use them to rationalize decisions already taken, or they rely on them to gain the upper hand on rivals. The interpretivist perspective also revealed how what seems to be objective accountability is, in reality, a highly subjective enterprise; it revealed that individuals frequently use them as weapons in the inevitable war of self-interest in the workplace. Management accounting systems not only reflect organizational reality, they are also used to construct a self-serving reality.

Next, we introduced the structuration framework which foregrounds how management accounting systems play a huge role in the production, reproduction, and alteration of an organization's fundamental social structures. Accounting systems not only provide a universal language for communicating financial meaning, they also reinforce the blueprints for relations of authority and dependency as well as the organization's codes for legitimate behavior and moral obligations. The major lesson here was that to ignore these dimensions is to ignore the essence of management accounting and control systems. Structuration theory precludes omission of power and morality considerations.

This was followed by considering four major critiques. Three of these addressed the problems and limitations of management accounting and control systems from the managerialist position. The other critique revealed how these systems emerged and took shape as part of the general disciplinary drive that appeared and spread during the modern epoch. It also opened up a space for a poststructuralist look at the perennial goal incongruence problem which, beyond doubt, is one of the most important elements in understanding the power and resistance aspects of control systems. While these critiques are disturbing, and it is all too easy to gloss over them, they provide us with essential material for developing a balanced and sophisticated intelligence of the inner workings of these systems.

Finally, we assessed, from a poststructuralist perspective, the recent proposal to empower employees by giving them total access to cost and management accounting information. Our analysis concurred wholeheartedly with the idea that remote-control management by the numbers will no longer do and that employees' concerns, jobs, and a stable society must be given priority over stock market winnings for top management and other shareholders. The empowerment theme was then taken to its radical extreme by calling for truly democratic organizations and real employee empowerment whereby employees, the people who produce the goods and services and the surplus wealth, truly run the show. In order for this to come about, we argued, employee accounting systems with their own discourse and discursive practices will be required.

This, of course, is what the poststructuralist position teaches us. Today's sociocultural world is neither inevitable nor immutable. Nor is it necessarily any better than previous ones. The way things are today will seem just as strange and foreign to future generations in, say, the twenty-second century as would Josiah Wedgwood's world of an untamed, slovenly, intemperate, and undisciplined work force were we to go back in time. Things were different then, they are different now, and they will be different in the future. So employee accounting systems are certainly not out of the question. They can be used, once developed, as a discourse to challenge and rebuke shareholders' and managers' accounting systems which currently hold sway. In the best of all Panglossian worlds, they may even be inevitable, at least for a while.

In presenting these frameworks and critiques I hope I have communicated my own sense of respect for the importance of each of the various models. While no one of them presents an absolute truth, it will no longer do to rely on simple solutions such as the belief that budget participation applied in the right circumstances, or that merely tightening or loosening existing controls as required, provide sufficient guidelines for designers and practitioners. The issues and problems encountered in practice are much more complex and important than that; they deserve sophisticated ways to understand them and to be looked at from multiple perspectives. I also hope this book might induce at least a few conversations beween proponents of specific frameworks. After all, we need differences of opinion to progress in this complex subject. We are all in this business together and, who knows, perhaps a new conceptual amalgam might be the happy result.

That is not to say good systems cannot be designed without these ideas; they can. What has always impressed me is that, in a loose way, management accounting and control systems executives operate with their own personal theories. And many of them put into place highly suitable and effective systems as we saw in the cases of Wedgwood, Empire Glass, Johnson & Johnson, General Motors, and du Pont. Practitioners do, in fact, look for patterns, they do form explanations, and they do make predictions, even if they have experience in only one or a few organizations. They have to.

What this book can do, however, is to help practitioners in formulating their theories. The various frameworks can be treated as a set of intellectual power tools, each with its own special purpose. The structuration framework in Chapter 10, for example, can be used to sensitize the analysis to the way accounting and control systems are always involved in relations of domination and morality. Likewise, the ideas in Chapter 6 bring to the forefront the relationship of control systems to strategy. While the frameworks in Chapter 7 show how management accounting systems perform according to the pattern of the component's technology or the pattern of its interdependencies. Chapter 8 indicates how accounting-based systems are complemented with other types of controls. And the critiques in Chapters 12 and 13 alert us not to ignore their darker side. Each of these frameworks can help us to be careful and humble in plying our trade of accountancy.

Beyond the models, the case studies, and the history of key ideas, I wanted to communicate my own sense of the importance of management accounting and control systems and my own sense of wonder at their many facets and complexities. It seems no exaggeration to say they are one of the most important social inventions of the last two centuries. They are vital to the workings of today's organizations and institutions. It is highly unlikely that enterprises such as the widely diversified meganational corporations or large-scale public sector organizations could have come into existence without them. They deserve to be understood in a sophisticated way. And that, of course, has been the central aim in writing this book.

References

Abernethy, M.A. and J.U. Stoelwinder (1991) Budget use, task uncertainty, system goal orientation and sub-unit performance: a test of the "fit" hypothesis in not-for-profit hospitals. *Accounting, Organizations and Society*, 105–120.
Ackoff, R.L. (1986) *Managing in Small Doses*. New York, John Wiley.
Alleman, R.H. (1985) Comptrollership at ITT. *Management Accounting*, 24–33.
Alchian, A. and H. Demsetz. (1972) Production, information costs and economic organization. *American Economic Review*, 777–795.
Amey, L.R. (1979) *Budget Planning and Control Systems*. London, Pitman.
———. (1986) *A Conceptual Approach to Management*. New York, Praeger.
Ansari, S.L. (1976) Behavioral factors in variance control: report on a laboratory experiment," *Journal of Accounting Research*, 189–211.
Ansari, S.L. and K.J. Euske. (1987) Rational, rationalizing, and reifying uses of accounting data in organizations. *Accounting, Organizations, and Society*, 549–570.
Anthony, P.D. (1977) *The Ideology of Work*. London, Tavistock.
Anthony, R.N. and J. Dearden. (1981) *Teacher's Guide to Accompany Management Control Systems*. Homewood, Ill., Irwin.
Anthony, R.N., J. Dearden and N.M. Bedford. (1989) *Management Control Systems*. Homewood, Ill., Irwin.
Anthony, R.N., J. Dearden and V. Govindarajan. (1992) *Management Control Systems*. Homewood, Ill., Irwin.
Anthony, R., J. Dearden and R. Vancil. (1965) *Management Control Systems: Cases and Readings*. Homewood, Ill., Irwin.
Antle, R. and G.D. Eppen. (1985) Capital rationing and organizational slack in capital budgeting. *Management Science*, 163–174.
Argyris, C. (1952) *The Impact of Budgets on People*. Ithaca, N.Y., Controllership Foundation, Cornell University.
Armstrong, P. (1987) The rise of accounting controls in British capitalist enterprises. *Accounting, Organizations and Society*, 415–436.
———. (1991) Contradiction and social dynamics in the capitalist agency relationship. *Accounting, Organizations and Society*, 1–25.
Arrow, K. (1964) Control in large organizations. *Management Science*, 1–36.
Arrington, C.E. and J.R. Francis. (1989) Letting the chat out of the bag: deconstruction, privilege, and accounting research. *Accounting, Organizations and Society*, 1–28.
Arrington, C.E. and A.G. Puxty. (1991) Accounting, interests, and rationality: a communicative relation. *Critical Perspectives on Accounting*, 31–58.
Baiman, S. (1982) Agency research in managerial accounting. *Journal of Accounting Literature*, 154–213.
———. (1990) Agency research in managerial accounting: a second look. *Accounting, Organizations and Society*, 341–371.
Baiman, S. and J.H. Evans, III. (1983) Pre-decision information and participative management control systems. *Journal of Accounting Research*, 371–395.

Barley, S.R. and G. Kunda. (1992) Design and devotion: surges in rational and normative ideologies of control in managerial discourse. *Administrative Science Quarterly*, 363–399.

Barnard, C.I. (1938) *The Functions of the Executive*. Cambridge, Mass., Harvard University Press.

Baudrillard, J. (1988) *Selected Writings*. Oxford, Polity Press.

Beaver, W.H. and J.S. Demski. (1974) The nature of financial accounting objectives: a survey and synthesis. Supplement to the *Journal of Accounting Research*.

Becker, S.W. and D. Green. (1962) Budgeting and employee behavior. *The Journal of Business*, 392–402.

Bedford, N.M. (1974) Discussion of opportunities and implications of the report on objectives of financial statements. *Studies on Financial Statement Objectives, 1974*, supplement to the *Journal of Accounting Research*, 15.

Bennis, W.G. (1966) *Changing Organizations*. New York, McGraw-Hill.

Best, S. and D. Kellner. (1991) *Postmodern Theory: Critical Investigations*. New York, Guilford Press.

Birnberg, J.G. (1992) Managerial accounting: yet another perspective. *Advances in Management Accounting*, 1–19.

Birnberg, J.G. and J.F. Shields. (1989) Three decades of behavioral accounting research in the United States. *Behavioral Research in Accounting*, 75–108.

Birnberg, J.G., M. Shields and W. McGhee. (1980) The effects of personality on subjects' information processing: a reply. *The Accounting Review*, 507–510.

Birnberg, J.G., I. Turopolec and S.M. Young. (1983) The organizational context of accounting. *Accounting, Organizations and Society*, 111–129.

Boland, R.S., Jr. and L.R. Pondy. (1983) Accounting in organizations: a union of natural and rational perspectives. *Accounting, Organizations and Society*, 223–234.

———. (1986) The micro dynamics of a budget-cutting process: modes, models, and structure. *Accounting, Organizations, and Society*, 403–422.

Bottomore, T. (1984) *The Frankfurt School*. London, Tavistock.

Bower, J. (1970) *Managing the Resource Allocation Process*. Boston, Mass., Division of Research, Graduate School of Business Administration, Harvard University.

Braverman, H. (1974) *Labor and Monopoly Capital: The Degradation of Work in the Twentieth Century*. (New York, Monthly Review Press.

Broadbent, J., R. Laughlin and S. Read. (1991) Recent financial and administrative changes in the NHS: a critical theory analysis. *Critical Perspectives on Accounting*, 1–29.

Brownell, P. (1981) Participation in budgeting, locus of control and organizational effectiveness. *The Accounting Review*, 844–860.

———. (1983) The role of accounting data in performance evaluation, budgetary participation and organizational effectiveness. *Journal of Accounting Research*, 456–472.

Brownell, P. and A.S. Dunk. (1991) Task uncertainty and its interaction with budgetary participation and budget emphasis: some methodological issues and empirical investigations. *Accounting, Organizations and Society*, 693–704.

Brownell, P. and M. Hirst. (1986) Reliance on accounting information, budgetary participation and task uncertainty, tests of a three-way interaction. *Journal of Accounting Research*, 241–249.

Brownell, P. and M. McInnes. (1986) Budgetary participation, motivation, and managerial performance. *The Accounting Review*, 587–600.

Bruns, W.J. and J.H. Waterhouse. (1975) Budgetary control and organization structure. *Journal of Accounting Research*, 177–203.

Bruns, W.J., Jr. and S.M. McKinnon. (1993) Information and managers: a field study. *Journal of Management Accounting Research*, Fall, 84–108.

Burchell, S., C. Clubb, A.G. Hopwood, T. Hughes and J. Nahapiet. (1980) The roles of accounting in organizations and society. *Accounting, Organizations and Society*, 5–27.

Burgstahler, D. and G.L. Sundem. (1989) The evolution of behavioral accounting research in the United States. *Behavioral Research in Accounting*, 75–108.

Burns, T. and G.M. Stalker. (1961) *The Management of Innovation*. London, Tavistock.

Burrell, G. and G. Morgan. (1979) *Sociological Paradigms and Organizational Analysis*. London, Heinemann.

Chandler, A.D., Jr. (1962) *Strategy and Structure*. Cambridge, Mass., M.I.T. Press.

———. (1977) *The Visible Hand: The Managerial Revolution in American Business*. Cambridge, Mass., Harvard University Press.

Chenhall, R. (1986) Authoritarianism and participative budgeting: a dyadic analysis. *The Accounting Review*, 263–272.

Chenhall, R. and D. Morris. (1986) The impact of structure, environment, and interdependence on the perceived usefulness of management accounting systems. *The Accounting Review*, 58–75.

———. (1991) The effect of cognitive style and sponsorship bias on the treatment of opportunity costs in resource allocation decisions. *Accounting, Organizations and Society*, 27–46.

Cherrington, D.J. and J.O. Cherrington. (1973) Appropriate reinforcement contingencies in the budgeting process. *Journal of Accounting Research: Empirical Research in Accounting: Selected Studies*, 225–253.

Chow, C.W., Cooper, J.C. and W.S. Waller. (1988) Participative budgeting: effects of a truth-inducing pay scheme and information asymmetry on slack and performance. *The Accounting Review*, 111–122.

Christensen, J. (1982) The determination of performance standards and participation. *Journal of Accounting Research*, 589–603.

Chua, W.F. (1986) Radical developments in accounting thought. *The Accounting Review*, 601–632.

———. (1988) Interpretive sociology and management accounting research: a critical review. *Accounting, Auditing & Accountability Journal*, 59–79.

Clark, J.M. (1923) *Studies in the Economics of Overhead Costs*. Chicago, The University of Chicago Press.

Clegg, S. and D. Dunkerley. (1980) *Organization, Class, and Control*. London, Routledge & Kegan Paul.

Collins, F. (1978) The interaction of budget characteristics and personality variables with budgetary response attitudes. *The Accounting Review*, 324–335.

Collins, F., P. Munter and D. Finn. (1987) The budgeting games people play. *The Accounting Review*, 29–49.

Cooper, D.J. (1980) Discussion of Towards a Political Economy of Accounting. *Accounting, Organizations, and Society*, 161–166.

Cooper, R., R.S. Kaplan, L.S. Maisel, E. Morrissey, and R.M. Oehm. (1992) *Implementing Activity-Based Cost Management: Moving From Analysis to Action*. Montvale, N.J., Institute of Management Accountants.

Covaleski, M.A. and M.W. Dirsmith. (1986) The budgeting process of power and politics. *Accounting, Organizations and Society*, 193–214.

———. (1988) The use of budgetary symbols in the political arena: an historically informed field study. *Accounting, Organizations and Society*, 1–24.

———. (1990) Dialectic tension, double reflexivity and the everyday accounting researcher: on using qualitative methods. *Accounting, Organizations and Society*, 543–573.

Covaleski, M.A., M.W. Dirsmith and S.F. Jablonsky. (1984) Traditional and emergent theories of budgeting: an empirical analysis. *Journal of Accounting and Public Policy*, 277–300.

Covaleski, M., M. Dirsmith and J. Michelman. (1993) An institutional theory perspective on the DR6 framework, case-mix accounting systems and health-care organizations. *Accounting, Organizations and Society*, 65–80.

Culbert, S.A. and J.J. McDonough. (1980) *The Invisible War: Pursuing Self Interests at Work*. New York, John Wiley.

———. (1985) *Radical Management: Power Politics and the Pursuit of Trust*. New York, The Free Press.

Cyert, R.M. and Y. Ijiri. (1974) Problems of implementing the Trueblood objectives report. Supplement to the *Journal of Accounting Research*, 29–42.

Cyert, R. and J.G. March. (1963) *The Behavioral Theory of the Firm*. Englewood Cliffs, N.J., Prentice Hall.

Daft, R.L. (1986) *Organizational Theory and Design*. St. Paul, Minn., West Publishing.

———. (1992) *Organizational Theory and Design*. St. Paul. Minn., West Publishing.

Daft, R.L. and N.B. Macintosh. (1978) A new approach to design and use of management information. *California Management Review*, 82–92.

———. (1981) A tentative exploration into the amount and equivocality of information processing in organizational work units. *Administrative Science Quarterly*, 207–224.

———. (1984) The nature and use of formal systems for management control and strategy implementation. *Journal of Management*, 43–66.

Dalton, G.W. and P.R. Lawrence. (1971) *Motivation and Control in Organizations*. Homewood, Ill., Irwin.

Dean, J. (1957) Profit performance measurement of division managers. *The Controller*, 423–428.

Dearden, J. (1960) Interdivisional pricing. *Harvard Business Review*, 117–126.

———. (1961) Problems in decentralized financial control. *Harvard Business Review*, 72–80.

———. (1987) Measuring profit center managers. *Harvard Business Review*, 84–88.

DeCoster, D.T. and J.P. Fertakis. (1968) Budget induced pressure and its relationship to supervisory behavior. *The Journal of Accounting Research*, 237–246.

Dent, J.F. (1990) Strategy, organization and control. *Accounting, Organizations and Society*, 3–26.

Dermer, J.D. (1973) Cognitive characteristics and the perceived importance of information. *The Accounting Review*, 511–519.

———. (1990) The strategic agenda: accounting for issues and support. *Accounting, Organizations and Society*, 67–76.

Derrida, J. (1976) *Of Grammatology*. Baltimore, Md., John S. Hopkins University Press.

———. (1978) *Writing and Difference*. London, Routledge & Kegan Paul.

Dirsmith, M.W. and S.F. Jablonsky. (1979) MBO, political rationality and information inductance. *Accounting, Organizations and Society*, 39–52.

Dopuch, N. (1993) A perspective on cost drivers. *The Accounting Review*, 615–620.

Dopuch, N. and S. Sunden. (1980) FASB's statements on objectives and elements of financial accounting: a review. *The Accounting Review*, 1–21.

Dreyfus, H.L. and P. Rabinow. (1983) *Michel Foucault: Beyond Structuralism and Hermeneutics*. Chicago, University of Chicago Press.

Driver, M.J. and T.J. Mock. (1975) Human information processing decision style theory and accounting information systems. *The Accounting Review*, 490–508.

Duncan, K. and K. Moores. (1989) Residual analysis: a better methodology for contingency studies in management accounting. *Journal of Management Accounting Research*, 89–103.

Dunk, S.A. (1989) Budget emphasis, budgetary participation and managerial performance: a note. *Accounting, Organizations and Society*, 321–324.

———. (1990) Budgetary participation, agreement on evaluation criteria and managerial performance. *Accounting, Organizations and Society*, 171–178.

———. (1992) The reliance on budgetary control, manufacturing process automation and production subunit performance: a research note. *Accounting Behavior and Organizations*, 195–204.

———. (1993) The effect of budget emphasis and information asymmetry on the relation between budgetary participation and slack. *The Accounting Review*, 400–410.

Earl, M.J. and A.G. Hopwood. (1980) From management information to information management. in H.C. Lucas, Jr., et al. *The Information Systems Environment*. Amsterdam, North Holland.

Eckel, L.G. (1976) Arbitrary and incorrigible allocation. *The Accounting Review*, 764–777.

Emmanuel, C. and D. Otley. (1985) *Accounting for Management Control*. Wokingham, UK, Van Nostrand Reinhold.

Emmanuel, C., D. Otley and K. Merchant. (1992) *Readings in Accounting for Management Control*. London, Chapman & Hall.

Engels, F. (1987) *The Condition of the Working Class in England*. New York, Penguin.

Ezzamel, M. (1992) *Business Unit & Divisional Performance*. London, Chapman & Hall.

Ezzamel, M. and H. Hart. (1987) *Advanced Management Accounting: An Organizational Emphasis*. London, Cassell.

Feldman, M.S. and J.G. March. (1981) Information in organizations as signal and symbol. *Administrative Science Quarterly*, 171–186.

Ferrara, W.L. (1960) Idle capacity as a loss—fact or fiction. *The Accounting Review*, 490–496.

———. (1961) Overhead costs and income measurement. *The Accounting Review*, 63–70.

———. (1963) Relevant costing—two points of view. *The Accounting Review*, 719–772.

———. (1990) The new cost management accounting: more questions than answers. *Management Accounting*, 48–52.

References

Fess, P.E. (1963) The relevant costing concept for income measurement: can it be defended? *The Accounting Review*, 723–732.
Fess, P.E. and W.L. Ferrara. (1961) The period cost concept for income measurement: can it be defended? *The Accounting Review*, 598–602.
Foucault, M. (1979) *Discipline and Punish: The Birth of the Prison*. New York, Vintage Books.
———. (1980) *The History of Sexuality Volume 1: An Introduction*. New York, Vintage Books.
Frank, W.G. (1990) Back to the future: a retrospective view of J. Maurice Clark's studies in the economics of overhead costs. *Journal of Accounting Research*, 153–160.
Fremgen, J.M. (1962) Variable costing for external reporting. *The Accounting Review*, 76–81.
Friedman, A. (1994) *Spider's Web: The Secret History of How the White House Illegally Armed Iraq*. New York, Bantam.
Galbraith, J. (1973) *Designing Complex Organizations*. Reading, Mass., Addison-Wesley.
———. (1977) *Organization Design*. Reading, Mass., Addison-Wesley.
Geneen, H. (1984a) The case for managing by the numbers. *Fortune*, October 1.
———. (1984b) *Managing*. New York, Avon.
Giddens, A. (1976) *New Rules of Sociological Analysis*. London, Hutchinson.
———. (1979) *Central Problems in Social Theory*. London, Macmillan.
———. (1984) *The Constitution of Society*. Cambridge, Polity Press.
Ginzberg, M.J. (1980) An organizational contingencies view of accounting and information systems implementation. *Accounting, Organizations and Society*, 369–382.
Goldratt, E.M. (1992) From cost world to throughput world in M.J. Epstein (ed.) *Advances in Management Accounting*. Greenwich, Conn., JAI Press.
Gordon, L.A. and D. Miller. (1976) A contingency framework for the design of accounting information systems. *Accounting, Organizations and Society*, 59–69.
Gordon, L.A. and K.J. Smith. (1992) Postauditing capital expenditures and firm performance: the role of asymmetric information. *Accounting, Organizations and Society*, 741–758.
Govindarajan, V. (1984) Appropriateness of accounting data in performance evaluation: an empirical examination of environment uncertainty as an intervening variable. *Accounting, Organizations and Society*, 125–135.
———. (1986) Decentralization, strategy, and effectiveness of strategic business units in multi-business organizations. *Academy of Management Review*, 844–856.
———. (1988) A contingency approach to strategy implementation at the business-unit level: integrating administrative mechanisms with strategy. *Academy of Management Journal*, 828–853.
Govindarajan, V. and J. Fisher. (1990) Strategy, control systems, and resource planning: effects on business unit performance. *Academy of Management Journal*, 259–285.
Govindarajan, V. and A.K. Gupta. (1985) Linking control systems to business unit strategy: impact on performance. *Accounting, Organizations and Society*, 51–66.
Greiner, L.E. (1972) Evolution and revolution as organizations grow. *Harvard Business Review*, 37–46.
Guess, R. (1981) *The Idea of a Critical Theory: Habermas and the Frankfurt School*. Cambridge, Cambridge University Press.
Gul, F.A. (1984) The joint and moderating effects of personality and cognitive style on decision making. *The Accounting Review*, 264–277.
Gupta, A.K. and V. Govindarajan. (1984a) Build, hold, harvest: converting strategic intentions into reality. *Journal of Business Strategy*, 34–47.
———. (1984b) Business unit strategy, managerial characteristics, and business unit effectiveness at strategy implementation. *Academy of Management Journal*, 24–41.
———. (1986) Resource sharing among SBUs: strategic antecedents and administrative implications. *Academy of Management Journal*, 695–714.
Gupta, M. (1993) Heterogeneity issues in aggregated costing systems. *Journal of Management Accounting Research*, Fall, 180–212.
Habermas, J. (1979) *Communication and the Evolution of Society*. Boston, Mass., Beacon Press.
Hall, R. (1962) The concept of bureaucracy: an empirical assessment. *American Journal of Sociology*, 32–40.
Handy, C. (1992) Balancing corporate power: a new federalist paper. *Harvard Business Reivew*, 59–72.

Harvey, D. (1990) *The Condition of Postmodernity: An Inquiry into the Origins of Cultural Change.* Oxford, Basil Blackwell.

Hayes, D.C. (1977) The contingency theory of management accounting. *The Accounting Review*, 22–39.

Hebdige, D. (1988) *Hiding in the Light.* London, Routledge.

Hedberg, B. and S. Jönsson. (1978) Designing semi-confusing information systems for organizations in changing environments. *Accounting, Organizations and Society*, 47–64.

Heilbroner, R.L. (1980) *Marxism: For and Against.* New York, W.W. Norton.

Herman, E.S. and N. Chomsky. (1988) *Manufacturing Consent: The Political Economy of the Mass Media.* New York, Pantheon.

Hiromoto, A. (1988) Another hidden edge—Japanese management accounting. *Harvard Business Review*, 22–26.

Hirst, M.K. (1981) Accounting information and the evaluation of subordinate performance: a situational approach. *The Accounting Review*, 771–784.

———. (1983) Reliance on accounting performance measures, task uncertainty, and dysfunctional behavior: some extensions. *Journal of Accounting Research*, 596–605.

Hirst, M.K. and J.R. Baxter. (1993) A capital budgeting case study: an analysis of a choice process and the roles of information. *Behavioral Research in Accounting*, 187–210.

Hofstede, G.H. (1967) *The Game of Budget Control.* Assen, Netherlands, Koninklijke Van Grocum.

Hopper, T.M. (1990) Social transformation and management accounting: finding the relevance in history in C. Gustafsson and L. Hassel (eds) *Accounting and Organizational Action.* Abo, Finland, Abo Academy Press.

Hopper, T.M. and P. Armstrong. (1991) Cost accounting, controlling labor and the rise of conglomerates. *Accounting, Organizations and Society*, 405–438.

Hopper, T.M. and N.B. Macintosh. (1993) Management accounting as disciplinary practice. *Management Accounting Research*, 181–216.

Hopper, T.M. and A. Powell. (1985) Making sense of research into the organizational and social aspects of management accounting: a review of its underlying assumptions. *Journal of Management Studies*, 429–465.

Hopper, T.M., J. Storey and H. Willmott. (1987) Accounting for accounting: towards the development of a dialectical view. *Accounting, Organizations and Society*, 437–456.

Hopwood, A. (1987) The archeology of accounting systems. *Accounting, Organizations and Society*, 207–234.

Hopwood, A.G. *An Accounting System and Managerial Behavior.* Hampshire, UK, Saxon House.

Horngren, C.T. and G.H. Sorter. (1961) Direct costing for external reporting. *The Journal of Accounting Research*, 84–93.

Horngren, C.T. and G.L. Sundem. (1990) *Introduction to Management Accounting.* Englewood Cliffs, N.J., Prentice Hall.

Horngren, C.T., G.L. Sundem, H.D. Teall and F.W. Selto. (1993) *Management Accounting.* Toronto, Prentice Hall.

Hoskin, K.W. and R.H. Macve. (1986) Accounting and the examination: a genealogy of disciplinary power. *Accounting, Organizations and Society*, 105–136.

———. (1988) The genesis of accountability: the West Point connections. *Accounting, Organizations and Society*, 37–74.

Hutcheon, L. (1989) *The Politics of Postmodernism.* London, Routledge.

Inoguchi, R., T. Nakajima and R. Pineau. (1958) *The Devine Wind.* Washington, Naval Institute Press.

Jameson, F. (1984) Postmodernism of the cultural logic of late capitalism. *New Left Review*, 52–92.

Jensen, M. and W.H. Meckling. (1976) Theory of the firm: managerial behavior, agency costs and ownership structure. *Journal of Financial Economics*, 305–360.

Johnson, H.T. (1992) *Relevance Regained: From Top-Down Control to Bottom-Up Improvement.* New York, The Free Press.

Johnson, H.T. and R. Kaplan. (1987) *Relevance Lost: The Rise and Fall of Management Accounting.* Boston, Mass., Harvard Business School Press.

Jönsson, S. and A. Grönlund. (1988) Life with a sub-contractor: new technology and management accounting. *Accounting, Organizations and Society*, 513–534.

References

Kanter, R. (1992) *The Challenge of Organizational Change: How Companies Experience It and Leaders Guide It.* New York, Maxwell Macmillan.

Kaplan, R.D. (1993) *The Romance of the Arabists.* New York, Macmillan.

Kaplan, R.S. (1983) Measuring manufacturing performance: a new challenge for managerial accounting research. *The Accounting Review*, 686–705.

———. (1993) Research opportunities in management accounting. *Journal of Management Accounting Research*, Fall, 1–14.

Kaplan, R.S. and A.A. Atkinson. (1989) *Advanced Management Accounting.* Englewood Cliffs, N.J., Prentice Hall.

Kenis, I. (1979) Effects of budgetary goal characteristics on managerial attitudes and performance. *The Accounting Review*, 707–721.

Khandwalla, P.N. (1972) The effect of different types of competition on the use of management controls. *Journal of Accounting Research*, 275–285.

Kilduff, M. (1993) Deconstructing organizations. *Academy of Management Review*, 13–31.

Kilmann, R.H. (1983) The costs of organization structure: dispelling the myths of independent divisions and organization-wide decision making. *Accounting, Organizations and Society*, 341–357.

Kim, K.K. (1988) Organization coordination and performance in hospital accounting information systems: an empirical investigation. *The Accounting Review*, 472–489.

Langton, J. (1984) The ecological theory of bureaucracy: the case of Josiah Wedgwood and the British pottery industry. *Administrative Science Quarterly*, 330–354.

Laughlin, R.C. (1987) Accounting systems in organizational contexts: a case for critical theory. *Accounting, Organizations and Society*, 479–502.

Loft, A. (1986) Towards a critical understanding of accounting: the case of cost accounting in the UK, 1914–1925. *Accounting, Organizations, and Society*, 137–139.

Lowe, E.A. and R.W. Shaw. (1968) An analysis of managerial biasing: evidence from a company's budgeting process. *The Journal of Management Studies*, 304–315.

Lowe, J. (1992) *The Secret Empire: How 25 Multinationals Rule the World.* Homewood, Ill., Business One Irwin.

Lukka, K. (1988) Budgetary biasing in organizations: theoretical framework and empirical evidence. *Accounting, Organizations and Society*, 281–301.

———. (1990) Ontology and accounting: the concept of profit. *Critical Perspectives on Accounting*, 239–261.

Lyotard, J. (1984) *The Postmodern Condition: A Report on Knowledge.* Manchester, UK, Manchester University Press.

Macdonell, D. (1986) *Theories of Discourse: An Introduction.* Oxford, Basil Blackwell.

Macintosh, N.B. (1981) A contextual model of information systems. *Accounting, Organizations and Society*, 39–53.

———. (1985) *The Social Software of Accounting and Information Systems.* Chichester, UK, John Wiley.

———. (1990) Annual reports in ideological role: a critical theory analysis in D.J. Cooper and T.M. Hopper (eds) *Critical Accounts*, London, Macmillan.

———. The profit manipulation phenomenon: a dialectic of control perspective. *Critical Perspectives on Accounting*, forthcoming.

Macintosh, N.B. and R.L. Daft. (1987) Management control systems and departmental interdependencies: an empirical study. *Accounting, Organizations and Society*, 40–61.

Macintosh, N.B. and R.W. Scapens (1990) Structuration theory in management accounting. *Accounting, Organizations and Society*, 455–477.

———. (1991) Management accounting and control systems: a structuration theory analysis. *Journal of Management Accounting Research*, 131–158.

Macintosh, N.B. and J.J. Williams. (1992) Managerial roles and budgeting. *Behavioral Research in Accounting*, 23–48.

March, J.G. and H. Simon (1958) *Organizations.* London, John Wiley.

Marx, K. (1944) *Economic and Political Manuscripts contained in Marx and Engels' Selected Works.* London, Lawrence and Wishart.

Mayo, E. (1933) *The Human Problems of an Industrial Civilization.* New York, Macmillan.

———. (1945) *The Social Problems of an Industrial Civilization*. Cambridge, Mass., Harvard University Press.

McGhee, W., M.D. Shields and J.G. Birnberg. (1978) The effects of personality on a subject's information processing. *The Accounting Review*, 681–697.

McKendrick, N. (1961) Josiah Wedgwood and factory discipline. *The Historical Journal*, 30–55.

———. (1970) Josiah Wedgwood and cost accounting in the industial revolution. *The Economic History Review*, 45–67.

McNair, C.J. and W. Mosconi. (1989) *Beyond the Bottom Line: Measuring World Class Performance*. New York, Dow Jones-Irwin.

Menzies, H.D. (1980) The ten toughest bosses. *Fortune*, April.

Merchant, K.A. (1985) Budgeting and propensity to create budgetary slack. *Accounting, Organizations and Society*, 201–210.

———. (1987) *Fraudulent and Questionable Financial Reporting: A Corporate Perspective*. Morristown, N.J., Financial Executives Research Foundation.

———. (1990) The effects of financial controls on data manipulation and management myopia. *Accounting, Organizations and Society*, 297–313.

Merchant, K.A. and J.-F. Manzoni. (1989) The achievability of budget targets in profit centers: a field study. *The Accounting Review*, 539–558.

Merchant, K. and J. Rockness. (1994) The ethics of managing earnings: an empirical investigation. *Journal of Accounting and Public Policy*, 79–94.

Merchant, K.A. and M.D. Shields. (1993) Commentary on when and why to measure costs *less* accurately to improve decision making. *Accounting Horizons*, 76–81.

Mia, L. (1988) Managerial attitude, motivation and the effectiveness of budget participation. *Accounting, Organizations and Society*, 465–476.

———. (1989) The impact of participation in budgeting and job difficulty on managerial performance and work motivation: a research note. *Accounting, Organizations and Society*, 347–358.

Milani, K. (1975) The relationship of participation in budget-setting, to industrial supervisor performance and attitudes: a field study. *The Accounting Review*, 274–284.

Miles, R.E. and C.C. Snow. (1978) *Organizational Strategy, Structure and Process*. New York, McGraw-Hill.

Miller, P. and T. O'Leary. (1987) Accounting and the construction of the governable person. *Accounting, Organizations and Society*, 235–265.

———. (1990) Making accounting practical. *Accounting, Organizations and Society*, 479–498.

Mintzberg, H. (1972) The myths of MIS. *California Management Review*, 92–97.

———. (1975) The manager's job: folklore and fact. *The Harvard Business Review*, 49–61.

———. (1979) *The Structuring of Organizations*. Englewood Cliffs, N.J., Prentice Hall.

Noreen, E. (1987) Commentary on Johnson and Kaplan's Relevance Lost. *Accounting Horizons*, 110–116.

———. (1991) Conditions under which activity-based cost systems provide relevant costs. *Journal of Management Accounting Research*, 159–168.

Norris, C. (1992) *Uncritical Theory*. Amherst, Mass., University of Massachusetts Press.

Onsi, M. (1973) Factor analysis of behavioral variables affecting budgetary slack. *The Accounting Review*, 535–548.

Otley, D.T. (1978) Budget use and managerial performance. *Journal of Accounting Research*, 122–149.

———. (1980) The contingency theory of management accounting: achievement and prognosis. *Accounting, Organizations and Society*, 413–428.

Otley, D.T. and A.J. Berry. (1980) Control, organization and accounting. *Accounting, Organizations and Society*, 231–244.

Otley, D.T. and F.J.B. Dias. (1982) Accounting aggregation and decision performance: an experimental investigation. *Journal of Accounting Research*, 171–188.

Ouchi, W.G. (1977) The relationship between organizational structure and organizational control. *Administrative Science Quarterly*, 25–113.

———. (1979) A conceptual framework for the design of organizational control mechanisms. *Management Science*, 833–848.

Parker, L.D., K.R. Ferris and D.T. Otley. (1989) *Accounting for the Human Factor.* Sydney, Prentice Hall.
Parsons, T. (1937) *The Structure of Social Action.* New York, McGraw-Hill.
Penno, M. (1984) Asymmetry of pre-decision information and managerial accounting. *Journal of Accounting Research,* 177–191.
Perrow, C. (1967) A framework for the comparative analysis of organizations. *American Sociological Review,* 194–208.
———. (1970) *Organizational Analysis: A Sociological Review.* Belmont, Calif., Wadsworth.
———. (1972) *Complex Organizations: A Critical Analysis.* Glenview, Ill., Scott, Foresman and Co.
———. (1986) *Complex Organizations,* 3rd ed. New York, Random House.
Posner, B.Z. and W.H. Schmidt. (1986) Values and the american manager: an update. *California Management Review,* 202–216.
Poster, M. (1990) *The Mode of Information: Poststructuration and Social Context.* Cambridge, Polity Press.
Preston, A., D. Cooper and R.W. Coombs. (1992) Fabricating budgets: a study of the production of management budgeting in the National Health Service. *Accounting, Organization and Society,* 561–594.
Puxty, A.G. (1993) *The Social & Organizational Context of Management Accounting.* London, Academic Press.
Reider, B. and G. Saunders. (1988) Management accounting education: a defense of criticisms. *Accounting Horizons,* 58–62.
Richardson, A.J. (1987) Accounting as a legitimating institution. *Accounting, Organizations and Society,* 341–356.
Roberts, J. (1990) Strategy and accounting in a U.K. conglomerate. *Accounting, Organizations and Society,* 107–126.
Roberts, J. and R.W. Scapens. (1985) Accounting systems and systems of accountability—understanding accounting practices in their organizational contexts. *Accounting, Organizations and Society,* 443–456.
Roethlisberger, F.J. and W.J. Dickson. (1939) *Management and the Worker.* Cambridge, Mass., Harvard University Press.
Ronen, J. and J.L. Livingston. (1975) An expectancy theory approach to the motivational impacts of budgets. *The Accounting Review,* 671–685.
Rose, M.A. (1991). *The Post-Modern & the Post-Industrial.* Cambridge, Cambridge University Press.
Roslender, R. (1992) *Sociological Perspectives on Modern Accounting.* London, Routledge.
Ryan, M. (1982) *Marxism and Deconstruction: A Critical Articulation,* Baltimore, Md., John S. Hopkins University Press.
Sampson, A. (1974) *The Sovereign State of ITT.* Greenwich, Conn., Fawcett.
San Miguel, J.G. (1976) Human information processing and its relevance to accounting: a laboratory study. *Accounting, Organizations and Society,* 357–373.
Sarup, M. (1993) *An Introductory Guide to Post-Structuralism and Postmodernism.* Athens, Ga, University of Georgia Press.
Scapens, R.W. (1985) *Management Accounting: A Review of Recent Developments.* London, Macmillan.
Schick, A.G., L.A. Gordon and S. Haka. (1990) Information overload: a temporal approach. *Accounting, Organizations and Society,* 199–220.
Schiff, M. and A.Y. Lewin. (1968) Where traditional budgeting fails. *Financial Executive,* 57–63.
———. (1970) The impact of people on budgets. *The Accounting Review,* 259–268.
Shank, J.K. (1981) *Contemporary Managerial Accounting: A Casebook.* Englewood Cliffs, N.J., Prentice Hall.
———. (1989) Strategic cost management: new wine or just new bottles. *Journal of Management Accounting Research,* 47–65.
Shank, J.K. and V. Govindarajan. (1989) *Strategic Cost Analysis: The Evolution from Managerial to Strategic Accounting.* Homewood, Ill., Irwin.
———. (1992) Strategic cost management: the value chain perspective. *The Journal of Management Accounting Research,* 179–197.
Simon, H.A. (1960) *The New Science of Management Decision.* New York, Harper & Row.

Simons, R. (1987) Accounting control systems and business strategy. *Accounting, Organizations and Society*, 357–374.

———. (1990) The role of management control systems in creating competitive advantage: new perspectives. *Accounting, Organizations and Society*, 127–143.

———. (1991) Strategic orientation and top management attention to control systems. *Strategic Management Journal*, 49–62.

———. (1992) The strategy of control. *CA Magazine*, March, 44–50.

Sloan, A.D., Jr. (1963) *My Years with General Motors*. New York, McFadden-Bartell.

Smart, B. (1985) *Michel Foucault*. London, Tavistock.

Smith, C., R. Whipp and H. Willmott. (1988) Case study research in accounting: methodological breakthrough or ideological weapon. *Advances in Public Interest Accounting*, 25–40.

Solomons, D. (1965) *Divisional Performance: Measurement and Control*. New York, Financial Executives Research Foundation.

Stedry, A. (1960) *Budget Control and Cost Behavior*. Englewood Cliffs, N.J., Prentice Hall.

Taylor, F.W. (1911) *The Principles of Scientific Management*. New York, Harper.

Thomas, A.L. (1969) The allocation problem in financial accounting theory. *Studies in Accounting Research No. 3*. American Accounting Association.

———. (1974) The allocation problem: part two. *Studies in Accounting Research No. 9*. American Accounting Association.

———. (1975a) Accounting and the allocation fallacy. *Financial Analysts Journal*, 37–41, 68.

———. (1975b) The FASB and the allocation fallacy. *The Journal of Accountancy*, 65–68.

———. (1978) Arbitrary and incorrigible allocations: a comment. *The Accounting Review*, 263–269.

Thompson, J.D. (1967) *Organizations in Action*. New York, McGraw-Hill.

Thompson, J.D. and A. Tuden. (1959) Strategies, structures and processes of organizational decision in J.D. Thompson, et al. *Comparative Studies in Administration*. Pittsburg, University of Pittsburg Press.

Thornton, D. (1988) Theory and metaphor in accounting. *Accounting Horizons*, 1–9.

Tichey, N.M. and S. Sherman. (1993) *Control Your Destiny or Someone Else Will*. New York, Doubleday.

Tinker, A.M. (1980) Towards a political economy of accounting: an empirical illustration of the Cambridge controversies. *Accounting, Organizations and Society*, 147–160.

Tinker, A.M., B.D. Merino and M.D. Neimark. (1982) The normative origins of positive theories: ideology and accounting thought. *Accounting, Organizations and Society*, 167–200.

Tinker, T. and M. Neimark. (1987) The role of annual reports in gender and class contradiction at General Motors: 1917–1976. *Accounting, Organizations and Society*, 71–88.

Wakefield, N. (1990) *Postmodernism: The Twilight of the Real*. London, Pluto Press.

Waller, W.S. (1988) Slack in participative budgeting: the joint effect of a truth-inducing pay scheme and risk preferences. *Accounting, Organizations and Society*, 87–100.

Walsh, E.J. and R.E. Stewart. (1993) Accounting and the construction of institutions: the case of the factory. *Accounting, Organizations and Society*, 783–800.

Waterhouse, J.H. and P.A. Tiessen. (1978) A contingency framework for management accounting systems research. *Accounting, Organizations and Society*, 413–428.

Watson, D.J.H. and J.V. Baumler. (1975) Transfer pricing: a behavioral context. *The Accounting Review*,, 466–474.

Weber, M. (1947) in T. Parsons, (ed.) *The Theory of Social and Economic Organization*. New York, The Free Press.

Weedon, C. (1987) *Feminist Practice & Poststructural Theory*. Oxford, Basil Blackwell.

Weick, K.E. (1979) *The Social Psychology of Organizing*. Reading, Mass., Addison-Wesley.

Weinwurm, E.H. (1961) The importance of idle capacity costs. *The Accounting Review*, 418–421.

Williams, J.J., N.B. Macintosh and J.C. Moore. (1990) Budget-related behavior in public sector organizations: some empirical evidence. *Accounting, Organization and Society*, 221–248.

Williamson, O.E. (1973) Markets and hierarchies: some elementary considerations. *American Economic Association*, 316–325.

Willmott, H.C. (1986) Unconscious sources of motivation in the theory of the subject: an exploration and critique of Giddens' dualistic models of action and personality. *Journal for the Theory of Social Behavior*, 105–121.

References

Woodward, J. (1965) *Industrial Organization: Theory and Practice*. London, Oxford University Press.
Wright, J., W. Winter, Jr., S. Zeigler and P. O'Dea. (1984) *Advertising*. Toronto, McGraw-Hill Ryerson.
Zimmerman, J.L. (1979) The costs and benefits of cost allocations. *The Accounting Review*, 504–521.

Name Index

n = note
t = table
f = figure

Abernethy, M.A., 111n3
Ackoff, R.L., 198, 202n
Alchian, A., 36–37
Alleman, R.H., 234n
Amey, L.R., 58n
Ansari, S.L., 153n2, 177n, 183n, 192n, 193, 213n30
Anthony, R.N., 15n, 20n, 21–22, 25–26, 67n, 69n, 97n, 99n, 102n11, 142n, 180n, 197n2, 198–99, 203, 239, 240
Argyris, Chris, 211–14, 247
Armstrong, P., 37n2, 41n5, 44, 217n
Arrington, C.E., 250n7, 252n
Arrow, K., 55, 55n2
Atkinson, A.A., 2n3, 205nn9–10
Antle, R., 213n32

Baiman, S., 37n3, 213n32
Barley, S.R., 210, 211n26, 249
Barnard, C., 198, 247, 247n4
Baudrillard, J., 251n9, 253
Baumler, J.V., 119n8
Beaver, W.H., 253n16
Becker, S.W., 213n27
Bedford, N.M., 20n, 69n, 10211, 240, 253n16
Bennis, W.G., 247
Bentham, Samuel, 228
Berry, A.J., 111n3
Best, S., 6n14
Birnberg, J.G., 111n3, 198n4, 213n28, 213n30
Boland, R.S., Jr., 153n2
Bottomore, T., 250n8
Bower, J., 153n3
Braverman, H., 39t, 43–44, 41n4, 43n, 209n22, 211n25
Broadbent, J., 250n7
Brothers Ltd., 101–02

Brownell, P., 11n4, 213n30, 213n32, 214n31
Bruns, W.J., Jr., 38–39, 111n2
Burchell, S., 150n, 169n2, 217n
Burgstahler, D., 213n28
Burns, T., 52
Burrell, G., 197n2, 211n25

Chandler, A.D., Jr., 90–93, 90n, 183n, 185
Chenall, R., 153n3, 119n8, 213n30
Cherrington, D.J., 213n27
Cherrington, J.O., 213n27
Chomsky, N., 251n14
Chow, C.W., 213n32
Christensen, J., 213n32
Chua, W.F., 3n7, 4n9
Clegg, S., 197n2, 211n25
Clubb, C., 150n, 169n2, 217n
Collins, F., 213n30, 213n32
Coombs, R.W., 207n13
Cooper D.J., 169n2, 207n13
Cooper, J.C., 213n32
Cooper, R., 207–08
Covaleski, M.A., 4n9, 153n2, 169n2, 207n13
Culbert, S.A., 156, 156n, 158, 160n
Cyert, R., 55, 253

Daft, R.L., 3n5, 75n, 87n1, 107n16, 111n2, 119n8, 123, 241n
Dalton, G.W., 15n, 16–17
Darwin, Charles, 31
Dean, J., 198–99
Dearden, J., 15n, 20n, 21–22, 25–26, 67n, 69n, 97n, 99n, 102n11, 142n, 158n, 180n, 198–99, 202, 203, 238, 240
DeCoster, D.T., 213n27
Demsetz, H., 36–37
Demski, J.S., 253n16
Dent, J.F., 87n2
Dermer, J.D., 87n2, 213n30
Derrida, J., 252n
Dias, F.J.B., 213n30

Dickson, W.J., 210
Dirsmith, M.W., 4n9, 153n2, 169n2, 207n13
Dopuch, N., 207n12, 253
Driver, M.J., 213n30
Duncan, K., 11n3
Dunk, S.A., 111nn4–5, 213nn31–32, 242n
Dunkerley, D., 197n2, 211n25

Earl, M.J., 150n
Eckel, L.G., 208n16, 216n
Emmanuel, C., 2n4, 119n8
Engels, Friedrich, 39t, 41–42, 41n7, 249–50, 250n6
Eppen, G.D., 213n32
Etruria, 13–15
Euske, K.J., 153n2, 193, 183n, 192n
Evans, J.H., III, 213n32
Ezzamel, M., 2n3, 138n

Feldman, M.S., 154, 207n13
Ferrara, W.L., 204, 208n17
Ferris, K.R., 2n4
Fertakis, J.P., 213n27
Fess, P.E., 208n17
Finn, D., 213n30
Fisher, J., 105n14
Foucault, M., 221–24, 228n, 229, 237, 239, 240n, 241–42, 249, 251n9, 252
Francis, J.R., 252n
Frank, W.G., 208n19
Fremgren, J.M., 208n17
Friedman, A., 159, 159n

Galbraith, J., 61, 58n
Geneen, H., 223, 228–38, 240, 242–43
Giddens, A., 188, 189, 169, 174, 176, 169n, 171n, 178n, 252
Ginzberg, M.J., 119n8
Goldratt, E.M., 207n12
Gordon, L.A., 57n, 102n10, 154n, 184n, 203
Govindarajan, V., 15n, 20n, 21–22, 25–26, 67n, 69n, 97n, 99n, 105nn13–14, 203, 204n, 239
Green, D., 213n27
Greiner, L.E., 65f, 62n
Grönlund, A., 247n2
Guess, R., 250n8
Gul, F.A., 213n30
Gupta, M., 105n14, 207n14

Habermas, J., 249–50, 251n12
Haka, S., 57n, 154n
Hall, R., 55n1
Handy, C., 247–48
Hart, H., 138n
Harvey, D., 251n9

Hayes, D.C., 111n2, 119n8, 123
Hebdige, D., 251n9
Hedberg, B., 152
Hegel, Georg, 42
Heilbronner, R.L., 41n6, 63n
Herman, E.S., 251n14
Hiromoto, A., 205n8
Hirst, M.K., 111n4, 213n31
Hobbes, Thomas, 30, 35–36, 39t
Hofstede, G.H., 214n
Hopper, T.M., 3n7, 41n5, 44–45, 217n, 221n1
Hopwood, A.G., 150n, 169n2, 213n29, 217n, 221n3, 240
Horngren, C.T., 2, 208n17, 240
Hoskin, K.W., 221n3
Hughes, T., 150n, 169n2, 217n
Hutcheon, L., 251n9

Ijiri, Y., 253
Inoguchi, T., 139n
International Telephone and Telegraph Company (ITT), 108–09, 228–38, 239, 242–43

Jablonsky, S.F., 153n2, 169n2
Jensen, M., 30
Johnson, Thomas, 44, 208, 245–46
Jönsson, S., 152, 247n2

Kanter, R., 247n3
Kaplan, R.S., 2n3, 44, 203–04, 205nn9–10, 207–08, 208nn14, 18
Kaplan, R.D., 159, 159n
Kellner, D., 6n14
Kenis, I., 213n27
Khandwalla, P.N., 111n1
Kilduff, M., 198n3
Kilmann, R.H., 119n8
Kim, K.K., 111n3
Kunda, G., 210, 211n26, 249

Langton, J., 11n
Laughlin, R.C., 250n7
Lawrence, P.R., 15n, 16–17
Lewin, A.Y., 177n, 213n32, 242n
Lincoln, Abraham, 248
Livingston, J.L., 215n
Loft, A., 221n3
Lowe, E.A., 213n32
Lowe, J., 176n, 238n
Lukka, K., 174n, 213n32
Lyotard, J., 251n9

Macdonnell, D., 251n9, 252
Macintosh, N.B., 2n4, 40, 45n, 75n, 101n, 107n15, 111n2, 114, 119n8, 122, 123,

Name Index 275

138n, 169n, 172f, 173f, 202n, 207n13, 213n29, 215n, 216, 217n, 221n1, 250n7
Macve, R.H., 221n3
Maisel, L.S., 207–08
Manzoni, J.-F., 242n
March, J.G., 55, 154, 198n3, 207n13
Marx, Karl, 39t, 41–42, 249, 251, 251nn12–13
Mayo, E., 209n23
McDonough, J.J., 156, 156n, 158, 160n
McGhee, W., 213n30
McInnes, M., 213n32
McKendrick, N., 11n
McKinnon, S.M., 38–39
McNair, C.J., 111n5
Mead, 39t
Meckling, W.H., 30
Menzies, H.D., 237
Merchant, K.A., 2n4, 45, 58n, 119n8, 202, 203, 208, 242n
Merino, B.D., 169n2
Merton, 39t
Mia, L., 111n4, 213n30
Michelman, J., 207n13
Milani, K., 213n27
Miles, R.E., 95n
Miller, D., 102n10
Miller, K.J., 184n
Miller, P., 197n1, 221n3
Mintzberg, H., 37, 114
Mock, T.J., 213n30
Moore, J.C., 119n8, 122, 123
Moores, K., 111n3
Morgan, G., 197n2, 211n25
Morris, D., 13n30, 119n8, 153n3
Morrissey, E., 207–08
Mosconi, W., 111n5
Munter, P., 213n30

Nader, Ralph, 188n, 192
Nakajima, T., 139n
Naphapiet, J., 150n, 169n2, 217n
Neimark, M., 5n11, 107n15, 169n2, 195
Noreen, E., 208n15, 208n21
Norris, C., 251n9

O'Dea, P., 145n
O'Leary, T., 197n1, 221n3
Oehm, R.M., 207–08
Onsi, M., 213n32, 242n
Otley, D.T., 2n4, 111n3, 119n8, 177n, 213n30, 214, 242n
Ouchi, W.G., 136n, 138n
Owen, Robert, 226–27, 234

Parker, L.D., 2n4
Parsons, T., 39t, 58

Penno, M., 213n32
Perrow, C., 4n10, 37n2, 111n1, 112, 211, 211n25
Pineau, T., 139n
Pondy, L.R., 153n2
Posner, B.Z., 253n17
Poster, M., 251n9
Powell, A., 3n7
Preston, A., 207n13
Puxty, A.G., 3n6, 5n12, 41n5, 45, 250n7

Read, S., 250n7
Reider, B., 208n20
Richardson, A.J., 174
Roberts, J., 94n, 174n
Rockness, J., 203
Roethlisberger, F.J., 210
Ronen, J., 215n
Rose, M.A., 251n9
Roslender, R., 3n6, 41n5
Ryan, M., 251n11, 252n

Sampson, A., 2n1, 232–38, 242–43, 108–09, 108n
San Miguel, J.G., 213n30
Sarup, M., 6n14, 251, 251n9
Saunders, G., 208n20
Scapens, R.W., 2n3, 37n2, 169n, 171n, 172f 173f, 207n13, 217n
Schick, A.G., 57n, 154n
Schiff, M., 177n, 213n32, 242n
Schmidt, W.H., 253n17
Selto, F.W., 240
Shank, J.K., 105n, 162n, 163–66, 204n
Shaw, R.W., 213n32
Sherman, S., 71–72, 71n10
Shields, J.F., 213n28
Shields, M.D., 208, 213n30
Simon, Herbert, A., 198, 198n3
Simons, R., 84, 95n, 98, 104–05, 104n
Sloan, Alfred, 1, 73, 184–92
Smith, Adam, 30–31, 35–36, 39t, 41, 43–44, 250n10, 251nn12–13
Smith, C., 195, 217
Smith, K.J., 203
Snow, C.C., 95n
Solomons, D., 201
Sorter, G.H., 208n17
Spencer, Herbert, 31, 39t, 58
Stalker, G.M., 52
Stedry, A., 84, 213n27
Steward, R.E., 227
Stoelwinder, J.U., 111n3
Storey, J., 45, 41n5
Sundem, G.L., 2, 213n28, 240
Sunden, S., 253

Sweezy, Paul, 41n4

Taylor, Frederick, 43, 191–92, 209–10
Teall, H.D., 240
Thomas, A.L., 208n16, 216n
Thompson, J.D., 111n1, 119n7, 127n, 150n
Thornton, D., 253n16
Tichey, N.M., 71, 71n10
Tiessen, P.A., 111n2
Tinker, T., 5n11, 107n15, 169n2, 195
Tuden, A., 150n
Turopolec, I., 111n3

Vancil, T., 199

Wakefield, N., 251n9
Waller, W.S., 213n32
Walsh, E.J., 227
Waterhouse, J.H., 111n2
Watson, D.J.H., 119n8

Weber, Max, 132, 136, 141, 251nn12–13
Wedgewood, Josiah, 11–15, 26, 46–48, 62, 84–85
Weedon, C., 251n9
Weick, K.E., 88n
Weinwurm, E.H., 208n17
Welch, Jack, 70–73
Wesley, John, 13–15, 47
Whipp, R., 195, 217
Williams, J.J., 40, 119n8, 122, 123
Williamson, O.E., 35n, 134n
Willmott, H., 45, 41n5, 171n, 195, 217
Winter, W., Jr., 145n
Woodward, J., 111n1
Wright, J., 145n

Young, S.M., 111n3

Zeigler, S., 145n
Zimmerman, J.L., 216n

Subject Index

Accounting, Organizations and Society, x
Activity based costing, 204–207
Agency, in social settings, 170–171, 190–192
Agency theory, 29–37, 46
   adverse selection (hidden information), 29–30
   asymmetric information, 35
   contract, 35–36
   incentive schemes, 34–35
   moral hazard (hidden action), 32
   Pareto optimal solution, 32
   self-interest, 30–31
   signalling, 33
   survival of the fittest, 31
   transaction cost, 36
Alienation, 42
Alignment, 161
Analyzer organization, 95, 100–102
Anthony, Robert, 198
Apollo Computers Co. case, 78–84, 178
Assessment and control typology, 127–132
   efficiency, instrumental, social tests, 128–131
   ends continuum, 127
   task instrumentality, 129
Asymmetric information, 35
Authoritarian control, 53
Autonomy crisis, 66, 79, 80
Avis Rent-a-Car, 238

Barnard, Chester, 198, 247
Baseball umpires, 153
Baysian decision theory, 33
Bentham, Samuel, 223
Big Three Auto companies, 131
British Petroleum, 247–248
Brothers, Ltd. case study, 101–102
Brown, Donaldson, 185–186
Budget participation, 212–213
Budget evaluation styles, 213–214
Bureaucratic controls, 132–133
   hierarchies, 133

   records and files, 134
   rules, 135
Bush, George, 159

Capitalism, 41, 42, 197, 246, 250, 251, 253, 254, 255
Cartesian philosophy, 200
Cellular principle (of control), 223, 239
Centralization, 85, 90, 92, 94, 247
Charismatic control, 135–136
Chrysler Corporation, 253–254
Churchill, Winston, 73, 136
Clan control, 138–140, 143, 166
Clark, J. Maurice, 208
Clinton, William (US President), 246
Clocking-in-system, 14
Close control style, 115
Closed-rational systems, 125–126, 147
Collegial control, 140–141, 143, 166
Command versus market control, 55–58, 77
   enforcement rules, 56
   goal congruence, 55
   omniscient management, 56
   operating rules, 56
Community Health Center case, 142–144
   see also Hyatt Hill Health Center case, 180–181
Comprehensive control style, 116–117
Comptroller's organization (ITT), 232–234
Contagion effect, 214
Control crisis, 66
Control styles typology, 132–141
   bureaucratic, 132–133, 143, 147
   charismatic, 135–136
   collegium, 140–141, 143, 144
   market, 136–137, 143, 147
   tradition (clan), 137–140, 143, 144
Corning Glass Co. case, 99–100
Critical theory (Habermas), 250

Darwin's law of survival of the fittest, 31

Davidson, Sidney, 208
Dean, Joel, 198
Dearden, John, 198
Decentralization, 57, 81, 83, 85, 94, 247–248
Deconstruction, 252–253
Defender organization, 95, 96–97, 104
de Gaulle, Charles, 136
Department of Defense (US), 107, 192–195
Desert Storm, 122
Deskilling, 42–43
Dialectic of control, 176–177
Discipline and control (Foucault) principles, 222–228
   disciplined mind, 225, 232–236
   efficient body, 223–225, 231–232, 239
   enclosure, 222–223, 230–231
   panopticism, 228
   resistance, 238, 240–243
Disciplining minds, 225–227, 232–237
Discourse analysis, 252–253
Discursive consciousness, 171, 190
Discursive formation, 221
Discursive practices, 222, 253
Divisionalization, 201
Domination structures, 175, 178, 180, 188, 194
Dressage, 224–225, 236, 239
Duality of structure, 171
du Pont Co. case, 90–94, 185
Durant, William (GM), 183

Effectiveness controls, 97–98
Efficiency controls, 96–97, 128, 143, 145, 147, 150
Efficient body principle, 223–225, 231–232
Einstein, Albert, 7, 130
Emerson, Waldo, 190
Empire Glass Co. case, 11, 15–20, 26, 46–48, 62, 85–86, 97
Employee accounting systems, 245–255
Empowerment, 73, 245–247
Enactment, 88
Enclosure principle, 222–223, 230–231
Enforcement rules, 56, 57, 83, 85
Engles, Friedrich, 41–42, 249, 250, 251
Enlightenment project, 200, 251
Environmental turbulence, 98
Examination, the, 227

Federalist governance, 247–248
Financial control chart, 186
Ford, Henry, 204
Ford Motor Co., 118, 204
Fragmenting, 160–161
Framing, 159–160

Frankford School of Social Philosophy, 249
Freud, Sigmund, 224
Freudian psychology, 221
Functions of the executive (Barnard), 198

Game spirit of budgeting, 214
Geneen, Harold, 229–238
General Appliance Corporation case, 67–68
General Electric case, 69–73, 107, 217
General Motors case, 188–194, 231
Gettysburg Address, 248
Ghandi, M., 73
Global conglomerates, 20, 109, 238
Goal congruence (incongruence), 45, 55, 57–58, 65, 198–199, 240–243
Gorbachev, general secretary USSR, 73, 178
Group dynamics, 211–212
Gulf War, 122, 159

Habermas, Jurgen, 249, 251
Harvard Business School, 231
Hawthorne study, 210
Hegel, Georg, 42
Herd metaphor of control, 52
Hierarchical supervision, 14, 113
Hierarchical surveillance, 225–226, 233–235
Highland clans (Scotland), 138
Historical-dialectic, 62–63, 65, 69, 70, 77, 81
Hitler, Adolf, 135, 159
Hobbes, Thomas, 30, 35
Human relations control rhetoric, 249
Human relations critique of management accounting, 209–216
Human resources plan, 15
Hyatt Hill Health Center case analysis, 180–181
   see also Community Health Center case, 142–144

Iacocca, Lee, 73, 253
Ideological control, 107–109
   strategy as, 107
Ideology, 197, 198, 251
Incentive schemes, 16, 34–35
Information processing model of control, 58–62, 79–80
Insect hive metaphor of control, 52
Instrumental assessment, 129, 143, 147
Interdependency and control, 119–123
   see technical rationality, interdependency and control
Interpretivist paradigm, 4, 149
Invisible war of self interest, 156–162, 165
Invisible hand of the market, 31, 41, 44, 200

Subject Index

Iraq, 123, 159
Israel, 159
ITT (International Telephone & Telegraph) case, 228–238

Johnson & Johnson case, 20–26, 46–48, 62, 85–86
Just-in-time, 216

Kamikaze pilots, 139–140
Knox, Walter & Thomas case, 144–147
Kuwait, 159

Labor process accounting paradigm, 41–45, 46–47
  alienation, 42
  class domination, 42
  commodification, 41, 42
  control of financial information, 44
  control of work process knowledge, 43
  control of wage labor, 41
  deskilling, 42–43
  invisible hand of the market, 41
  manager's ambiguous status, 44
  scientific management, 43
  specialization of labor, 41
Lateral relations, 60–61, 62
Leadership crisis, 65
Legitimation structures, 174, 178, 187–188, 193–194
Lincoln, Abraham, 248
Logos, 42

Managed (manipulated) profits, 201–203, 246
Management accounting and control systems (defined), 2
Managerial class, 44–45, 211
Managerialism, 197–198, 249
Managing by the numbers, 231–236, 245–246
Maneuver, the, 224
Market control, 57–58, 81, 85, 136–137
Master budget, 57
Matrix management, 54, 60
McNamara, Robert, 118, 193
Mechanistic and organic control, 52–54, 79, 85
  authoritarian commands, 53
  external circumstances, 32
  matrix organization, 54
  politics and careerism, 54
  social technology, 52
Metanarratives, 251
Military schools, 226

Modernity, 6, 251
Multinational corporations, 1, 26, 57, 176, 259
Mussolini, Benito, 135, 159

Nader, Ralph, 192
Nerve center model, 37–40, 47
  decision role, 38
  dissemination role, 39
  informational role, 38
  interpersonal role, 38
  spokesman role, 39
  strategy-maker role, 39
Normalizing sanctions, 226–227, 236, 240

Objective accountability war at work, 156–162, 165–166
  alignment, 161
  coping strategies–framing, fragmenting, playing it both ways, 159–161
  disorientation, 161
  inputs, activities, outputs, impact accountability, 157–158, 165
Omniscient information, 56, 57
One-man show, 90
Ontological security, 171, 190
Open-natural systems, 125–126, 147
Operating rules, 56, 57, 83, 85
Operation work out, 71
Organic control
  see Mechanistic and organic control
Organic structures, 53–54, 81
Owen, Robert, 226–227

Panopticism, 228, 237
  accounting as, 240
Pareto optimal, 31
Participation in budgeting and accounting systems, 51, 212–213, 215–216, 246–247
Partitioning, 222
Path-goal theory of financial controls, 215
Perestroika, 178
Perot, Ross, 246
Phases (historical) model of control, 62–69
  autonomy and direction, 63–64
  crises (leadership, autonomy, control, red-tape, spirituality), 64–69
  dynamic tension, 63
  General Electric case illustration, 63–73
  historical-dialectic analysis, 62
  phases (creativity, direction, delegation, coordination, collaboration), 64–69
Plato's master-slave example, 63
Playing it both ways, 161

Postmodernism, 251–253
Postmodernist paradigm, 6, 240–242
Poststructuralist, 252, 258
Power resources (authoritative, allocative), 175, 180
Practical consciousness, 171
Private School Costing System case, 162–166
Product life cycle model of control, 105–107
Profit planning and control reports, 16
Prospector organization, 95, 97–98, 104
Prospects control style, 117–118

Radical-humanist paradigm, 5, 89, 250
Radical-structuralist paradigm, 5–6
Rational control rhetoric, 249
Reactor organization, 95, 102–104
Records and files, 134
Red tape crisis, 67, 70
Relevance lost (management accounting critique), 203–209
   ABC, 205–209
   concerns, 207–209
   strategic cost management, 204–207
   value chain analysis, 204–205
Remote control management, 245–246
Resistance to control systems, 212, 214, 238, 240–243
Results control style, 116
Rockerfeller, John D., 191
Rommel, Field Marshall, 136
Rules, 14, 133–134
Running blind organization, 95, 103

Saudi Arabia, 159
Scientific management, 43, 44, 191, 239
Self-contained units, 60, 77, 81, 83
Signification structures, 172–173, 180, 185–187, 193
Simon, Herbert, 198
Slack resources, 59–60, 81, 242
Sloan, Alfred D., Jr, 73, 184–192
Smith, Adam, 30, 41, 251
Social technology, 52, 54, 77, 79
Social tests, 129–131
Socially-constructed, 149
Spencer, Herbert, 30, 31, 58
Spiritual crisis, 69, 71
Standardization, control by, 121–122
Star Wars, 107
Stock options, 254
Stratagem, 89
Strategic cost management, 204–207
Strategic planning, 12–13
Strategy and control, 87–110
   at General Motors, 184–185
   defender, prospector, analyzer, reactor, 88, 96–104
   global versus business unit, 94
   programmed versus interactive use of controls, 104–105
Strategy (competitive types), 105–107
   build, differentiate, hold, harvest, divest, 105–109
   defined, 89
Structural functionalist paradigm, 4, 123
Structuration theory (Giddens), 170–179
Structuration theory of management accounting, 169–181
   agency, defined, 170–171
   allocative and authoritative resources, 175, 188
   dialectic of control, 176–177, 190
   dimensions of structuration, 172–176, 185–189, 193–194
   discursive and practical consciousness, 171, 190
   duality of structure, 171, 190
   ontological security, 171, 190
   routine and crisis (critical) situations, 177–178, 189–190
   structuration, 171
   structures, defined, 170
Symbolic use of accounting systems, 154–156
   defensive, 154
   offensive, 155
   symbolic posturing, 155

Tandem Computers case, 107–108, 109
Taylor, Frederick, 43, 191
Technical rationality, interdependence and control, 119–123
   feedback, 122–123
   long-linked, pooled, and reciprocal interdependence, 119–123
   planning, measurement, coordination, 120
   serial, mediating, and intensive technical rationality, 119–123
   standardization, 121–122
Technology, defined, 112–113
   types of, 113–114
Technology framework of control, 111–119
   control styles — close, results, comprehensive, prospects, 113–118
   craft, programmable, professional, research technologies, 113–114
   technology defined, 112
Tight control, 90
Timetable, 223
Townsend, Robert, 238, 243

Subject Index

Tradition control, 137–140
Transaction cost economics, 36
Transamerica Finance Co. case, 74–78
Transcendental signifiers, 251
Uncertainty and accounting and control systems, 150–153
    actual uses (answers, ammunition, rationalization), 152–153
    ideal uses (answers, computation, inspiration, learning), 151–152

Uncertainty and scorekeeping, 127
    defined, 127
    efficiency tests, 128, 143, 147
    instrumental tests, 129, 143, 147
    social tests, 129–131, 143
United Nations, 122, 141, 159
Universal National case, 102–104

Value chain analysis, 204–205
Verdi, Giuseppe, 130
Vertical information processing, 60, 77, 83
Vietnam War, 118, 159

Weapons Repair Accounting System case, 192–195
Weber, Max, 251
Wedgwood Potteries case, 11–15, 26, 46, 47, 62, 84–85
Welsch, Jack, 70, 72, 73